# The Malaysia-Japan Model on Technology Partnership

Khairuddin Ab. Hamid • Osamu Ono
Anas Muhamad Bostamam • Amy Poh Ai Ling
Editors

# The Malaysia-Japan Model on Technology Partnership

## International Proceedings 2013 of Malaysia-Japan Academic Scholar Conference

Springer

*Editors*
Khairuddin Ab. Hamid
Universiti Malaysia Sarawak
Sarawak, Malaysia

Osamu Ono
Meiji University
Tokyo, Japan

Anas Muhamad Bostamam
Sony Corporation
Tokyo, Japan

Amy Poh Ai Ling
The University of Tokyo
Tokyo, Japan

ISBN 978-4-431-54438-8 ISBN 978-4-431-54439-5 (eBook)
DOI 10.1007/978-4-431-54439-5
Springer Tokyo Heidelberg New York Dordrecht London

Library of Congress Control Number: 2014933781



Printed on acid-free paper

Springer is part of Springer Science+Business Media (www.springer.com)

# Foreword

The Malaysia–Japan Academic Scholar Conference (MJASC) 2013 was organized by researchers and student committee members from the Look East Policy (LEP), which was convened successfully at Meiji University. The LEP was announced by Malaysia's fourth prime minister, Tun Dr. Mahathir Mohamad, on 8 February 1982, and to date more than 15,000 Malaysians have studied or trained at more than 60 universities in Japan. The most important aspect of the Look East Policy is development of a value system, a work culture and ethic that are compatible globally. This is the key to our both nation's success. Malaysia also has a strong bilateral relationship with Japan, and this significant event for LEP was marked by gathering students and policy makers as well as their Japanese counterparts at MJASC 2013, where papers in various fields of study and research were presented.

The first version of this event was held last year: the Malaysia–Japan Academic Scholar Seminar (MJASS) 2012. MJASS 2012 made a positive impact on the significance of the Look East Policy, especially on the relationship between Malaysia and Japan in terms of human capital and the development of both countries. MJASS 2012 also provided an opportunity for Malaysian and Japanese students to gather and gain knowledge from the keynote speeches, the panel discussions, and the oral and poster presentations.

The Malaysia–Japan Academic Scholar Seminar (MJASS) is now called the Malaysia–Japan Academic Scholar Conference (MJASC) as a more suitable name for an international conference. MJASC provides a unique platform for a diverse community of constituents involved in the Look East Policy as a medium to exchange knowledge and information about the latest research. The MJASC program includes keynote lectures, panel discussions, poster sessions, and research presentations—a good platform to promote the gathering and sharing of knowledge. Thus, MJASC will concentrate on being a major source of information that compiles the research and studies of Malaysian and Japanese students here in Japan.

Hiromi Naya
Former President of Meiji University

# Preface

We were honored to welcome the attendees of the Malaysia Japan Academic Scholar Conference (MJASC) 2013 which was held at the Surugadai Campus of Meiji University, Tokyo, Japan, November 8–9, 2013.

First and foremost, allow us to take this opportunity to express our gratitude to the committee members of MJASC 2013 for their dedication and hard work in the realization of the conference. We would also like to thank all Malaysian and Japanese academicians who shared their research work at MJASC 2013. Such contributions allowed them not only to share their research knowledge and expertise, but at the same time to create more opportunities for academicians in Malaysia and Japan to collaborate in the future.

We believe that this was the best time to hold a conference in the fast-growing field of technology research. The proceedings that came out of the conference will be valuable both to contributors and to readers who have common interests in nanoscience engineering, mechanical engineering, electrical and electronic engineering, computer science, information technology, and related areas. These proceedings are a source of research findings for Malaysia and Japan specifically and for other countries in general, among researchers, those in industry sectors, and government policy makers. They will serve as a useful resource–reference and platform to reflect the significance of the Look East Policy outcomes and products.

We are confident that these proceedings are of high quality among peer-reviewed publications. We would like to thank our colleagues on the editorial board, the first authors, and panel reviewers who have contributed to these conference proceedings, and the publisher Springer for their support.

Finally, we believe that these proceedings provide the reader with a great intellectual experience, thus providing the best platform for the future development and collaboration of Malaysia and Japan.

| | |
|---|---|
| Sarawak, Malaysia | Khairuddin Ab. Hamid |
| Tokyo, Japan | Osamu Ono |
| Tokyo, Japan | Anas Muhamad Bostamam |
| Tokyo, Japan | Amy Poh Ai Ling |
| December 2013 | |

# Contents

**Part I Science and Technology**

**The Investigation of Zinc-Rich Paint (ZRP) Behavior in NaCl Solution by Electrochemical Methods** . . . . . 3
Azizul Helmi Bin Sofian and Kazuhiko Noda

**An Investigation on Performance of 2D Simulated Fuel Cells Using Taguchi Method** . . . . . 9
Teck-Ming Ng and Nooryusmiza Yusoff

**Application of Correction Line in Floated Particle Size Distribution Measurement Using the Buoyancy Weighing-Pan Method** . . . . . 23
M.P. Khairunnisa, Y. Ohira, M. Shimadzu, and S. Yamanaka

**Fine Pitch Wire Bonding with Insulated Cu Wire: Free Air Ball and Ball Bond** . . . . . 35
Leong HungYang, Yap BoonKar, Navas Khan, Mohd Rusli Ibrahim, and L.C. Tan

**Effect of Red and Blue Lights on Photomorphogenesis in *Brassica chinensis*** . . . . . 49
Nurul Najwa Ani, Ahmad Nizar Harun, Saiful Farhan M. Samsuri, and Robiah Ahmad

**Quiet Standing of Time-Delay Stability with Derivative Feedback Control** . . . . . 59
Fitri Yakub, Ahmad Zahran Md. Khudzari, and Yasuchika Mori

**Social Determinants of Psychiatric Morbidity Among Malaysian Children** . . . . . 71
Ali Adli, YT Ooi, YL Tan Isaac, WY Ho Katrina, HM Chai, Seet Ruijia, KW Teh, and KH Wong

**Modelling Community Response for Ecological Significance Based on Laboratory Simulations of Variable Copper Exposure** . . . . . . . . . . 85
Khun C. Tan and Carolyn Oldham

**Part II Mechanical and Manufacturing Engineering**

**Application of Integrated Fuzzy-AHP for Design Concept Evaluation: A Case Study on Mold Design Selection** . . . . . . . . . . 101
Faiz Mohd Turan and Badrul Omar

**Numerical Modelling of Dry and Wet Steam Flows in Converging-Diverging Nozzles** . . . . . . . . . . 115
Hasril Hasini, Norhazwani Abd Malek, and Mohd. Zamri Yusoff

**How Quality Engineering in Japan Affects Global Product Performance** . . . . . . . . . . 125
Rozzeta Dolah and Zenichi Miyagi

**Developing a Heuristic Robotics Integrated Production Model for Economic Production Quantity in Mass Customization Manufacturing** . . . . . . . . . . 139
S.E.S. Bariran, K.S.M. Sahari, and A.H. Mahlouji

**Systematic Way for Problem Identification of Car Braking System by Using TRIZ** . . . . . . . . . . 147
N.F. Mesbah and S.M. Yusof

**Part III Social Sciences and Humanities**

**High-Performance Work System and Safety Performance** . . . . . . . . . . 157
Dayang Nailul Munna Abg Abdullah and Vanissa Karupaiah

**Determinant Factors of Insurance Demand by SMEs in Malaysia** . . . . . . . . . . 167
Junainah Junid, Siti Zahrah Buyong, and Khuzaimah Akasah

**The Perceptions of Malaysians in a Japanese Company: A Case Study in Shimano Components (Malaysia) Sdn. Bhd.** . . . . . . . . . . 177
Nor Fyadzillah Mohd Taha

**A Conceptual Model Towards Information Security Culture in Health Informatics** . . . . . . . . . . 187
Noor Hafizah Hassan and Zuraini Ismail

**Media Conversion of Paralinguistic and Nonlinguistic Speech Information into Animated Texts for Foreign Language Learning** . . . . . 197
Nur Syafikah Binti Samsudin and Kazunori Mano

**Analysis of Pancreatic Tissue Injury Caused by Ultrasonically Activated Devices** . . . . . 205
Bik Ee Lau, Tao Gao,, and Hideki Hayashi

**The Integration of the Kano Model and SERVQUAL into the House of Quality for Developing Occupational Safety and Health Training Program** . . . . . 213
Mohd Saiful Izwaan Saadon, Zainol Mustafa, and Khairul Za'im Kamarulzaman

**An Urgent Need for Policy Change: Reform in Curriculum and Practice Through Values-Based Pedagogies and Cooperative Learning in the Curriculum in Malaysia, Japan and Thailand** . . . . . 225
Suma Parahakaran

**Part IV Electronic and Computer Engineering**

**The Effects of Doping Bismuth on the Thermoelectric Properties of $Zn_4Sb_3$** . . . . . 237
Afiqa Binti Mohamad and Atsumi Hisao

**Study of Transmission and Reflection Characteristics of Microstrip Line During Application of Noise Suppression Sheet by Adhesive** . . . . . 243
Lim Yang Wei, Kyota Otsuka, and Takanobu Ohno

**An Improved VEPSO Algorithm for Multi-objective Optimisation Problems** . . . . . 253
Kian Sheng Lim, Salinda Buyamin, Anita Ahmad, Sophan Wahyudi Nawawi, Zuwairie Ibrahim, Faradila Naim, Kamarul Hawari Ghazali, and Norrima Mokhtar

**Electrical and Photoluminescence Study of Undoped $CuGaSe_2$ Single Crystal Thin Film** . . . . . 265
Sathiabama Thiru, Miki Fujita, Atsushi Kawaharazuka, and Yoshiji Horikoshi

**Challenges in Adopting Cloud Computing for Automation Process Controller in Manufacturing** . . . . . 271
Kean Wei Tan

**Face Recognition Human–Machine Comparison Under Heavy Lighting** . . . . . 283
Evgeny V. Mozgunov

**Design of Passive Mark for Mobile Robot Navigation Purpose** . . . . . 291
Khairul Salleh Mohamed Sahari and Kavinthran Devarajan

**Changes of Coastline: A Study Case of Carey Island-Morib Coast, Selangor, Malaysia** . . . . . 301
Nor Lita Fadilah Rameli and Mokhtar Jaafar

**Assessing the Condition of Buried Pipe Using Ground-Penetrating Radar (GPR)** . . . . . 311
S.W. Wahab, D.N. Chapman, C.D.F. Rogers, K.Y. Foo, S.W. Nawawi, and K.H. Abas

**Evaluation of Palm Decanter Cake as an Alternative Fish Feed** . . . . . 321
Ying Ping Chang and Jun Hoeng Chuah

**Human Head Pose Estimation and Its Application in Unmanned Aerial Vehicle Control** . . . . . 327
Chun Fui Liew and Takehisa Yairi

**Japan and US Smart Grid Effort? A Case Study** . . . . . 337
Amy Poh Ai Ling

**Threat to Individuals' Privacy in the Era of Social Networking: A Case Study of Facebook** . . . . . 345
Chai Yeun Hoe

**Part V Green Technology**

**Modelling of the Output Power from a Grid-Connected Photovoltaic System Using the Artificial Neural Network** . . . . . 355
Shahril Irwan Sulaiman, Sulaiman Shaari, and Ahmad Maliki Omar

**Basic Research on the $TiO_2$ Electrode and the Coloring Effect of Dye-Sensitized Solar Cell** . . . . . 363
Arini Nuran, Matsutake Daiki, Mohamad Norsyafiq, and Akira Fujiki

**An Experimental Study on Bioclimatic Design of Vertical Greenery Systems in the Tropical Climate** . . . . . 369
Badrulzaman Jaafar, Ismail Said, Mohd Nadzri Md Reba, and Mohd Hisham Rasidi

**The Adoption of Green Manufacturing Practices in Furniture Firms** . . . . 377
Puteri Fadzline Tamyez, Norzanah Mat Nor, and Syed Jamal Abdul Nasir

**Part VI Invited Papers**

**Institute for Research, Development and Innovation (IRDI) of the International Medical University (IMU), Malaysia** . . . . 387
Wan-Loy Chu, Eng-Lai Tan, Stephen Ambu, Chee-Onn Leong, Vishna Devi Nadarajah, Patricia Kim-Chooi Lim, Shew-Fung Wong, Geok-Lin Khor, James Michael Menke, and Joon-Wah Mak

**The Roles of the Institute of Nano Electronic Engineering (INEE), UniMAP in NCER: From Nanotechnology Perspectives** . . . . 397
U. Hashim, E.N. Elliazir, Jasni M. Ismail, A.R. Ruslinda, and M.K. Md. Arshad

**OER in Action: Case Study of Wawasan Open University** . . . . 409
Vighnarajah

**Lead Initiatives on Teaching and Learning in Malaysian Higher Education Institutions** . . . . 415
Rosna Awang-Hashim, Nena Padilla-Valdez, Norhafezah Yusof, and Tengku Faekah Tengku Ariffin

# MJASC 2013 Committee

## International Advisory Board Member

Y. Bhg. Datuk Ibrahim Ahmad
Director General,
Majlis Amanah Rakyat (MARA)

Y.Bhg. Dato' Mohd Tajudin Don
Director, Human Capital Development Division,
Public Service Department of Malaysia

Prof. Datuk Dr. Khairuddin Ab. Hamid
Professor,
Malaysia University of Sarawak.

Prof. Dr. Rugayah Mohamed,
Deputy Vice Chancellor,
Malaysia National Defense University

Prof. Dr. Osamu Ono,
Deputy Dean of Graduate School,
Meiji University, Japan.

**Chair :**

Dr. Amy Poh Ai Ling

**Vice Chair :**

Noor Afiza Razali

**Secretary :**

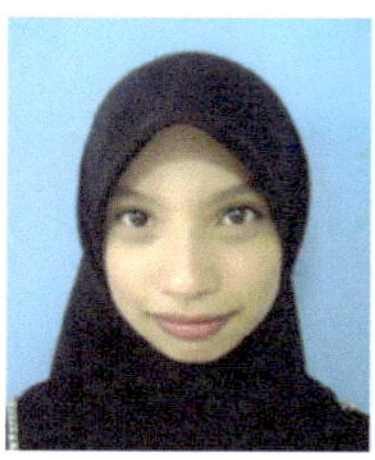

Nurul Adni Ahmad Ridzuan

**Treasurer :**

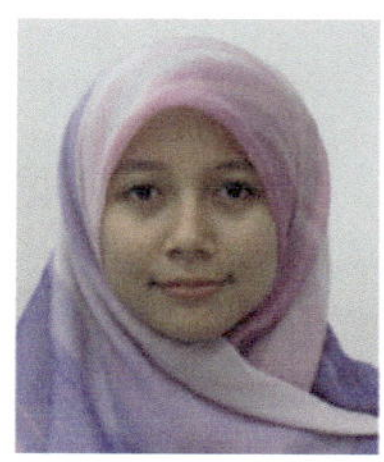

Nur Afiqah Adilla Binti Nawi

Aime Aqelah

**Lead Reviewer :**

Dr. Anas Muhamad Bostamam

**Publication :**

Nur Iqlimma Aishya Binti Alumi

Siti Maisarah binti Muhammad Yusop

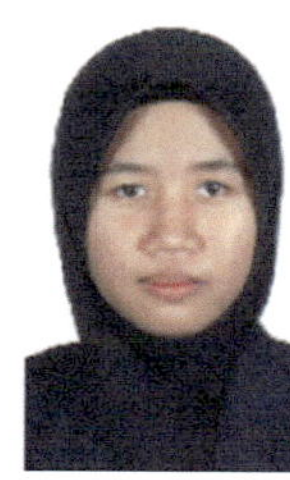

Nur Syafiqah Nadia binti Johari

Khairul khalil Ishak

Siti Nor Khuzaimah binti Amit

Crystal Ting Ching Ching

Jeffrey Tan

**Secretariat :**

Sarah Rashiqah Binti Abdul Rashid

Calvin Ong Kah Yee

Nurul Sheeda Suhaimi

**Protocol :**

Rozzeta bt. Dolah

Nor Fariza Afifi bt Wahab

Sathia Bama

Siti Norhidayah binti Che Mohd Yusoff

Dr. Haslina Rani

**Publicity :**

Abul Hasan Johari B Ahmad Tajudin

Ahmad Muzammir bin Muhammad

Sarah Rashiqah Binti Abdul Rashid

Akmal bin Ahmed

**IT :**

Muhammad Khairi Faiz Ahmad Hairuddin

Omar Shafiq Bin Suhaimi

**Program Book :**

Azizul Helmi bin Sofian

**Program Book :**

Abdul Rifqi Al-Abqary bin Abdul Walid

# Part I
# Science and Technology

# The Investigation of Zinc-Rich Paint (ZRP) Behavior in NaCl Solution by Electrochemical Methods

**Azizul Helmi Bin Sofian and Kazuhiko Noda**

## 1 Introduction

Corrosion is a serious problem that industries around the world are facing every day. Protection of metals from corrosion is a serious concern. Metals are protected from corrosion not just to reduce the financial cost but also to prevent serious injuries and several fatalities. As a result, corrosion protection is important and its significance can be distinguished in three areas. The first area of implication is economics, comprising of the objective of reducing material losses; the second area of implication is the revision life and safety of operations while the third is the management of the metal resources [1, 2].

Zinc primers or paints have been used in various industries for over half a century. These coatings can be considered to be practical coatings to virtually any environment demanding long-term corrosion protection [3, 4]. In a comparative evaluation of inorganic or organic zinc coating, it was established that only metallic zinc is capable of providing galvanic protection [5].

Zinc-rich paints can be used as a single-coat primer for temporary protection or as a primer in a multilayer system for long-term protection. The protective mechanism of single-coat zinc-rich paint has been studied for zinc-rich epoxy polyamide and zinc-rich ethyl silicate paints [6]. The life span and residual life of the zinc-rich primer coatings were evaluated with electrical measurement and are shown by [7].

A.H.B. Sofian (✉)
Shibaura Institute of Technology, 3-7-5 Toyosu, Koto-ku, Tokyo 135-8548, Japan

Universiti Malaysia Pahang, Lebuhraya Tun Razak, Kuantan, Pahang 26300, Malaysia
e-mail: na11102@shibaura-it.ac.jp

K. Noda
Shibaura Institute of Technology, 3-7-5 Toyosu, Koto-ku, Tokyo 135-8548, Japan

K. Ab. Hamid et al. (eds.), *The Malaysia-Japan Model on Technology Partnership: International Proceedings 2013 of Malaysia-Japan Academic Scholar Conference*,
DOI 10.1007/978-4-431-54439-5_1, © Springer Japan 2015

# 2 Experimental

## 2.1 *The Preparation of Samples*

The test specimens were pure iron (Nilaco Corporation, Japan) with purity of 99.5 % with 10 × 20 × 0.5 mm dimensions were used as a metallic substrate. The metal substrates were polished with emery papers till #1000. Abraded metal substrates were then cleansed with ethanol using ultrasonic bath and dried for 15 min prior to spray application. Commercial ZRPs with zinc content of 74 % and 96 % (Atomic Paint Co. Ltd., Japan) were applied onto the metal substrate as per manufacturer's instructions and dried at ambient temperature for 1–2 h under controlled condition in a laboratory. The metal substrates were coated with 1, 3, and 5 layers respectively. All edges and rear sides of the painting area were sealed with resin to avoid corrosion of the unprotected area of metallic substrates during immersion in salt solution. These metal substrates were cured according to the resin supplier before the tests.

## 2.2 *Potentiodynamic Polarization Measurement*

Potentiodynamic polarization measurements were carried out for coated substrate using the FRA5022 model. Measurements were carried out by applying a potential of 1.0 V in both positive and negative direction. 0.5 M NaCl was the electrolyte medium and platinum electrode and Ag/AgCl as counter and reference electrodes, respectively. Electrochemical measurements were carried out on 20 $mm^2$ area. All experiments were conducted at ambient temperature. The electrolyte solution was made from 0.5, 1.0 M NaCl () and bi-distilled water. The potentiodynamic current-potential curves were obtained by changing the electrode potential automatically from −1.0 to 1.0 V at a scan rate of 1.0 $mVs^{-1}$. EIS measurements were carried out in a frequency range from 100 kHz to 10 mHz with an amplitude of 5 mV peak to peak using ac signals at open circuit potential.

# 3 Results and Discussion

## 3.1 *Polarization Measurements*

Figure 1 presents typical cathodic and anodic polarization curves recorded for metallic substrates with different thicknesses in 0.5 M NaCl solution at a scan rate of 1.0 $mVs^{-1}$ at 25°C. It is obvious from Fig. 1 that the corrosion potential for all coated samples is more positive than that marked in the case of 96 % ZRP-5.

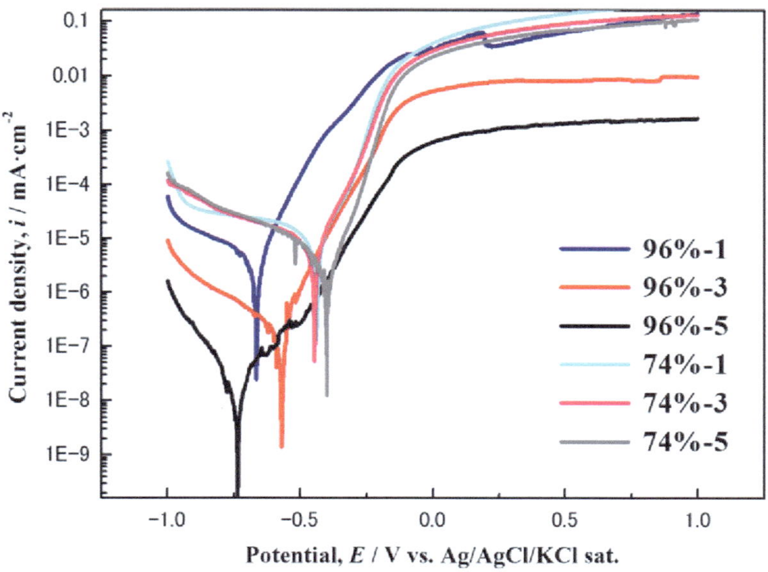


**Fig. 1** Polarization cathodic curves of coated metallic substrate with ZRP in 0.5 M NaCl solution

It is observed that the corrosion current density, $I_{corr}$, values of ZRP coatings gradually decrease as the coating thickness increases up to 5 layers and the 96 % ZRP-5 has the lowest value of $I_{corr}$.

The samples coated with 96 % ZRP-5 have the most negative $E_{corr}$ value with respect to the zinc contents. In this case, 74 % ZRP-5 shows a corrosion potential of −0.45 V versus Ag/AgCl which indicates that the 74 % ZRP-5 is noble toward the electrochemical corrosion with good barrier effect instead of galvanic protection compared to the 96 % ZRP coated sample with $E_{corr}$ value of −0.76 V versus Ag/AgCl.

### *3.2 Electrochemical Impedance Spectroscopy EIS*

The impedance behavior of coated metallic substrates in 0.5 M NaCl solution with various thicknesses of two different zinc concentrations, 74 % and 96 % ZRP, was studied. Measurements were conducted at the respective corrosion potentials at 25°C, and results obtained are depicted in Fig. 2a. The recorded impedance spectra are characterized by two relaxation times. The first one corresponded to a high-frequency (HF) capacitive loop depressed and deformed on its left area, which is typical of a charge transfer, while the second one could be explained as an inductive loop in the capacitive plane. When the thickness of the coating increased, a new loop is weakly initiated and becomes well defined at 96 % ZRP-3 layer coating. Samples coated with 96 % ZRP-1 show the lowest impedance resistance which

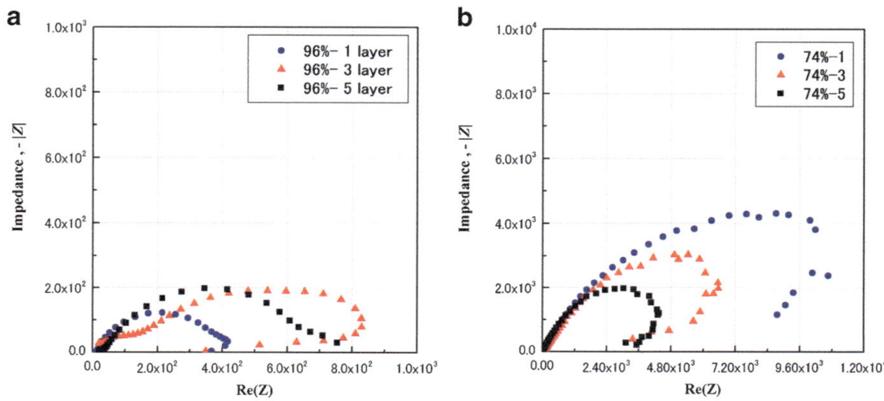


**Fig. 2** EIS Nyquist (**a**) spectra of 96 % ZRP and (**b**) 74 % ZRP with various thicknesses in 0.5 M NaCl solution

indicated that even with less coating, galvanic protection can be achieved, while 96 % ZRP-3 and ZRP-5 layers may provide better protection, not just galvanic protection but also barrier effect. The presence of thicker coating may facilitate a strong affinity of zinc particle toward the metallic substrate. Therefore, a metallic substrate with higher coating thickness shows a higher impedance resistance where barrier effect comes first and then the galvanic protection takes place.

However, one has to consider that metallic substrates coated with 74 % ZRP (Fig. 2b) brought a shift of impedance resistance to 9.6 KΩ, a bigger digit compared to samples coated with 96 % ZRP. Therefore, the influence of zinc contents of ZRP on the number of coating thickness observed on an EIS diagram was also studied at 0.5 M. The appearance of two capacitive loops can be observed from the Nyquist plot. The increase in the size of the capacitive loop with the deduction of coating thickness shows that 74 % ZRP provides barrier effect on the metallic substrate. The barrier is probably related to epoxy from the paint which acts as a barrier surface on the metallic substrate. According to Cachet et al. [8] the deformation of the left part of the HF capacitive loop is associated to the existence of pores on the surface of the coated metallic substrate with the substrate surface being inactive.

## *3.3 Conclusions*

By employing the EIS technique with polarization measurements, it was shown that superior galvanic protection efficiency was achieved on metallic substrate coated with 96 % ZRP. However, 74 % ZRP also exhibits very good corrosion protection effect which comes from the barrier protection of epoxy resin in paint formulation. Low concentrations of NaCl (<1 M) solution were found not to be adequate mean to initiate the weakest process to confirm the porous state of coated metallic substrate. The above measurements of coated metallic substrate with ZRP with crosscut scribe confirmed the superior healing or corrosion protection when zinc content of the coating is 96 % ZRP-3,5 layers. Galvanic protection competency of ZRP depends not only on the quantity of zinc particles used in the formulation but also on the affinity of zinc particles on metallic substrate. However, the zinc coatings were shown to provide adequate corrosion protection to the metallic substrate in NaCl solution and the effect of pores of the size was not particularly marked.

## References

1. Kruba L, Stucker P, Schuster T (2005) Less metal, more protection. Eur Coat J 10:38–43
2. Eliasson J (2003) Prevention and management of marine corrosion. In: Intertanko presentation, Lloyd's list events conference (economics of coatings/corrosion protection of ships selecting the correct type of anticorrosion protection for underwater application on new buildings) April 2 and 3, 2003, London
3. Brodd RJ, Leger VE, Bard AJ (eds) (1976) Encyclopedia of electrochemistry of the elements, vol 6. Marcel Decker, New York, p 35
4. Munger CG, Vincent LD (1999) Corrosion prevention by protective coatings, 2nd edn. NACE, USA
5. Karakasch N (2001) Zinc coating review-2000 and beyond. Corrosion Manage 10(1):3
6. Armas RA, Real SG, Gervasi C, Di Sarli AR, Vilche JR (1992) Zinc-rich paints on steels in artificial seawater by electrochemical impedance spectroscopy. Corrosion 48:379
7. Jagtap RN, Nambiar R, Hassan SZ, Malsche VC (2007) Predictive power for life and residual life of the zinc rich primer coatings with electrical measurement. Prog Organic Coat 58:253–258
8. Cachet C, Saidani B, Wiart R (1992) The behavior of zinc electrode in alkaline electrolytes II. A kinetic analysis of anodic dissolution. J Electrochem Soc 139:644–654

# An Investigation on Performance of 2D Simulated Fuel Cells Using Taguchi Method

**Teck-Ming Ng and Nooryusmiza Yusoff**

## 1 Introduction

The fuel cell is a device that converts the chemical energy from a fuel into electricity through a redox reaction with oxygen or other oxidizing agent. Hydrogen is the most common fuel; natural gas and alcohols such as methanol are frequently used. Fuel cells are considered as green, reliable, and highly efficient power generation technology in this century. Among various types of fuel cells, proton exchange membrane fuel cells (PEMFCs) and direct methanol fuel cells (DMFCs) are commonly used because of low weight, high current density, easy design, and low $CO_2$ emission. These are the most promising power sources for transportable electronic devices and transportation applications [1].

Fuel cells are used for primary and backup power for commercial, industrial, and residential buildings [2]. The energy efficiency of a fuel cell is generally between 40 and 60 % or up to 85 % if waste heat is captured for use. It is much higher than a combustion engine, which is usually 30 % efficient.

Lately, there has been a growing concern about acid rain and greenhouse gas releases which have made renewable technologies an attractive option [3]. Continuing work to meet increasing energy demand and also to preserve the global environment, the development of fuel cell systems with readily available fuels, high efficiency, and minimal environmental impact is urgently required. But the problem rising now is how to maximize the performance in terms of current density (A/m$^2$) produced by PEMFC and DMFC.

In fact, it can be maximized by manipulating operating parameters and design system variables simultaneously. These variables/parameters can be classified into two which are processed variable and design variable in fuel cells. Process variable

T.-M. Ng (✉) • N. Yusoff
Department of Chemical Engineering, Universiti Teknologi PETRONAS, Bandar Seri Iskandar, 31750 Tronoh, Perak, Malaysia
e-mail: teckming.ng@petronas.com.my

K. Ab. Hamid et al. (eds.), *The Malaysia-Japan Model on Technology Partnership: International Proceedings 2013 of Malaysia-Japan Academic Scholar Conference*, DOI 10.1007/978-4-431-54439-5_2, © Springer Japan 2015

includes the direction flow of reactants, pressure and velocity of reactants' flow, operating temperature, structure design, and so on, while design variable includes channel length and width, type and porosity of membrane, and other design variations of a fuel cell as well. Hence, PEMFC and DMFC were simulated in this work to identify the best operating condition to deliver the best performance of a fuel cell system.

The objective of this work is to maximize the current density of a fuel cell by investigating the effects of fuel and oxidant concentration, membrane properties, and direction flow of reactants in fuel cells by using Taguchi method.

Taguchi method for design of experiment (DOE), analysis of mean (ANOM), and analysis of variance (ANOVA) is applied to measure the significance of 4 factors, to determine the optimum configurations of 4 factors, and to estimate and validate optimal current density. Taguchi method has been effectively applied in many engineering disciplines. There are four parameters that need to be investigated with two levels in this work. These parameters include fuel concentration at anode channel, oxidant concentration at cathode channel, directional flow of reactants, and membrane properties. So, $L_8$ of Taguchi method has been chosen in this work. Initially, there are $2^4$ which are 16 experiments needed to be carried out for each fuel cell. But by using Taguchi method, the number of experiments for each fuel cell is 8. So the amount of experimentation can be reduced to half; thus, it saves time and resource at all.

## 2 Literature Review

There are numerous categories of fuel cells, but they all comprise of an anode (negative side), a cathode (positive side), and an electrolyte that permits electron to travel between the two sides of the fuel cell [4]. A catalyst oxidizes the fuel at anode, usually hydrogen or other hydrogen compound, converting fuel into a positively charged ion and a negatively charged electron. The electrolyte is a substance specifically designed so ions can pass through it, but the electrons cannot. Electrons are drawn from the anode to the cathode through an external circuit, producing direct current electricity. Chemical redox reactions occur at the interfaces of the three different sectors. The net result of the reactions is that fuel is consumed, water or carbon dioxide (DMFC) is formed, and an electric current is produced, which can be used to power electrical devices as indicated in Fig. 1.

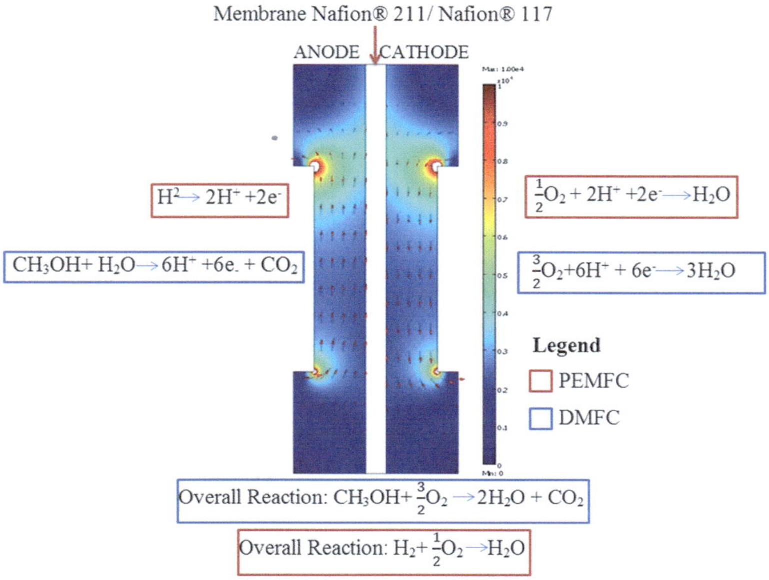


**Fig. 1** Redox reaction with stoichiometry equations

# 3 Modeling and Methodology

## *3.1 Proton and Electron Transport Balance*

A conductive media DC application mode describes the potential distributions in the three subdomains, as shown in Fig. 2 by using the following equations (a = anode, m for membrane, and c = cathode):

$$\nabla \cdot (\kappa_{s,\mathrm{eff}} \nabla \varphi_s) = 0 \ \text{ in } \ \Omega a \tag{1}$$

$$\nabla \cdot (\kappa_{m,\mathrm{eff}} \nabla \varphi_m) = 0 \ \text{ in } \ \Omega m \tag{2}$$

$$\nabla \cdot (\kappa_{s,\mathrm{eff}} \nabla \varphi_s) = 0 \ \text{ in } \ \Omega c \tag{3}$$

Charge transfer current density expression is generally described by using the Butler–Volmer electrochemical kinetic expression as a boundary condition. For the electrolyte potential equation, this results in a condition where the inward normal ionic current densities at the anode and cathode boundaries, $i_a$ and $i_c$, are specified as in below:

$$i_e = L_{\mathrm{act}}(1 - \varepsilon_{\mathrm{mac}}) j_{\mathrm{agg},e} \tag{4}$$

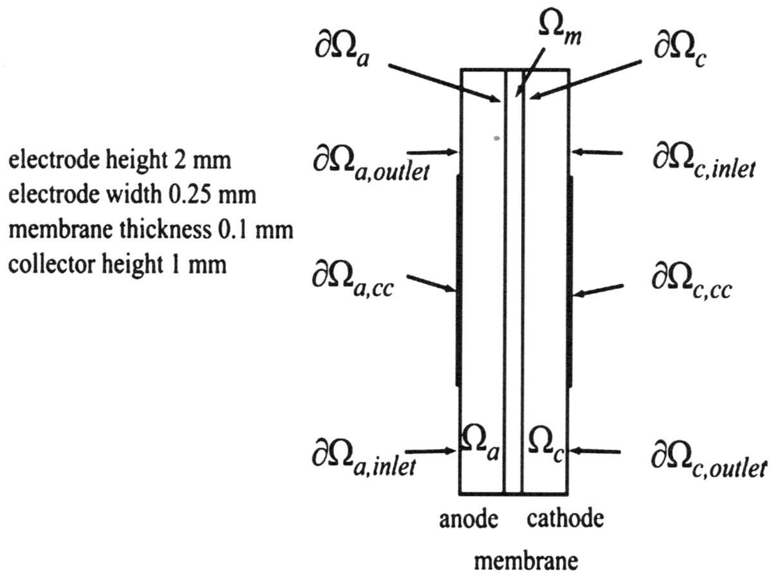


**Fig. 2** Geometry models with subdomain labels

## 3.2 *Modeling for Anode and Cathode*

The current density can be expressed analytically by solving a combination of the Maxwell–Stefan diffusion equation and the Butler–Volmer electrode kinetic equation for agglomerate with constant electric and ionic potentials. The resulting equations for the current density in the anode and cathode are [6]

$$j_{\mathrm{agg,e}} = -6n_{\mathrm{e}}F\left(\frac{D_{\mathrm{agg}}}{R_{\mathrm{agg}}^{2}}\right)(1-\lambda_{\mathrm{e}}\coth\lambda_{\mathrm{e}})\beta_{\mathrm{e}} \tag{5}$$

The dissolved $H_2$ and $O_2$ at the surface of the agglomerates are related to the molar fractions of the respective species in the gas phase through Henry's law. The concentration of the species is partial pressure of the species divided by Henry's constant. Henry's constant for oxygen is 769.23 L/atm./mol, while 1,282.05 L/atm./mol is for hydrogen:

$$C_{\mathrm{agg,H_2}} = \frac{p_{\mathrm{H}}x_{\mathrm{H}}}{K_{\mathrm{H}}} \tag{6}$$

$$C_{\mathrm{agg,O_2}} = \frac{p_{\mathrm{O}}x_{\mathrm{O}}}{K_{\mathrm{O}}} \tag{7}$$

The potential difference between the cathode and anode current collectors corresponds to the total cell voltage. The total cell voltage as the boundary condition at the cathode current collector is as follows:

$$\varphi_s = 0 \text{ in } \partial\Omega_{a,cc} \text{ (current collector at anode)} \tag{8}$$

$$\varphi_s = V_{cell} \text{ in } \partial\Omega_{c,cc} \text{(current collector at cathode)} \tag{9}$$

## 3.3 Porous Media Transport Balance

Darcy's law was being applied to specify the fluid flow through a porous medium. The gas velocity is given by the continuity equation as below:

$$\nabla \cdot (\rho u) = 0 \text{ in } \Omega_a \text{ and } \Omega_c \tag{10}$$

Darcy's law for porous media states that the gradient of the pressure, the viscosity of the fluid, and the structure of the porous media determine the velocity:

$$u = kp/\eta \cdot \nabla p \tag{11}$$

Combined with these boundary conditions, Darcy's law determines the gas flow velocity and preserves the total mass conservation in the anode and cathode gas backing. The main assumptions used in the modeling are as follows [5]:

1. Ignore the formations of the $CO_2$ bubbles and water vapor.
2. Membranes are fully hydrated.
3. Methanol in DMFC is fully consumed at the interface of the cathode membrane and the cathode catalyst layer.
4. The flow in the electrolyte channel is laminar flow.
5. The fuel cell is isothermal and operates at steady state.

## 3.4 Design of Experiment

A symmetric 2D simulation flow diagram is shown in Fig. 3. In this work, 2D models were simulated in COMSOL Multiphysics version 3.5a. First of all, fuel cell geometry has been drawn in COMSOL. A set of equations involved in this 2D simulation has been chosen such as Darcy's law (fluid flow through a porous medium), Maxwell–Stefan mass transport equation (diffusion in multicomponent systems), Henry's law (at constant temperature, the amount of gas that dissolves in a liquid is directly proportional to the partial pressure of the gas), and Butler–Volmer kinetics electrochemical kinetic expression as a boundary condition. Next, constants, domain equations, and boundary conditions involved in this simulation were specified in Fig. 2. This model consists of 3 domains which are an anode (Ωa), a membrane (Ωm), and a

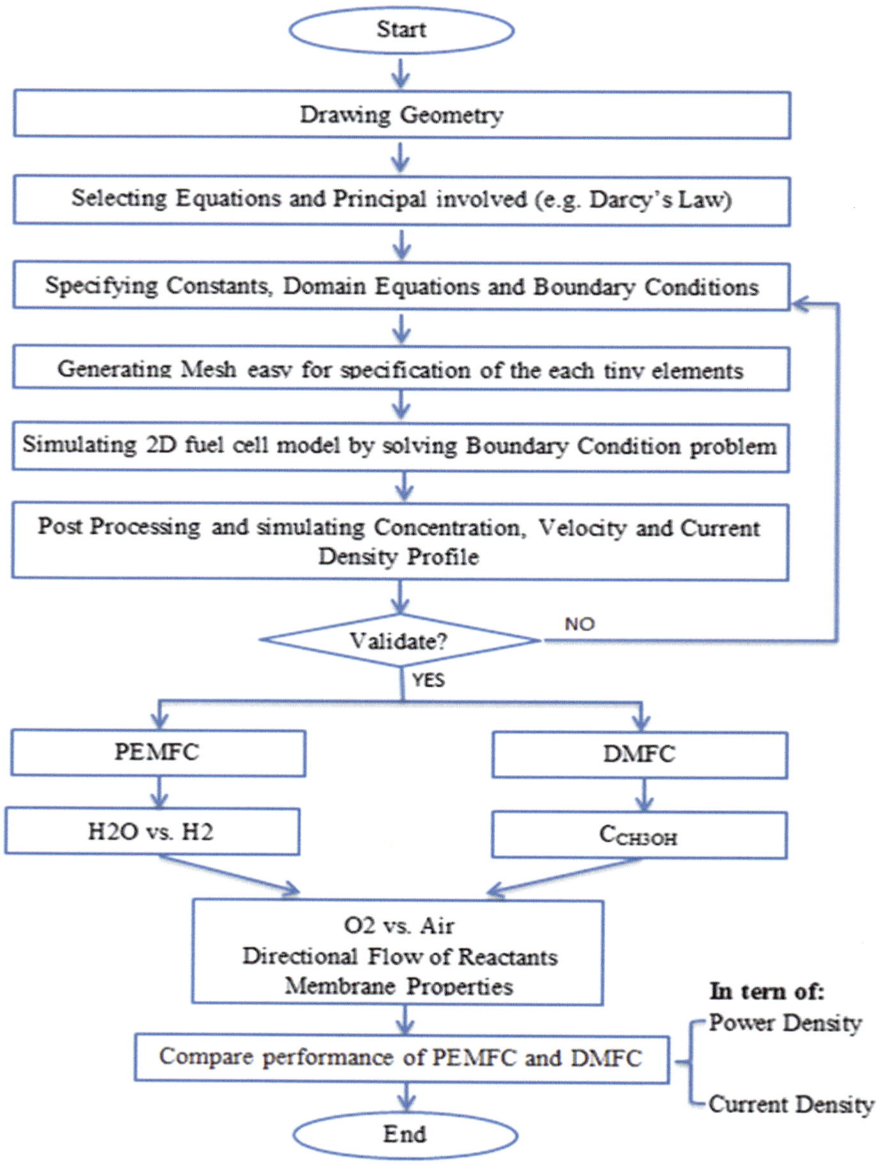


**Fig. 3** Methodology for 2D fuel cell simulation

cathode (Ωc). Each of the porous electrodes is in contact with a gas distributor, which has an inlet channel (∂Ωa, inlet), a current collector (∂Ωa, cc), and an outlet channel (∂Ωa, outlet). The same notation is used for the cathode side.

A mesh model has been generated easily for the specification of each tiny element in this mesh. In this work, the maximum element size for membrane is 50 μm and for both anode and cathode is 10 μm.

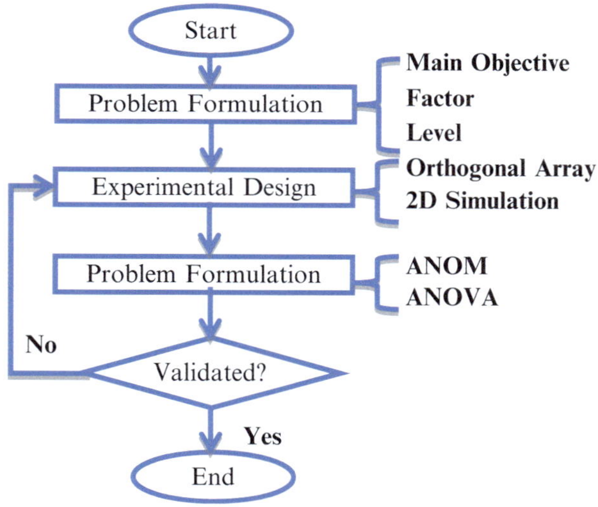


**Fig. 4** Design of experiments by using Taguchi method

In this work, PEMFC and DMFC models have been simulated and validated by experimental data. It means that these two models were validated and can work in real life. If these two fuel cell models were not validated, the developer has to go back to the early stage, whereby specifying constants, domain equations, and boundary conditions over again. The next step was running some case studies by manipulating some parameters on fuel cells.

The implementation of Taguchi method can be illustrated as shown in Fig. 4. At the top of that is the formulation of a problem which requires defining a main objective. In this work, the main objective is to maximize the performance of PEMFC and DMFC by using Taguchi method. Furthermore, it consists of factors and levels of experiment. Controllable factors A to D were set up in an $L_8$ $(2^4)$ orthogonal array. The next step is designing experiments by using orthogonal array. There are four controllable factors at two levels. An $L_8$ array which consists of eight rows and four columns was chosen. The row is representing simulation run, and column represents number of factors in this work. It should be noted that in order to take enough response of all four variables toward object function, only eight simulation runs are needed to be done for each fuel cell.

Step 3 is analyzing the results of simulation. Table 1 shows Taguchi orthogonal array used in this simulation. In this work, there were two statistical tools that have been used which were ANOM and ANOVA. ANOM is an analysis of mean which compares means and variances across several groups of result data, while analysis of variance, ANOVA, is a particular form of statistical hypothesis testing heavily used in the analysis of experimental data. A statistical hypothesis test is a method of making decisions using data. Finally, step four is a validation of experimental result. For each run in Lg orthogonal array, it will yield the highest current density of the particular fuel cell. A preliminary visualization of trend of each factor's average contribution at all 2 levels can be made through response plot. Response plot is used to identify the optimal design configuration for validating the result obtained from simulation.

**Table 1** Taguchi orthogonal array

| | Factor | | | |
|---|---|---|---|---|
| Experiment | A | B | C | D |
| 1 | 1 | 1 | 1 | 1 |
| 2 | 1 | 1 | 2 | 2 |
| 3 | 1 | 2 | 2 | 1 |
| 4 | 1 | 2 | 1 | 2 |
| 5 | 2 | 1 | 2 | 1 |
| 6 | 2 | 1 | 1 | 2 |
| 7 | 2 | 2 | 1 | 1 |
| 8 | 2 | 2 | 2 | 2 |

# 4 Results and Discussions

## *4.1 Data Gathering and Analysis for Fuel Cells*

From Fig. 5, PEMFC has better performance (max 3,600 $A/m^2$) than DMFC (max 3,025 $A/m^2$) whereby the current density difference $\Delta J_e$ is 575 $A/m^2$ in optimum operating configuration $\mathbf{A_2B_2C_2D_2}$.

## *4.2 Analysis of Mean (ANOM)*

The analysis of mean (ANOM) is a graphical method for comparing a group data by means to define if any one of the data varies significantly from the overall mean. ANOM is a type of multiple comparison method. For PEMFC, factor B ranks number 1, while factor A ranks number 1 for DMFC. It means that factor B is an important factor for PEMFC and factor A is an important factor for DMFC as shown in Table 2 below.

## *4.3 Analysis of Variance (ANOVA)*

In statistics, analysis of variance (ANOVA) is a collection of statistical data and their associated method, in which the variance in a particular manipulated variable is separated into components attributable to different variations.

Factors are assigned to experimental units by a combination of randomization and blocking to ensure the validity of the results by using Taguchi orthogonal array. Factor B (oxidant concentration) affects the most in PEMFC performance, while factor A (methanol concentration) gives the most impact on DMFC performance. From ANOM plot, $\mathbf{A_2B_2C_2D_2}$ is the best configuration for PEMFC and DMFC

| Experiment | A | B | C | D | PEMFC Current Density (A/m$^2$) | PEMFC Power Density (W/m$^2$) | DMFC Current Density (A/m$^2$) | DMFC Power Density (W/m$^2$) |
|---|---|---|---|---|---|---|---|---|
| 1 | 1 | 1 | 1 | 1 | 2800 | 1960 | 1263 | 884 |
| 2 | 1 | 1 | 2 | 2 | 3420 | 2394 | 1650 | 1150 |
| 3 | 1 | 2 | 2 | 1 | 3430 | 2401 | 1271 | 890 |
| 4 | 1 | 2 | 1 | 2 | 3325 | 2376 | 1635 | 1145 |
| 5 | 2 | 1 | 2 | 1 | 2636 | 1845 | 2300 | 1610 |
| 6 | 2 | 1 | 1 | 2 | 3036 | 2125 | 2750 | 1925 |
| 7 | 2 | 2 | 1 | 1 | 3050 | 2135 | 2360 | 1652 |
| 8 | 2 | 2 | 2 | 2 | 3600 | 2520 | 3025 | 2117.5 |

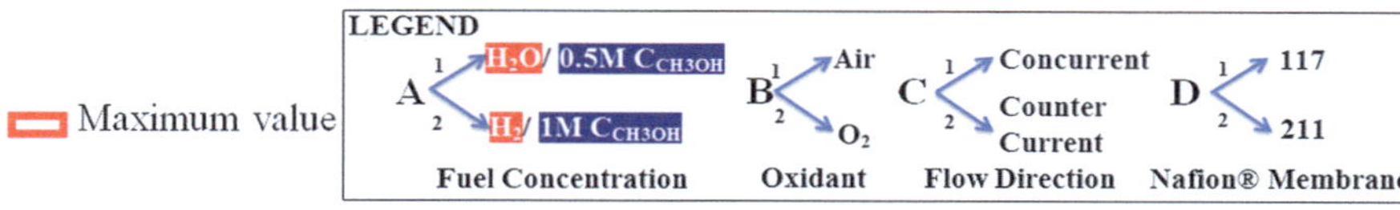


**Fig. 5** Simulation result in Taguchi table

**Table 2** ANOM table for PEMFC and DMFC

| | PEMFC | | | | DMFC | | | |
|---|---|---|---|---|---|---|---|---|
| Level | A | B | C | D | A | B | C | D |
| 1 (A/m$^2$) | 3,080.50 | 2,973 | 3,052.75 | 2,979 | 1,454.75 | 1,990.75 | 2,002 | 1,798.5 |
| 2 (A/m$^2$) | 3,243.75 | 3,351.25 | 3,271.50 | 3,345.25 | 2,608.75 | 2,072.75 | 2,061.5 | 2,265 |
| Mean | 3,162.13 | 3,162.13 | 3,162.13 | 3,162.13 | 2,031.75 | 2,031.75 | 2,031.75 | 2,031.75 |
| Effect | 163.25 | 378.25 | 218.75 | 366.25 | 1,154 | 82 | 59.5 | 466.50 |
| Rank | 4 | 1 | 3 | 2 | 1 | 3 | 4 | 2 |

as shown in Table 3. It gives the maximum current density for both these fuel cells. Higher fuel and oxidant concentration accelerate redox reaction in fuel cells, counter-current flow of reactants maximizes the transfer rate of heat and mass, and by using Nafion® 211 membrane, there is more hydrogen ion that can pass through the membrane to combine with oxygen to form water and produce electricity.

## 4.4 *Validation of Taguchi Method*

For each case study in this project, there are eight combinations of experimental runs for each fuel cell. By simulation design, only one will yield the highest current density. Initial visualization of trends of each factor contribution at all levels is possible through a main effect plot. Here, average values of factor k $\bar{x}_{kl}$ are plotted against all 2 levels in this project. The effect plot may be used to detect optimum design configuration for the purpose of verifying the results. An additional experiment is run to compare both experimental and calculated outputs (current density).

**Table 3** ANOVA table for PEMFC and DMFC

| | PEMFC | | | | DMFC | | | |
|---|---|---|---|---|---|---|---|---|
| Level | **A** | **B** | **C** | **D** | **A** | **B** | **C** | **D** |
| 1 | 6,662.64 | 35,768.27 | 11,962.89 | 33,534.77 | 332,929 | 1,681.00 | 885.06 | 54,405.56 |
| 2 | 6,662.64 | 35,768.27 | 11,962.89 | 33,534.77 | 332,929 | 1,681.00 | 885.06 | 54,405.56 |
| Sum | 13,325.28 | 71,536.53 | 23,925.78 | 67,069.53 | 665,858 | 3,362.00 | 1,770.13 | 108,811.13 |
| Sum of square | 106,602.3 | 572,292.3 | 191,406.25 | 536,556.3 | 5,326,864 | 26,896 | 14,161 | 870,489 |
| DOF | 1 | 1 | 1 | 1 | 1 | 1 | 1 | 1 |
| Variance of factor 1 | 106,602.3 | 572,292.3 | 191,406.25 | 536,556.3 | 5,326,864 | 26,896 | 14,161 | 870,489 |
| % Contribution | 7.577334 | 40.67878 | 13.6052385 | 38.13865 | 85.38817 | 0.431135 | 0.23 | 13.95 |
| Rank | 4 | 1 | 3 | 2 | 1 | 3 | 4 | 2 |

The calculated optimum current density $x_{\mathrm{opt}}$ is obtained as shown in Table 4 by summing up global mean $\overline{x}$ with maximum deviations of average of 4 factors over all 2 levels, $\overline{x}_{\mathrm{kl}}$, from its corresponding average $\overline{x}_{\mathrm{k}}$:

$$x_{\mathrm{opt}} = \overline{x} + \left[\sum\nolimits_{k=1}^{K} \max(\overline{x}_{kl}) - x_k\right]; \ \forall l = 1, \ldots, L, \tag{12}$$

where

$$\overline{x}_k = \overline{x}; \ \forall k = 1, \ldots, K \tag{13}$$

## 5 Conclusion

In conclusion, the main objectives of the project have been successfully achieved. These models enable us to view in two-dimensional and study the effects of fuel and oxidant concentration, reactants' flow direction, and membrane properties over the full range of operating current densities and performance. There is a good agreement between the results of the 2D simulated models and the experimental data in the validation section in Sect. 4. In this work, oxidant concentration in the cathode channel has much influence on current and power density of PEMFC, while fuel concentration in anode channel is the main effect that affects DMFC's performance.

## Nomenclatures

| | |
|---|---|
| $c_{\mathrm{i,\ agg}}$ | Concentrations of agglomerate surface (mol/m$^3$) |
| $c_{\mathrm{i,\ ref}}$ | Reference concentrations (mol/m$^3$) |
| $D_{\mathrm{agg}}$ | Agglomerate gas diffusivity (m$^2$/s) |
| $D_{\mathrm{ij}}$ | Maxwell–Stefan diffusivity matrix (m$^2$/s) |
| $E$ | "a" (anode) or "c" (cathode) |
| $F$ | Faraday's constant (C/mol) |
| $i_{\mathrm{0a/c}}$ | Exchange current densities (A/m$^2$) |
| $j_{\mathrm{agg,\ a/c}}$ | Current densities in agglomerate model |
| $K$ | Henry's constant (Pa·m$^3$/mol) |
| $kp$ | Electrode's permeability (m$^2$) |
| $L_{\mathrm{act}}$ | Active layer's thickness (m) |
| M | Concentration (mol/L) |
| $n_e$ | "Charge transfer" number (1 is anode, 2 is cathode) |
| $p$ | Pressure (Pa) |
| $R$ | Gas constant |
| $R_{\mathrm{agg}}$ | Agglomerate radius (m) |
| $S$ | Specific area of the catalyst (m) |

**Table 4** Validation table for PEMFC and DMFC

| | PEMFC | | | | DMFC | | | |
|---|---|---|---|---|---|---|---|---|
| Level | **A** | **B** | **C** | **D** | **A** | **B** | **C** | **D** |
| Maximum | 3,243.75 | 3,351.25 | 3,271.5 | 3,345.25 | 2,608.75 | 2,072.75 | 2,061.5 | 2,265 |
| Mean | 3,162.125 | 3,162.125 | 3,162.125 | 3,162.125 | 2,031.75 | 2,031.75 | 2,031.75 | 2,031.75 |
| Max–Mean | 81.625 | 189.125 | 109.375 | 183.125 | 577 | 41 | 29.75 | 233.25 |
| $X_{opt}$ | 3,725.375 | | | | 2,912.75 | | | |

% Difference of optimum current density (if <20 %, acceptable) PEMFC $= \frac{3,725.375-3,600}{3,725.375} \times 100\% = 3.365\%$ DMFC $= \frac{3,025-2,912.75}{2,912.75} \times 100\% = 3.854\%$
PEMFC and DMFC are validated and acceptable by using $\mathbf{A_2B_2C_2D_2}$ configuration in Taguchi method

| | |
|---|---|
| $T$ | Temperature (K) |
| $\mathbf{u}$ | Velocity (m/s) |
| $\mathbf{u}$ | Gas velocity (m/s) |
| w | Weight fraction |
| x | Mole fraction of the species |
| $\varepsilon_{mac}$ | Porosity (the macroscopic porosity) |
| $\eta$ | Gas viscosity (Pa/s) |
| $\kappa_{m}$, $\kappa_{m,\ eff}$ | Membrane ionic conductivity (S/m) |
| $\kappa_{s,\ eff}$ | Effective electronic conductivity (S/m) |
| $\rho$ | Density of the gas phase (kg/m$^3$) |
| $\varphi_m$ | Potential (V) in the membrane phase |
| $\varphi_s$ | Potential (V) in the solid phases |

## References

1. Smirnov A, Burt A, Celik I (2006) Multi-physics simulations of fuel cells using multi-component modeling. J Power Sources 158(1):295–302
2. Dyer CK (2002) Fuel cells for portable applications. J Power Sources 106(1–2):31–34
3. Colpan CO, Funga A, Hamdullahpur F (2012) 2D modeling of a flowing-electrolyte direct methanol fuel cell. J Power Sources 209:301–311
4. Siegel C (2008) Review of computational heat and mass transfer modeling in polymer-electrolyte-membrane (PEM) fuel cells. J Energy 33(9):1331–1352
5. Feng L, Cai W, Li C, Zhang J, Liu C, Xing W (2012) Fabrication and performance evaluation for a novel small planar passive direct methanol fuel cell stack. Fuel 94:401–408
6. Lottin O, Antoine B, Colinart T, Didierjean S, Maranzana G, Moyne C, Ramousse J (2009) Modeling of the operation of polymer exchange membrane fuel cells in the presence of electrodes flooding. Int J Thermal Sci 48(1):133–145

# Application of Correction Line in Floated Particle Size Distribution Measurement Using the Buoyancy Weighing-Pan Method

**M.P. Khairunnisa, Y. Ohira, M. Shimadzu, and S. Yamanaka**

## 1 Introduction

Floating particles' size distribution measurements have been measured using the microscopic or the photographic method and the laser diffraction/scattering method [1–3]. We have looked at the size distribution measurement of floating particles in a different manner [11, 15–17]. Settling particles are measured using various methods. One typical method is the sedimentation balance method [4]. Theoretically, the sedimentation balance method can be applied to floating particles that include bubbles and liquid droplets.

In this study, we experimentally investigate the applicability of the floated balance method on measurements of floating particles' size distribution; then the correction line will be applied. Floating particles include micro bubbles, liquid drops, and solid particles, which are smaller than the liquid density. In terms of easy handling and availability of standard samples, we used hollow glass beads as the solid particles. The floated particle size cannot be measured by using any other method except using the laser diffraction/scattering method which is expensive. We have developed the buoyancy weighing-pan method which is the economically way to measure floated particle size [7–9, 12–14, 16].

Correction line will be applied to the particle-size distribution that is obtained from the buoyancy weighing-pan method result. This research had been used recently in farming industries and material engineering.

M.P. Khairunnisa (✉) • Y. Ohira • M. Shimadzu • S. Yamanaka
Division of Applied Sciences Graduate School of Engineering, Muroran Institute of Technology, 27-1 Mizumoto, Muroran 050-8585, Japan
e-mail: khairunnisa88@gmail.com; ohira@mmm.muroran-it.ac.jp

K. Ab. Hamid et al. (eds.), *The Malaysia-Japan Model on Technology Partnership: International Proceedings 2013 of Malaysia-Japan Academic Scholar Conference*, DOI 10.1007/978-4-431-54439-5_3, © Springer Japan 2015

## 2 Theory

Let us assume the particles are uniformly dispersed in suspension [10]. The schematic diagram of particle floating is shown in Fig. 1. As shown in Fig. 1 the volume of the submerged weighing pan $V$ depended on the particles between the top of the weighing pan and the bottom of the pan in the suspension. The initial density of suspension $\rho_{S0}$ and the particle mass in the weighing pan $m_0$ in suspension at $t = 0$ is given by the following equations [1]:

$$\rho_{so} = \frac{m_o}{v} + \left(1 - \frac{m_o}{v_{\rho_p}}\right)\rho_L \tag{1}$$

$$M_0 = V\rho_{\mathrm{Pan}} - V\rho_{S0} = V(\rho_{\mathrm{Pan}} - \rho_{S0}), \tag{2}$$

where the liquid density is $\rho_L$, the particle density is $\rho_P$, and the initial concentration of suspension is $m$ (kg–solid/m$^3$–suspension). As shown in Fig. 1, the concentration of suspension $m$ decreases with time, because large particles have floated. The density of suspension $\rho_S$ and the apparent mass of the pan $M$ in suspension at $t = t$ is given by the following equations:

$$\rho_S = \frac{m}{V} + \left(1 - \frac{m}{V\rho_P}\right)\rho_L, \tag{3}$$

$$M = V\rho_{\mathrm{Pan}} - V\rho_S = V(\rho_{\mathrm{Pan}} - \rho_S) \tag{4}$$

As shown in Fig. 1c, the concentration of suspension $m$ is finally zero, because small particles also have floated. The final density of suspension $\rho_{SF}$ and the final

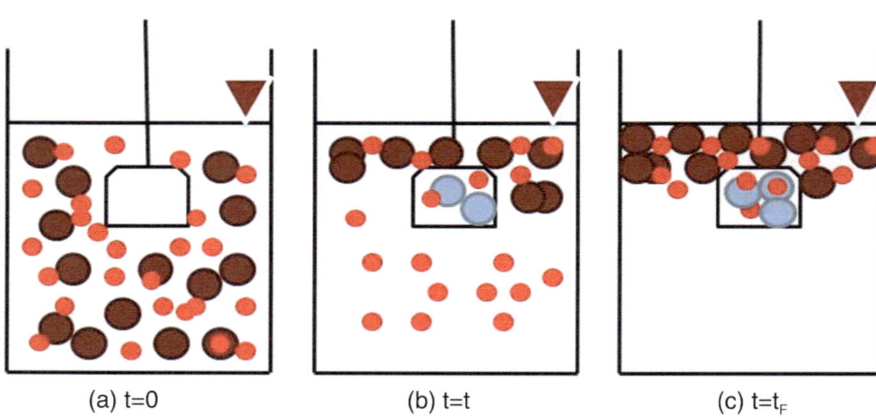


**Fig. 1** Schematic diagram of particle floating

apparent mass of the weighing pan $M_F$ in suspension at $t = t_F$ are given by the following equations:

$$\rho_{SF} = \frac{m_F}{V} + \left(1 - \frac{m_F}{V\rho_P}\right)\rho_L, \tag{5}$$

$$M_F = V\rho_{Pan} - V\rho_{SF} = V(\rho_{Pan} - \rho_{SF}) \tag{6}$$

Equation (10) shows the mass balance of particles in suspension:

$$m = m_F\int_{x_i}^{x_{max}} f(x)dx + m_F\int_{x_{min}}^{x_i} \frac{v(x)t}{h} f(x)dx, \tag{7}$$

where $v(x)$ is the floating velocity of the particle and $f(x)$ is the mass frequency of the particle size $x$. Differentiating Eq. (11) with respect to time $t$, we obtain

$$\frac{dm}{dt} = 0 + m_F\int_{x_{min}}^{x_i} \frac{v(x)}{h} f(x)dx \tag{8}$$

From Eqs. (7) and (8)

$$m = m_F\int_{xmin}^{xi} \frac{v(x)}{h} f(x)dx + \left(\frac{dm}{dt}\right)t, \tag{9}$$

$$m_R = m_F\int_{xmin}^{xi} \frac{v(x)}{h} f(x)dx, \tag{10}$$

And from Eqs. (9) and (10), we obtain Eq. (11)

$$m = m_R + \left(\frac{dm}{dt}\right)t \tag{11}$$

The apparent mass of the submerged weighing pan $M$ is given by Eq. (4). It gradually decreases from $M_0$ to $M_F$. The volume and the density of the submerged weighing pan are constant values. Differentiate Eq. (6) with respect to the time $t$, and during this time $m_0$ is equal to 0. From that we obtain

$$\rho_{S0} = m_0 = 0, \tag{12}$$

$$V\rho_S = m, \tag{13}$$

$$\frac{dM}{dt} = -\frac{dm}{dt}, \tag{14}$$

$$M = m_0 - m, \tag{15}$$

Therefore, from Eqs. (4) and (14), we obtain

$$M = M_R + \left(\frac{dM}{dt}\right)t, \tag{16}$$

$$M_R = m_0 - m_R, \tag{17}$$

$$M_R = R \tag{18}$$

The value of $M_R$ is calculated from the tangent line based on Eq. (18). The cumulative mass percentage oversize is

$$D = (1 - R) = 1 - m_R = 1 - m_F \int_{x_{min}}^{x_i} \frac{v(x)t}{h} f(x)dx \tag{19}$$

$$D = \left[1 - \frac{M_0 - M_R}{M_0 - M_T}\right] \times 100 \tag{20}$$

Particle size $x$ is given by the following equation using Stokes formula where g is the gravitational acceleration, $\mu_L$ is the viscosity of dispersion liquid containing the dispersant , and $\rho_L$ is the dispersion liquid density:

$$x = \sqrt{\frac{18\mu_L v(x)}{g(\rho_L - \rho_P)}} \tag{21}$$

The floating velocity of the particles $v(x)$ is calculated using Eq. (22):

$$v(x) = \frac{h}{t} \tag{22}$$

The particle-size distribution of the suspended particles is prepared by calculating the particle size at each time and then plotting the corresponding mass percentage oversize. Thus, the theory of the buoyancy weighing-pan method is similar to that of sedimentation balance analysis.

# 3 Experiments

## 3.1 Experimental Apparatus

Figure 2 schematically illustrates this experiment diagram, meanwhile Fig. 3 shows the photograph of the weighing pan that was made from stainless steel. The analytical balance (GR-300: A&D Co., Ltd., minimum readout mass 0.1 mg)

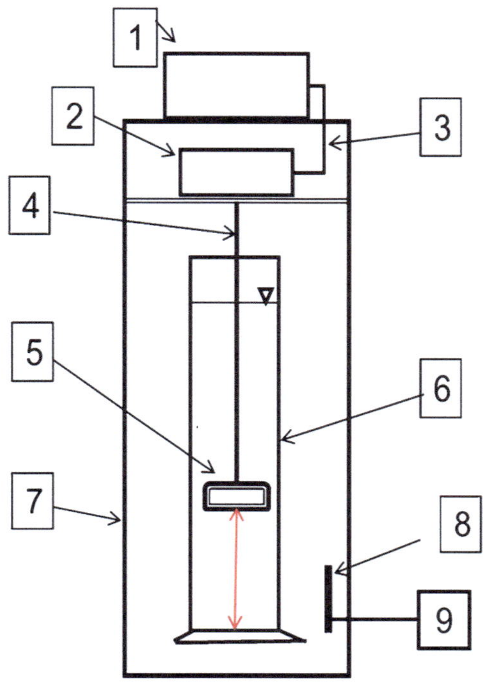


**Fig. 2** Schematic diagram of experimental apparatus for the buoyancy weighing-pan (BWP) method. (*1*) Personal computer, (*2*) analytical balance, (*3*) RS-232C cable, (*4*) hanging wire, (*5*) weighing pan, (*6*) measuring cylinder, (*7*) thermal insulation vessel, (*8*) heating panel, (*9*) controller

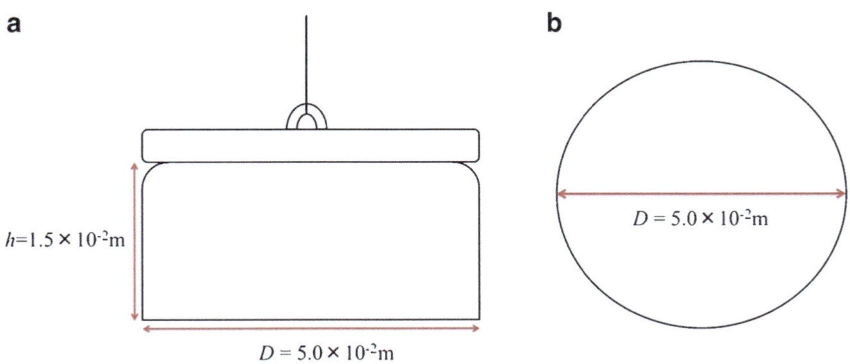


**Fig. 3** (**a**) Weighing pan for the BWP method. (**b**) Weighing pan for the BWP method

had a below-balance-weighing hook for hanging measurement and was equipped with an RS-232C interface. The measuring cylinder that was used in this research was 1 L. During the experiment a personal computer was connected to an analytical balance with RS-232C, and the data was collected in a 3-second interval.

### 3.2 Sample Particle

The sample materials that were used in this experiment were hollow glass beads K37, or also known as Glass Bubbles K37, which was standardized by Japan Industrial Standards (JIS). The particle-size distributions were measured using the laser diffraction/scattering method (MT3300EX, Nikkiso Co., Ltd.) and the micrograph method (SZX7, Olympus).

### 3.3 Experimental Conditions and Method

All suspensions had a solid concentration of 10 kg/m$^3$ (ca. 1 wt.%). NaHMP ((sodium hexametaphosphate), Kanto Chemical Co., Inc.) for calcium carbonate was used as dispersant and ion-exchanged water was used as dispersion liquid. Both room temperature and liquid temperature were approximately 298 K. Each suspension was initially stirred for 10 min and the measuring time was 2 h. In this work, all the particles were in the laminar flow region.

To prepare the suspension, 1,000 mL of ion-exchanged water and the particles to be tested were mixed in a glass cylinder. This experiment was using the hanging wire, which did not extend due to the weight of the weighing pan; a weighing pan was hung from the analytical balance. After thoroughly stirring the suspension, the weighing pan was set in suspension, and this was recorded as $t = 0$. The measuring data, which consisted of time $t$ and the corresponding apparent mass of the submerged weighing pan $M$, were recorded on a personal computer. After the measurements, we calculated the PSD of the tested particles based on the aforementioned theory.

## 4 Results and Discussion

### 4.1 Glass Bubbles K37

Figure 4 shows the change in the apparent mass of the pan $M–M_0$ with time when Glass Bubbles 37 particles were used [5, 6]. The apparent mass of the pan linearly decreased to approximately 900 s, at which point the Glass Bubbles K37 particles floated above the pan, and the apparent mass of the pan became constant $M_\infty$. The change in the apparent mass was due to the change in the buoyant mass against the pan as well as particles floated.

Figure 5 shows the PSDs that are obtained from the laser diffraction/scattering method, the microscopic method, and the BWP method. After we compare the

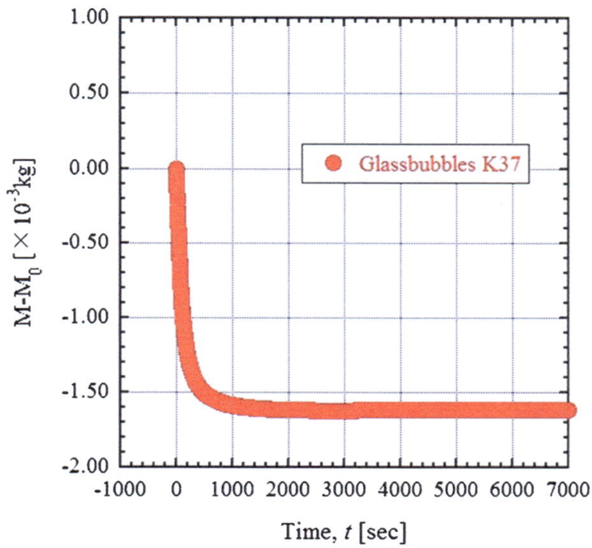


**Fig. 4** Apparent mass of the pan as a function of time (Glass Bubbles K37)

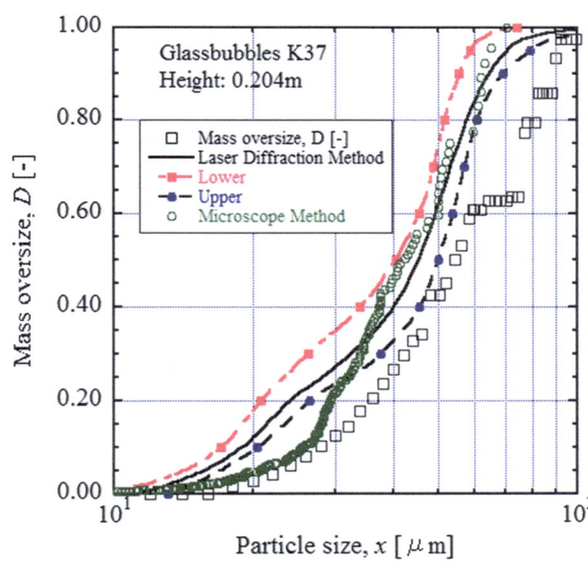


**Fig. 5** Particle distributions of the Glass Bubbles K37

BWP method result and the laser diffraction/scattering method and the microscopic method, the PSDs that determine using the BWP method overall bigger 1 μm than the laser diffraction/scattering method. By comparing the PSD lines that were obtained from this method, both of them had the same graph curve, which means the BWP method can measure the PSDs of Glass Bubbles K37.

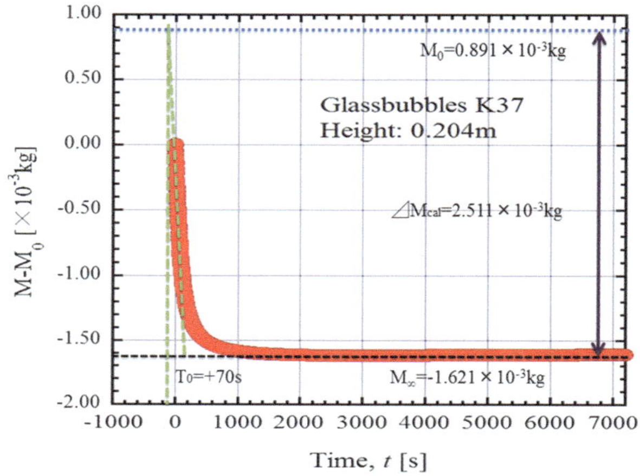


**Fig. 6** Apparent mass of the pan as a function of time (Glass Bubbles K37)—after correction

## 4.2 *Correction Line*

We assume there are time lags between the real 0s and experiment 0s. We need to calculate the real 0s to get the precise experiment data. From the theory that is given in the theory section, we can calculate the real 0s.

The final apparent mass $M_F$ for the Glass Bubbles K37 also can be interpreted as below:

$$M_F = M_{0r} + \frac{m_F}{\rho_P}(\rho_L - \rho_P) \tag{23}$$

$$\Delta M_{cal} = M_{0r} - M_F, \tag{24}$$

The real 0s can be calculated by inserting the value of $M_F$, $m_F$, $\rho_P$, and $\rho_L$. The value of needed value is shown below:

$$M_{0r} = M_F - \frac{m_F}{\rho_P}(\rho_L - \rho_P), \tag{23'}$$

$$m_F = \frac{\pi}{4} D^2 h C_0, \tag{25}$$

where $M_F$ (from Fig. 4) $= 1.621 \times 10^{-3}$ kg, $m_F = 1.482 \times 10^{-3}$ kg, $D = 0.05$ m, $h = 0.204$ m, and $C_0 = 3.7$ kg/m$^3$.

Based on Eq. (23′) $M_0$ is equal to $2.511 \times 10^{-3}$ kg. From that we can calculate the $\Delta M_{cal}$ using Eq. (24).

Figure 6 shows the apparent mass of the pan as a function of time after the correction was made when the Glass Bubbles K37 was used. Figure 7 shows the up-close apparent mass of the pan as a function of time after the correction was made. Both of these graphs show that the real 0s is $T_0 = +70$s.

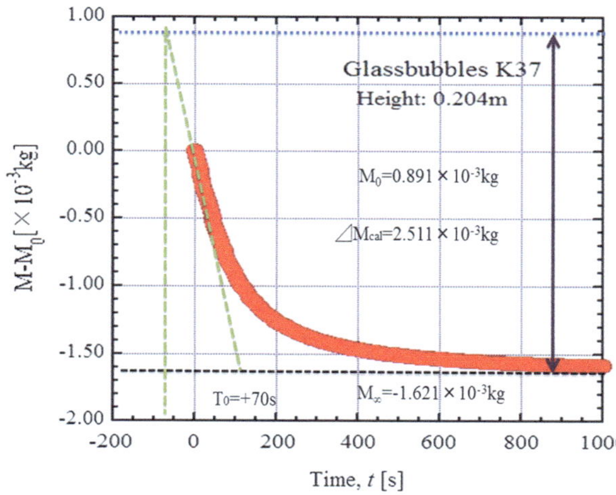


**Fig. 7** Up-close graph for apparent mass of the pan as a function of time (Glass Bubbles K37)—after correction

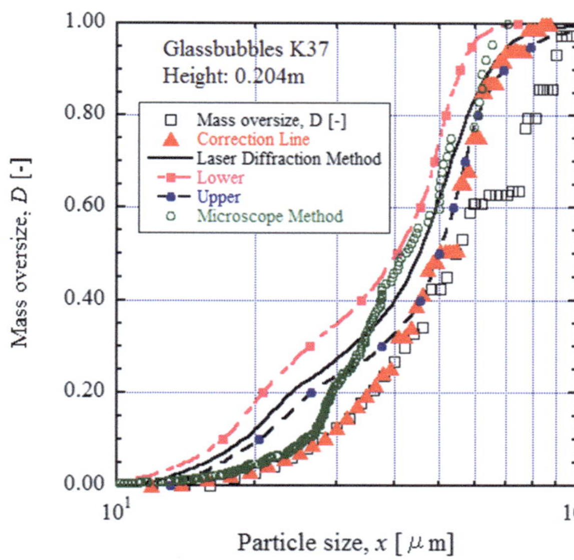


**Fig. 8** Particle distributions of the Glass Bubbles K37—after correction

Then, to search the correction line, we need to put this $T_0 = +70\text{s}$ value into Eq. (26):

$$x = \sqrt{\frac{18\mu_L h}{g(\rho_L - \rho_P)(T_0 + t)}}, \tag{26}$$

After the calculation, we can get the correction line as shown in Fig. 8. The line before correction and after correction shows a big difference, where the particle size was different from 0.1 to 4 μm. The laser diffraction/scattering method and microscopic method gave results similar to that of the BWP method. Hence, the BWP method can measure the PSD of the Glass Bubbles K37.

## 5 Conclusions

We experimentally investigated the applicability of the BWP method to measure particle-size distribution. This method is easy, useful, and economical. The results led to the following conclusions:

1. The particle-size distributions of the Glass Bubbles K37 can be measured using the BWP method.
2. The precision of particle-size distribution after adding the correction line is comparable to that obtained by a laser diffraction/scattering method.
3. The buoyancy weighing-pan method is a suitable method to measure floated particle size in Malaysia because of the easy handling of the machine and its economic benefits.

**Acknowledgments** The authors would like to thank K. Nakano and R. Tambun for their experimental assistance.

## Nomenclatures

| | |
|---|---|
| $C$ | Cumulative mass percentage oversize |
| $D(x)$ | Mass oversize of particle size $x$ |
| $f(x)$ | Mass frequency of the particle size $x$ |
| $g$ | Gravitational acceleration (m/s$^2$) |
| $h$ | Submerged length of weighing pan (m) |
| $M$ | Apparent mass of weighing pan at $t = t$ (kg) |
| m | Concentration of suspension |
| $t$ | Time (s) |
| $u$ | Superficial velocity (m/s) |
| $V$ | Volume of the weighing pan (m$^3$) |
| $v(x)$ | Floating velocity of particle size $x$ (m/s) |
| $W_{\mathrm{B}}$ | Buoyant mass of the submerged weighing pan in the suspension (kg) |
| $x$ | Particle size (m) |
| $\mu_{\mathrm{L}}$ | Liquid viscosity (Pa s) |
| $\rho_{\mathrm{L}}$ | Liquid density (kg/m$^3$) |
| $\rho_{\mathrm{P}}$ | Particle density (kg/m$^3$) |
| $\rho_{\mathrm{Pan}}$ | Weighing-pan density |
| $\rho_{\mathrm{S}}$ | Density of suspension (kg/m$^3$) |

### Subscripts

| | |
|---|---|
| cal | Calculation |
| max | Maximum |
| min | Minimum |
| 0 | Initial $t = 0$s |

| | |
|---|---|
| F | Final |
| L | Liquid |
| S | Suspension |
| r | Real |
| R | Mass oversize of particle size |

## References


1. Allen T (1990) Particle size measurement, 4th edn. Chapman and Hall, London, pp 345–355
2. Society of Chemical Engineering of Japan. Chemical engineering handbook, 5th edn. Maruzen, Tokyo, pp 224–231
3. Obata E, Watanabe H (1986) Measurement of particles sizes by fluidization, vol 4, Encyclopedia of fluid mechanics. Gulf Publishing, Houston, pp 221–236
4. Stanly-Wood NG, Obata E, Takahashi H, Ando K (1990) Liquid fluidisation curve. Powder Technol 60:61–70
5. Oden S (1925) The size distribution of particles in soils and experimental methods of obtaining them. Soil Sci 19:1–35
6. Fukui K, Yoshida H, Shiba M, Tokunaga Y (2000) Investigation about data reduction and sedimentation distance balance method. J Chem Eng Jpn 33:393–399
7. Arakawa M, Shimomura G, Imamura A, Yanazawa N, Yokoyama T, Kaya N (1984) A new apparatus for measuring particle size distribution based on centrifugal sedimentation. J Soc Mater Sci Jpn 33:1141–1145
8. Obata E, Watanabe H, Endo N (1982) Measurement of particles size and size distribution of particles by fluidization. J Chem Eng Jpn 15:23–28
9. Obata E, Watanabe H, Mukaida K, Akiyoshi M, Ando K (1986) Measurement of particle size distribution by numerical values of fluidization curve. Kagaku Kougaku Ronbunshu 12:619–621
10. Obata E, Takahashi H, Akiyoshi M, Ando K, Watanabe H (1990) Measurement of particle size by superficial velocity and pressure drops across liquid-solid fluidization in high Reynolds number. J Soc Powder Technol 27:301–307
11. Obata E, Ohira Y, Ohta M (2009) New measurement of particle size distribution by a buoyancy weighing-bar method. J Powder Technol 196(2):163–168
12. Obata E, Ohira Y, Ohta M (2006) New measurement of the particle size distribution by buoyancy weighing method. In: Proceedings of the 5th world congress on particle technology," Yuichi Ohira, Florida, USA, CD-ROM
13. Nakano K, Tambun R, Shimadzu M, Ohira Y, Obata E (2011) Application of the buoyancy weighing-bar method to floating size distribution measurement. Abstract of joint seminar on environmental science and disaster mitigation research, Muroran, pp 88–89
14. Khairunnisa MP, Tambun R, Shimadzu M, Ohira Y, Obata E (2012) Floating particle size distribution using the floated balance method. In: Abstract of joint seminar on environmental science and disaster mitigation research, Muroran, pp 115–116
15. Khairunnisa MP, Tambun R, Ohira Y, Obata E (2012) Measurement of velocity mean particle size of all the particle within five minutes by buoyancy weighing-bar method. In: Hydraulic measurement & experimental method conference 2012 (HMEM 2012) Snowbird, Utah, USA

16. Khairunnisa MP, Ohira Y, Obata E (2012) Effect of height and suspension concentration in the floated balance method. In: Malaysia-Japan academic scholar conference 2012 (MJASC2012) Waseda University, Tokyo, Japan
17. Khairunnisa MP, Tambun R, Shimadzu M, Ohira Y (2013) Particle size distribution measurement of the hollow glass beads by the buoyancy weighing-bar method. In: Abstract of joint seminar on environmental science and disaster mitigation research, Muroran, Japan, pp 89–90

# Fine Pitch Wire Bonding with Insulated Cu Wire: Free Air Ball and Ball Bond

**Leong HungYang, Yap BoonKar, Navas Khan, Mohd Rusli Ibrahim, and L.C. Tan**

## 1 Introduction

In semiconductor packaging industry, wire bonding is still the main interconnect technology to form electrical connection between the chips and substrates [1]. However, as the market trend is moving for lower cost and more I/O with smaller bond pad pitch devices, this poses challenges in wire bond technology. Increasing gold prices have pushed Cu wire to become the choice for wire bonding interconnects for high-density applications such as communication and networking products. However, Cu wire bonding poses many challenges due to its harder physical property compared to gold wire and its susceptibility to corrosion [2].

Insulated Cu wire technology is receiving increased attention in the semiconductor industry due to the flexibility in wire bonding design rule offered by the insulation layer. There is no more concern for wire-to-wire shorts. Longer wire, crossed wire, touched wire, and simpler wire looping is possible if insulated wire is used. Insulated wire brings wire bonding interconnection to high-density devices which have multiple tiers of wires to accommodate more than 1,000 wires per package. The technology can be adopted easily as it uses existing wire bonding infrastructure. Cu wire with suitable electrical insulation prevents electrical shorts between the wires and also allows complex looping for multiple die applications [3]. Previous research works were focused on insulated Au wire only. Weak stitch formation which leads to tail short and stoppages during wire bonding is one of the key challenges. Capillary with roughen tip surface was proposed to improve the insulated wire stitch bond [4].

L. HungYang (✉) • Y. BoonKar
Centre for Microelectronic and Nano Engineering, COE, Universiti Tenaga Nasional, KM7 Jalan Ikram-Uniten, 43009 Kajang, Selangor, Malaysia
e-mail: hung_yang@hotmail.com

N. Khan • M.R. Ibrahim • L. Tan
Fresscale Semiconductor Malaysia SDN BHD, Free Industrial Zone Sungai Way, 47300 Petaling Jaya, Selangor, Malaysia

K. Ab. Hamid et al. (eds.), *The Malaysia-Japan Model on Technology Partnership: International Proceedings 2013 of Malaysia-Japan Academic Scholar Conference*, DOI 10.1007/978-4-431-54439-5_4, © Springer Japan 2015

Ball shear, wire pull, and stitch pull for Cu wire are higher compared to Au wire due to the hardness of Cu wire [5]. The common issues with Cu wire bonding are increased pad structure damage and excessive Al splash. Excessive Al splash on bond pad during Cu wire bonding may cause wire-to-wire short between adjacent bond pads. This poses higher risk to Cu wire bonding especially for pitch wire bonding and sensitive bond pad with low-k or ultra low-k devices [6].

Besides the above-mentioned challenges, insulation on the wire will pose process challenges such as residue-free FAB and good ball bond joints. This study aims to focus and understand on the behavior of the free air ball and ball bond formation of insulated Cu wire compared with Cu wire.

## 2 Experimental

The key materials include 0.8 mil insulated Cu wire and a commercial Cu wire with 4 N purity. Samples were bonded via a commercial wire bonder, equipped with Cu kit to supply forming gas (95 % $N_2$ 5% $H_2$) with gas flow 0.6–0.8 L/min. The package type is a thermally enhanced BGA with fine bond pad pitch size. The aluminum bond pad is 1.2 μm in thickness. A suitable capillary with roughened surface for the 0.8 mil wire was used in this study.

### 2.1 Insulation Characteristics

Insulated wire is the core wire covered with a layer of organic coating material coated by the supplier to prevent wire-to-wire short which is shown in Fig. 1. The organic material is supplier's proprietary information. The melting point is around 300 °C.

The coating thickness was verified by ion milling. Five sets of measurements were taken using FESEM (field emission scanning electron microscopy).

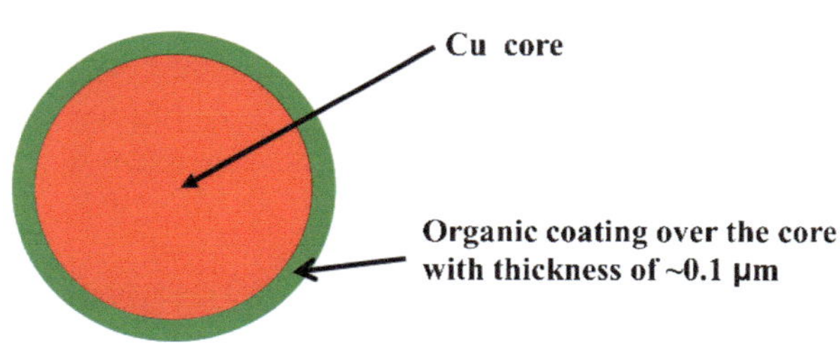


**Fig. 1** Insulated wire cross-section image

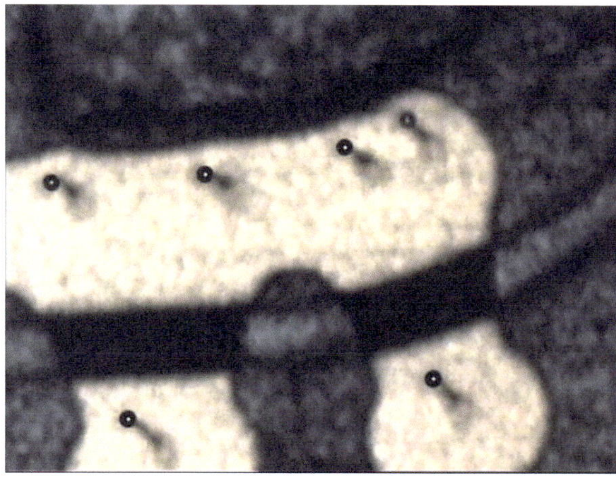

**Fig. 2** Optical image of FAB forming on substrate

## 2.2 FAB Characteristics

FAB is formed during electronic flame off (EFO) firing. When high potential energy is transferred from the electrode to the wire, a spark is induced which melts the wire to form a spherical ball due to surface tension. A good FAB determines the quality of the ball bond formation. To study the FAB, a FAB farming program was developed to bond a large quantity of FAB on the substrate for inspection as shown in Fig. 2. EFO parameter settings were benchmarked to the Cu wire.

We defined a list of criteria for a good Cu wire free air ball (FAB) in wire bonding. Round, spherical, and consistent size must be achieved for Cu wire FAB. Clean and residue-free FAB is an additional criterion for insulated Cu FAB. To study this, we used the below methods:

1. Low-power and high-power optical scope inspection
2. SEM (scanning electron microscope) imaging
3. FIB (focus ion beam) imaging

An experiment was carried out to compare the FAB size with a set of EFO settings between Cu wire and insulated Cu wire. Each setting, a total of 30 measurements, was taken on the FAB size. EFO time was a constant while the EFO current varied from 40, 45, 50, 55 to 60 mA.

## 2.3 Ball Bond

The bonding parameters were optimized to ensure the Cu ball is sticking on bond pad and at the same time not too strong that leads to bond pad damage. Ball bond strength was accessed by destructive ball shear test and wire pull test at time zero

(T0 h) using Dage 4000 series. The ball shear test is used to measure the integrity of the ball bond through horizontal shear force. Wire pull test is performed by placing the hook at the first kink of the ball bond. The minimum limit of acceptable ball shear is 4.5 g while wire pull test is 3.5 g.

Bonded Cu wire and insulated Cu wire samples were also subjected to isothermal aging at 225 °C for 4.5 h in nitrogen gas-purged oven. The purpose of the short time of thermal aging is to expose the wire-bonded package under a certain level of thermal stress. It acts as an early indicator to determine the bondability and reliability of the first bond recipe creation. Ball shear and wire pull tests are the key responses.

Al remnant and Al splash are also key responses during recipe development. Al splash is the difference between the measurements of the ball diameter of Y and Y1 direction. Y is the measurement of the ball diameter when the high-power scope was focused on the ball while Y1 is the measurement of the same ball diameter when the high-power scope was focused on the outer ring of the ball at the Y direction. A total of 16 readings was taken. The Zygo 3-D measurement tool was applied to determine the amount of Al remnant for the bonded balls. The requirement of Al remnant must be more than 0.05 μm of the Al original thickness. The Zygo measurement was performed on two units (six ball bonds per unit) for both wires.

Cratering test was performed using phosphoric acid etching solution on six full bonded samples each for Cu and insulated Cu to ensure no bond pad cratering. All the bonded wires were etched off, and optical inspection was performed on the bond pad for both wires.

Six samples of Cu ball and insulated Cu ball bonded samples were baked under the temperature 225 °C for 2 h to accelerate the IMC growth. Using acid nitrate etching solution, the IMC coverage was inspected on the bond pad and calculated using image software tool. The ball bond quality is justified and compared in detail between the two wires based on the key responses mentioned above.

# 3 Results and Discussion

## 3.1 *Insulation Characteristics*

Based on the FESEM image in Fig. 3, we clearly observed the insulation layer coating as a different color tone compared to the bare Cu wire. The average coating thickness was in a range of 100–140 nm which correlates to the thickness information provided by the supplier. The measurement results were consistent at five location points.

**Fig. 3** FESEM image of coating layer thickness measurement

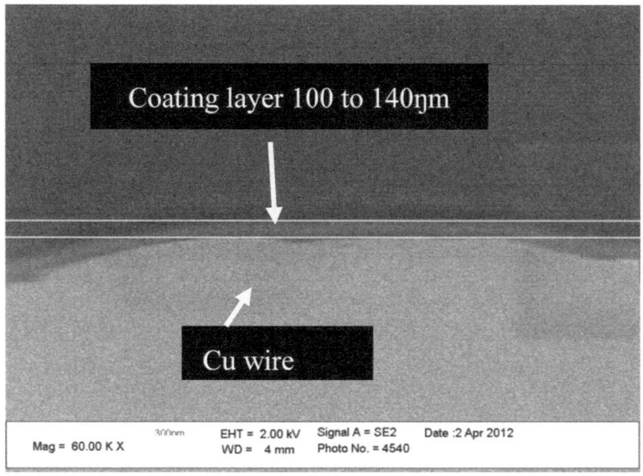


**Table 1** EFO current and EFO time for Cu wire and insulated Cu wire

| EFO setting | Cu wire | Insulated Cu wire |
|---|---|---|
| EFO current (mA) | A | 0.8A |
| EFO time (μs) | B | B |

## 3.2 *FAB Characteristics*

### 3.2.1 FAB Roundness and Residue Analysis

Low-Power and High-Power Optical Scope Inspection

Table 1 shows the optimized EFO settings for insulated Cu wire and Cu wire, respectively. We observed that insulated Cu wire needs 20 % lower EFO current compared to bare Cu wire to form a similar FAB size.

Both FAB were consistently round and no surface abnormality observed as shown in Fig. 4. The size of the FAB was measured using high-power optical scope.

SEM Imaging

Thorough inspection was carried out via SEM imaging to confirm the FAB quality. The insulated Cu FAB has consistent size and roundness, demonstrating that the EFO parameters are suitable. During the EFO firing, the thin insulation was burned off exposing the Cu for the ball bond formation. A previous study on insulated Au wire FAB by Robert Lyn et al. [6] showed that the FAB has watermelon stripe pattern showing that partial organic coating still remained on the FAB surface. With the same approach, we were not able to observe the color tone difference between

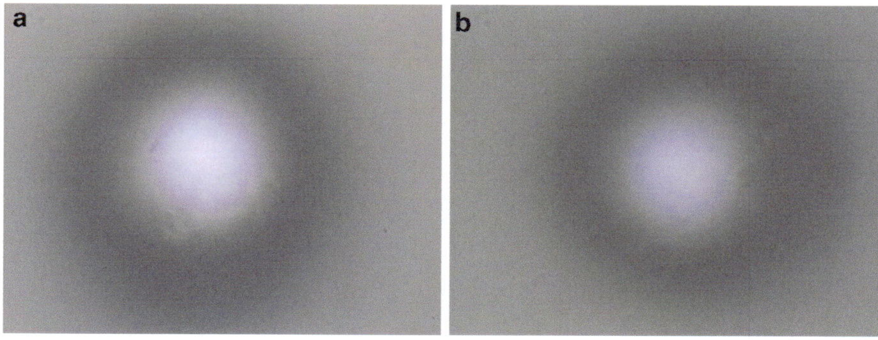


**Fig. 4** Optical image of FAB for (**a**) bare Cu wire and (**b**) insulated Cu wire

**Fig. 5** Backscatter electron imaging of insulated Cu FAB

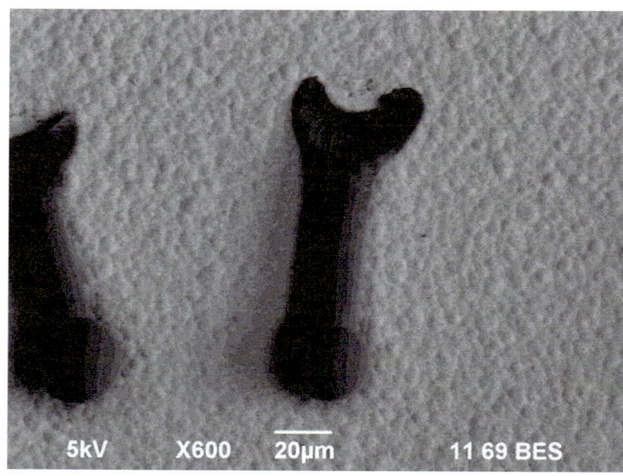


the insulation layer and bare Cu wire on this insulated Cu FAB even with backscatter electron mode imaging. The reason could be that the organic coating was too thin. The electron beam penetrated deeper into the Cu wire. As a result, only Cu element was shown in Fig. 5.

## FIB Imaging

In Fig. 6, the Cu grain structure was clearly seen on the FAB surface of Cu wire and ball neck or heat-affected zone (HAZ) surface. However, due to the presence of the insulation layer on the core wire, there is no visible Cu grain structure on the insulated Cu wire FAB. A very thin remnant of insulation in watermelon stripe was seen on the FAB of insulated Cu wire in Fig. 7. The insulated Cu FAB was clean without any organic residue.

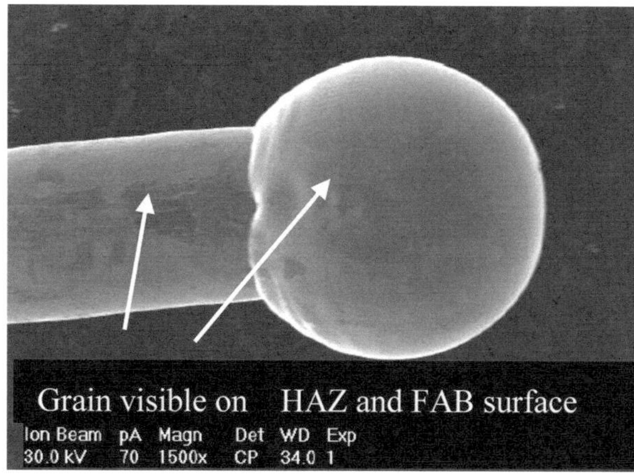


**Fig. 6** FAB formation of Cu wire, showing visible Cu grain structure without any "watermelon" stripe pattern

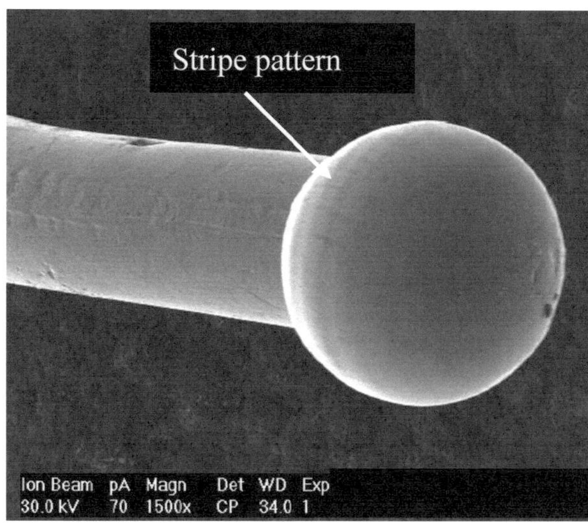


**Fig. 7** FAB formation of insulated Cu wire, showing thin lines of "watermelon" stripe pattern

### 3.2.2 Energy to Form a FAB

Figure 8 is the graph of FAB size versus a set of EFO currents. We observed that the Cu wire FAB size is consistently smaller than insulated Cu wire for all the EFO current settings. It is clearly shown in Fig. 9 under the same bare Cu EFO setting that the insulated Cu FAB is larger than the Cu bare wire FAB. Insulated Cu wire consumed around ~20 % lower energy to form a FAB compared to Cu wire. At low EFO current of 40 and 45 mA, no proper FAB was able to form with Cu wire. From the microscope, we observed that the Cu wire merely melted, but the energy was not enough to form a FAB. It is clearly shown that organic coating covered on the Cu wire could affect the FAB size. The heat loss during the FAB formation could be

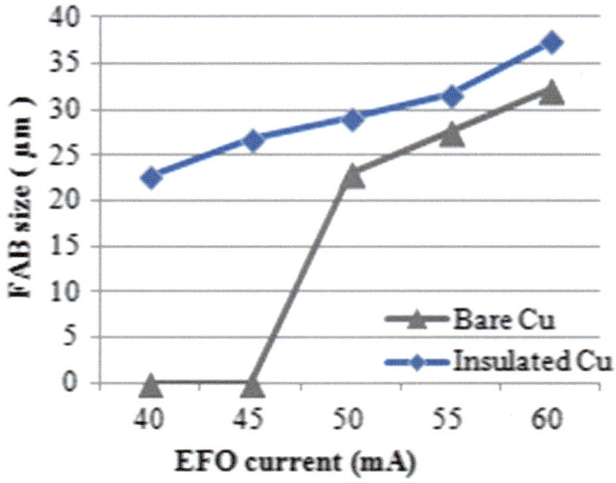


**Fig. 8** FAB size versus EFO current for Cu wire and insulated Cu wire, EFO time = constant

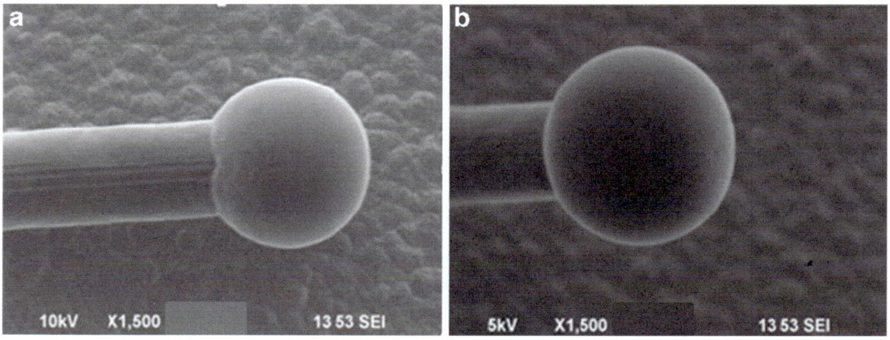


**Fig. 9** SEM image of FAB at same EFO setting: (**a**) bare Cu wire and (**b**) insulated Cu wire

lesser due to the effect of the organic coating covering the wire which requires less energy compared to the Cu wire to form a FAB. The organic coating constricts the heat at the tip of the wire [7].

## 3.3 *Ball Formation*

### 3.3.1 Bonding Parameters

Cu wire recipe was taken from the production line as the baseline parameters. It was found that contact power, contact force, bond power, bond time, bond power, and bond force were the minimum parameters to prevent NSOP during auto bonding for the insulated Cu wire bonding. Bond power and bond force are the significant factors for the desired ball bond height (BBH) and ball bond diameter (BBD) in this study. These two parameters were selected for further DOE to create the first bond recipe.

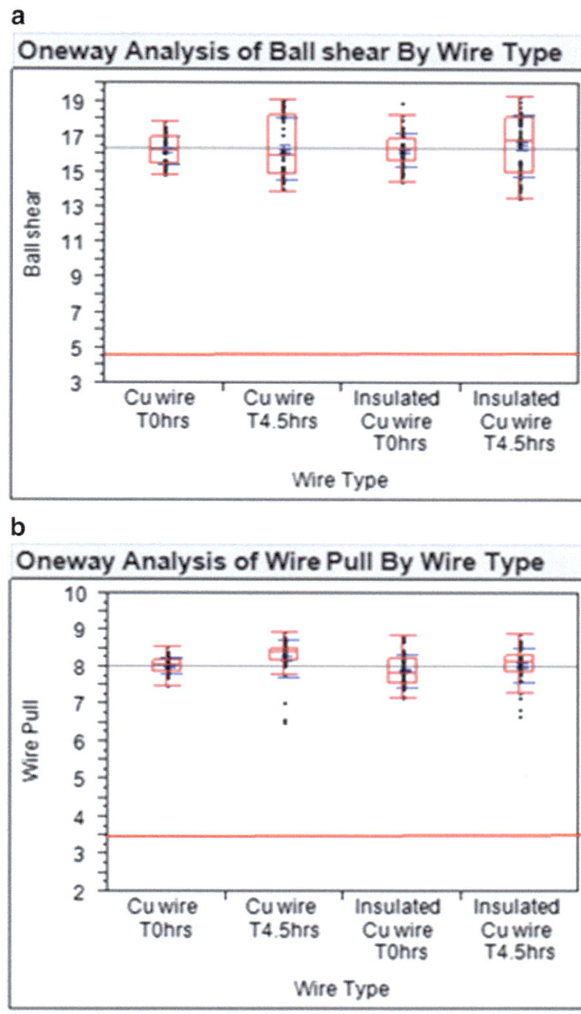


**Fig. 10** Ball bond strength comparison between Cu wire and insulated Cu wire: (**a**) ball shear test and (**b**) wire pull test

We found that insulated wire samples require similar bonding parameters except the bond power value which is 0.7 times lower than the Cu wire samples to get the desired ball thickness and size. The ball shape of insulated Cu is no different from Cu ball bond. Therefore, insulated Cu wire sample requires less demanding ball bond parameters than Cu wire sample.

### 3.3.2 Mechanical Test Results

The ball shear test and wire pull measurements passed the minimum requirement. Ball shear test and wire pull test results for samples subjected to T0 h and T4.5 h at 225 °C for both wires are shown in Fig. 10a, b, respectively. Ball shear strength is

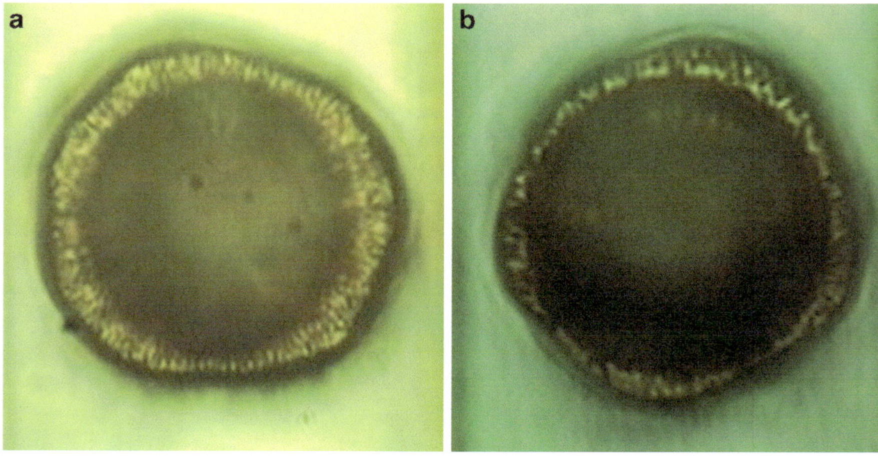


**Fig. 11** Optical photo of bonded ball shape: (**a**) Cu wire and (**b**) insulated Cu wire

comparable between Cu ball bond and insulated Cu ball bond at T0 h and after thermal aging. No ball lift and lifted metal was observed. The shear mode of insulated Cu bonded ball is through Al bond pad, similar to Cu bonded ball.

There are no significant differences on wire pull strength for both the wire samples subjected to T0 h and T4.5 h at 225 °C. Wire pull break modes for both wires all break at neck. No ball lift or metal lift observed at T4.5 h at 225 °C, but there were a few wires which gave slightly lower wire pull value after thermal aging.

### 3.3.3 Al Splash

Figure 11 shows the optical images of the ball bond and Fig. 12 shows the optical images of the ball bond focusing on Al splash. The SEM images of Cu and insulated Cu wire ball bond are shown in Fig. 13. Figure 14 shows the boxplot of the Al splash measurements for both wires. In this study, insulated Cu wire ball bond has 31 % lesser splash compared to Cu wire ball bond.

The average Al remnant for Cu wire samples is 0.23 μm, while the Al remnant for insulated Cu wire is 0.1 μm. The minimum spec for allowable Al remnant is 0.05 μm. The Al remnant for insulated Cu bonded ball is 43 % lower than that of Cu bonded ball. This indicates that the insulated Cu ball is softer although the insulated Cu wire has applied lower bond power.

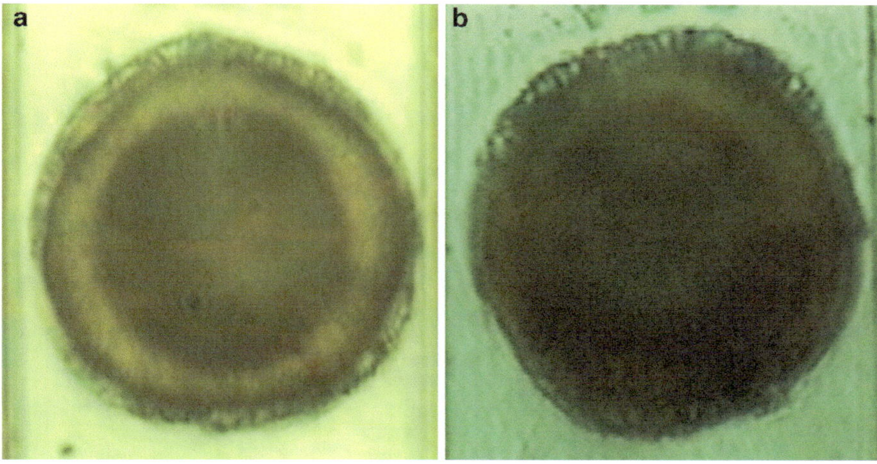


**Fig. 12** Optical photo of bonded ball focused on Al splash: (**a**) Cu wire and (**b**) insulated Cu wire

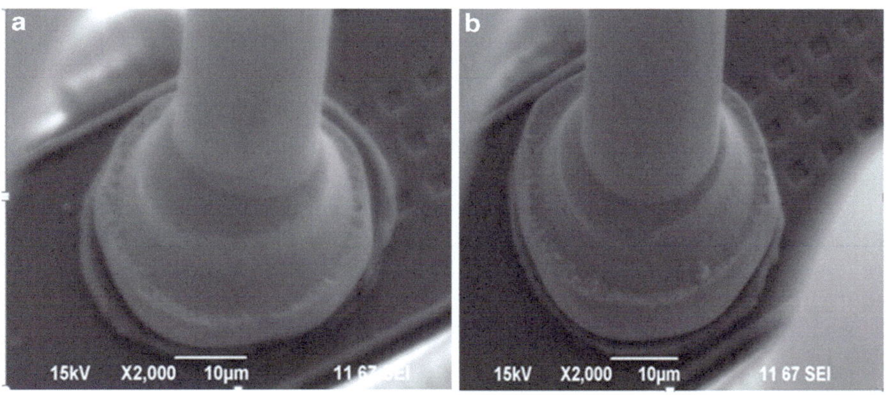


**Fig. 13** SEM photo of bonded ball: (**a**) Cu wire and (**b**) insulated Cu wire

**Fig. 14** Al splash of bare Cu wire and insulated Cu wire

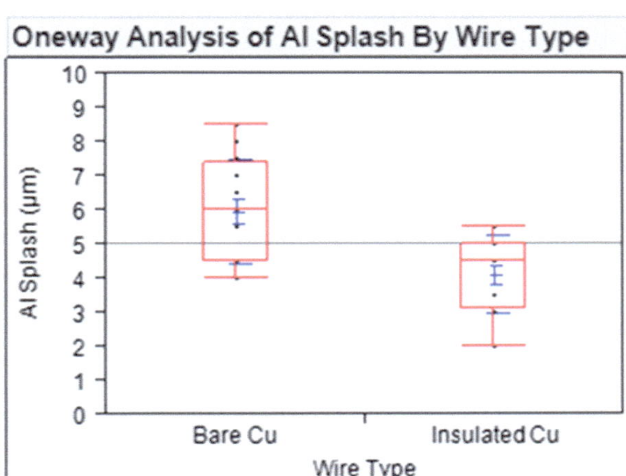

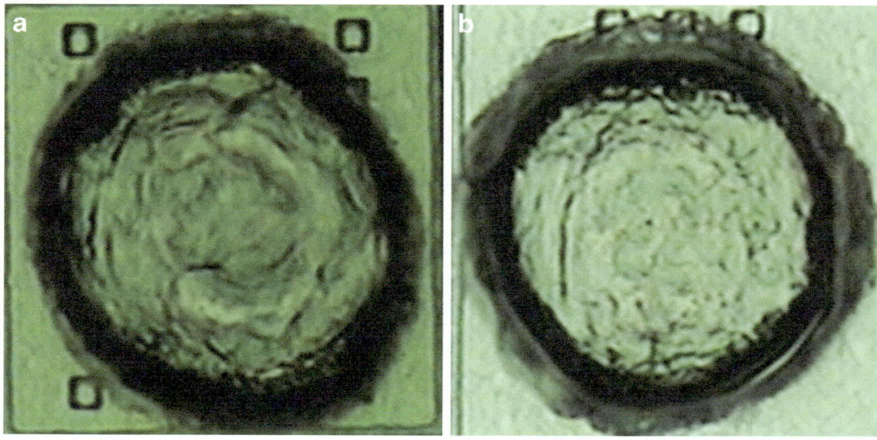


**Fig. 15** IMC coverage of (**a**) Cu wire ball bond and (**b**) insulated Cu wire ball bond

### 3.3.4 IMC Test and Cratering

Uniform and larger surface of the bonded ball has to form IMC with the aluminum pad for a reliable joint. When a new wire material is introduced, a check on the IMC coverage is critical. Based on the IMC coverage test, insulated Cu bonded ball has >80 % IMC coverage, comparable to Cu bonded ball. Figure 15 shows the IMC coverage for Cu and insulated Cu, respectively. Both ball bonds have relatively similar IMC coverage pattern.

Cratering test was performed, and no bond pad damage and peeling were observed for Cu wire samples and insulated Cu wire samples. Therefore, the ball bonding recipe was considered good.

## 4 Conclusion

A detailed process characterization on the 0.8 mil insulated Cu wire FAB and ball bond formation was successfully established. Insulated Cu wire ball bonding has comparable performances as the current Cu wire which is currently running in production line. As a summary:

1. Spherical, consistent shape, and residue-free FAB of insulated Cu wire were obtained using forming gas.
2. Insulated Cu wire consumed ~20 % lesser energy than Cu wire to form a FAB.
3. Insulated Cu wire samples require less demanding parameters for ball bond formation.
4. The bond strength of insulated Cu wire is comparable to that of Cu wire.
5. Insulated Cu wire-bonded ball has lesser Al splash.

The insulated wire technology offers a lot of potential for high-density devices applications. However, in-depth study works are required before high-volume production. Other areas which are currently under development are stitch bonding performance and reliability study which will be published in the next paper.

## References

1. Harman G (2010) Wire bonding in microelectronics, 3rd edn. McGraw-Hill, New York
2. Ibrahim MR et al (2006) The challenges of fine pitch copper wire bonding in BGA packages. In: 31st international conference on electronics manufacturing and technology, Petaling Jaya, pp 347–353
3. Lyn R et al (2007) Assembly using X-Wire™ insulated bonding wire technology. SEMICON, Singapore
4. Harun F et al (2007) Wirebonding insulated wire and capillary therefor. U.S. Patent 726,230
5. Kid WB et al (2011) Metallurgical bond integrity of C45 ultra fine pitch with 18 μm copper wire. In: IEEE regional symposium on micro and nanoelectronics (RSM), Kota Kinabalu, pp 236–240
6. Leng EP et al (2010) A study on fine pitch Au and Cu WB integrity vs Ni thickness of Ni/Pd Au bond pad on C90 low K wafer technology for high temperature automotive. In: Proceedings of the 34th electronic manufacturing technology symposium, pp 1–7
7. Uno T (2011) Enhancing bondability with coated copper bonding wire. Microelectron Reliab 51:88–96

# Effect of Red and Blue Lights on Photomorphogenesis in *Brassica chinensis*

**Nurul Najwa Ani, Ahmad Nizar Harun, Saiful Farhan M. Samsuri, and Robiah Ahmad**

## 1 Introduction

Red and blue light-emitting diodes (LEDs) are necessary for plant development and physiology. Plant morphogenesis and differentiation of plant tissue and cell were controlled by photoperiod and light quality and quantity [33]. Several studies have shown that the combination of different light resulted in many positive effects on growth, development, nutrition, appearance, and the edible quality of plants [11, 14, 23, 28]. The quality of light which refers to the color or wavelength that reached a plant's surface was strongly influencing plant growth and development [19]. The major energy sources for photosynthetic $CO_2$ assimilation in plants were from red (R) and blue (B) lights. There are studies that examined photosynthesis of higher plants in the reaction of light spectra. It is well known that R and B ranges have a maximum reaction for light spectra [6, 20]. Previous studies show that LED light has strong effect on several plant growth, such as maize [12], grape [38], banana [10], strawberry [35], potato [17, 31], Chrysanthemum [1, 13, 21, 22], *Withania somnifera* [26], Cymbidium [41], Eucalyptus [34], Phalaenopsis orchids [43], Zantedeschia [18], Lilium [27], Spathiphyllum [36], *Rehmannia glutinose* [11, 14], and *Euphorbia milii* [7].

Past studies have proven the combination of R and B lights in controlled environments as a lighting source for the effective production of many plant species, including *Brassica chinensis* [3, 9, 15, 22, 37, 40, 45, 46]. LED lights are widely used for plant photomorphogenesis and growth at present. No research is available on the effects of RB lights on the photomorphogenesis in *B. chinensis* grown in hydroponic system in controlled environment with different photoperiod under low light intensity which is constant at 100 μmol $m^{-2}$ $s^{-1}$ photon flux density.

N.N. Ani (✉) • A.N. Harun • S.F.M. Samsuri • R. Ahmad
Universiti Teknologi Malaysia Kuala Lumpur, Jalan Semarak,
54100 Kuala Lumpur, Malaysia
e-mail: nurulnajwa.ani@gmail.com

K. Ab. Hamid et al. (eds.), *The Malaysia-Japan Model on Technology Partnership: International Proceedings 2013 of Malaysia-Japan Academic Scholar Conference*,
DOI 10.1007/978-4-431-54439-5_5, 

In order to apply the findings to *B. chinensis* quality and production, it is important to investigate the effect of combination of RB with low intensity under different photoperiod when provided as the sole source of light. Therefore, the hypothesis of this study was that plants would grow better under RB LED pulse lighting (continues 1 h lights, 15 min dark in a day) compared to RB LEDs (12 h lights, 12 h dark). The final goal of the research was to develop a year-round and rapid production system for fresh, high-quality, pesticide-free, and economically feasible hydroponic *B. chinensis* that is produced close to the final retail market.

# 2 Material and Methods

## 2.1 Plant Material and Growth Conditions

Seeds of *B. chinensis* were germinated in sponge cubes (3 × 3 cm) and hydroponically grown for 14 days in an environmentally controlled growth room. The temperature was at constant 20 °C under a light intensity 100 μmol $m^{-2}$ $s^{-1}$ photon flux density (PFD) for 12 h under RB lights. Uniform-sized seedlings of *B. chinensis* at the 3-leaf stage were individually raised in a polystyrene foam cube, then mounted into a Styrofoam plate with 15 holes, and placed in a container (55.5 × 42 × 13 cm) filled with complete nutrient solution in a control environment room. The nutrient solution was renewed every week and adjusted to pH 6 and an electrical conductivity of 2.5 mS $cm^{-1}$. The air temperature, relative humidity, and $CO_2$ levels for all treatments were respectively maintained throughout the experiment at 22/20 °C (day and night), 65 % and 400 ppm under a light intensity 100 μmol $m^{-2}$ $s^{-1}$ PFD.

## 2.2 Light Treatments

Treatments with T1 (12 h lights,12 h dark) and T2 (1 h lights, 15 min dark in a day) consisted a ratio of 16:4 combination of red (R) and blue (B) light source. The peak emissions of the B (454 nm) and R (660 nm) LEDs closely coincide with the absorption peaks of chlorophylls *a* and *b*, and the reported wavelengths are at their respective maximum photosynthetic efficiency [30]. The same light intensity expressed as photosynthetic PFD of 100 μmol $m^{-2}$ $s^{-1}$ was measured daily above the plant canopy and maintained by adjusting the distance of the LEDs to the plant canopy. Plants were harvested at 30 days after transplant.

## 2.3 Plant Growth Measurements

Measurements included plant height (PH), number of leaf, plant fresh weight (FW), plant dry weight (DW), moisture content, and leaf area (LA). PH and leaf number were recorded on the plants in four replicates every week. Measurements included plant fresh weight (FW), plant dry weight (DW), moisture content, and leaf area (LA). Plant tissue samples were dried in a drying oven for 48 h at 65 °C before weighing. The LA ($cm^2$) of every plant was measured by an LA meter (LI-3100, LI-COR).

## 2.4 Chlorophyll (chl)

To examine chlorophyll content, chlorophyll was extracted from leaves of five plants at a similar position for both treatments. Leaves were weighed to 0.1 g (fresh weight, FW). The sample is added to the bottle and mixed with 20 mL of 80 % acetone, closed using aluminum foil, and left for 1 week until the leaf turned white. Optical density was measured with a UV 3101PC scanning spectrophotometer at 663 nm for chlorophyll *a* and at 645 nm for chlorophyll *b* [48]. Concentrations of chl *a* and chl *b* were determined from the following equations [29]:

$$\text{Total chlorophyll (mg/L)} = 20.2\ D_{645} + 0.02\ D_{663}$$
$$\text{Chlorophyll}\,a = 12.7\ D_{663} + 2.69\ D_{634}$$
$$\text{Chlorophyll}\,b = 22.9\ D_{645} + 0.02\ D_{663}$$

## 2.5 Gas Exchange Measurements

Measurements of net photosynthesis (μmol $CO_2$ $m^{-2}$ $s^{-1}$), leaf stomatal conductance (mol $H_2O$ $m^{-2}$ $s^{-1}$), and transpiration rate (mol $H_2O$ $m^{-2}$ $s^{-1}$) of 20 different leaves per treatment were monitored using a Portable Photosynthesis System Li-6400XT (LICOR, USA). To assess the trade-off between $CO_2$ uptake and water loss, instantaneous water-use efficiency (WUE) was calculated as ratio between photosynthetic rate and transpiration rate (μmol $CO_2$/μmol $H_2O$). Diurnal measurements of gas exchange were taken from 0900 h to 1500 h on the fifth youngest fully expanded leaf of four plants in each replicate for T1 and T2. Statistical assessment was done on gas exchange parameters at between 1100 and 1200 h, which was presumed to be the diurnal period when photosynthetic rates would be maximal [8].

## 2.6 *Statistical Analysis*

Statistical analyses were conducted with statistical product and service solutions for Windows, version 16.0 (SPSS). All measurements were evaluated for significance by an analysis of variance (ANOVA) followed by the least significant difference (LSD) test at the $p < 0.05$ level [23].

# 3 Results

## 3.1 *Plant Growth and Morphology and Pigment Contents*

Results of the photomorphogenesis measurements of *B. chinensis* are influenced by two photoperiod light treatments shown in Table 1, and plants showed distinct growth responses to T1 and T2. Plants FW and DW were the greatest when grown under T2 treatment as compared to T1. The LA decreased in the order of plants grown under T1 and the parameters under T2 were significantly higher than under T1 light. In addition, a normal appearance with plant height (PH) and number of leaf of the *B. chinensis* plants were observed. However, plants grown under T1 looked small or even severely dwarfed. Bigger plant sizes with higher moisture contents were also produced under T2, with the mean of fresh plant weight of 152.78 g and dry weight of 5.76 g and moisture contents of 96.22 %.

Chl *a* contents of *B. chinensis* leaves in both treatments were higher than the respective chl *b* contents. However, significant differences were observed in pigment contents (chl *a*, chl *b*, and total chl) regardless of the photoperiod light treatment (Table 1).

### 3.1.1 Leaf Photosynthesis Rate

The diurnal mean leaf photosynthesis rate of *B. chinensis* under RB (16:4) LED in different photoperiods is shown in Fig. 1. Plants grown under T2 gave a higher mean value of 3.557 μmol $CO_2$ $m^{-2}$ $s^{-1}$ compared to T1 with a value of 0.909 μmol

**Table 1** Influence of treatment on fresh weight (FW), dry weight (DW), leaf area (LA), chlorophyll *a* (chl *a*), chlorophyll *b* (chl *b*), and total chlorophyll at 30 days after sowing

| | Parameter | | | | | | | | |
|---|---|---|---|---|---|---|---|---|---|
| Treatment | FW (g) | DW (g) | Moisture content (%) | PH (cm) | Number of leaf | LA ($cm^2$) | Chl *a* | Chl *b* | Total Chl |
| T1 | 52.78$^a$ | 2.11$^a$ | 95.99$^a$ | 3.59$^a$ | 10.20$^a$ | 2.34$^a$ | 12.07$^a$ | 9.06$^a$ | 7.99$^a$ |
| T2 | 152.50$^b$ | 5.76$^b$ | 96.22$^b$ | 3.74$^b$ | 11.35$^b$ | 4.03$^b$ | 10.57$^b$ | 7.69$^b$ | 6.79$^b$ |

*Note*: Plants were subjected to two different photoperiods under RB (16:4) lights. Figures with the same letter superscript within columns are not statistically different using LSD test at $P < 0.05$ probability level

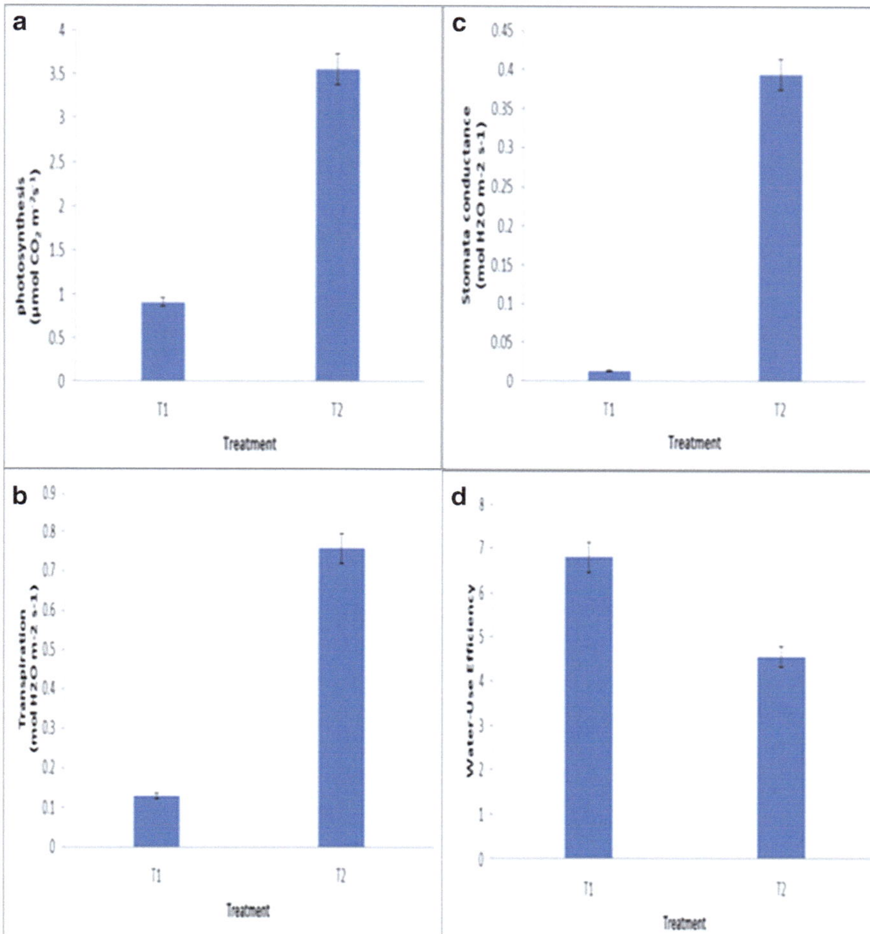


**Fig. 1** *Brassica chinensis* gas exchange measurement under RB (16:4) LED in different photoperiods were observed for 30 days, and the parameters measured are photosynthesis (**a**), transpiration rates (**b**), leaf stomatal conductance (**c**), and water-use efficiency (WUE) (**d**). Data for *B. chinensis* were hydroponically cultured and represent the mean ± standard error of 4 replicate plants. *Vertical bars* indicate the value of standard error

$CO_2$ $m^{-2}$ $s^{-1}$. Leaf of plant exposed to T1 (12 h light and 12 h dark) and T2 (continues 1 h light and 15 min dark in a day) showed that there were significant ($p < 0.05$) differences among both treatments.

### 3.1.2 Leaf Transpiration Rate

The diurnal mean leaf transpiration rate of *B. chinensis* under RB (16:4) LED in different photoperiods is shown in Fig. 1. Similar to photosynthesis, T2 depicted higher transpiration rate when compared to counterparts within the time frame of the study. The result of the statistical comparison between treatment means in 30 days shows that there were significant ($p < 0.05$) differences among both treatments.

### 3.1.3 Water-Use Efficiency (WUE)

The diurnal mean leaf WUE of *B. chinensis* under various levels of photoperiods is shown in Fig. 1. The T1 plants recorded higher WUE values than its T2 counterparts thus implying that photoperiod influenced the WUE of plants substantially. Plants grown under T2 gave lower values of WUE 4.54 compared to T1 with values of 6.79. The result from the comparisons of treatment means for 30 days showed that there were significant ($p < 0.05$) differences among both treatments.

### 3.1.4 Leaf Stomatal Conductance

The diurnal mean leaf stomatal conductance of *B. chinensis* under different treatments is depicted in Fig. 1. All plants under T2 conditions gave a higher value of 0.39 mol $H_2O$ $m^{-2}$ $s^{-1}$ than those in T1. Regardless of treatments, the stomatal conductance values were relatively higher for T2 plants which also contributed to higher photosynthesis of the plants. This was due to the impact of stomatal opening which maintained photosynthetic efficiency without much considerable change in photoperiod. The outcome of the statistical analysis of treatment means for 30 days in T1 and T2 revealed that there were significant ($p < 0.05$) differences.

## 4 Discussion

*B. chinensis* is widely grown in Malaysia, and its production is very important, both economically and commercially. The spectral quality of lights is the constant intensity and quantity of different photoperiods emitted by a light source and perceived by photoreceptors within a plant. Plant yields and quality are the result of interactions of various environmental factors under which plants are grown.

The present study examined the effects of different photoperiod light spectral conditions on the yield and quality of *B. chinensis* grown under the same environmental conditions. The FW and DW was comparatively greater in plants grown under T2 than under T1. These results indicate that the relationship between FW and DW will slightly affect the moisture contents that give a higher result (Table 1). The greater FW is likely associated with the greater LA achieved under these conditions. The larger leaf allowed greater light interception, which may have led to the significant increase in biomass [23]. The induction of the shade-avoidance syndrome requires the perception of the spectral changes associated with shade, rather than changes in total light quality [16]. The increase in plant height and enhancement of number of leaf as a consequence of pulse light and long photoperiod found in T2. However, there are significant different shown between T1 and T2 according their growth and will be defined as T2 have a greater growth and photomorphogenesis impact due to the additional photoperiod that is given in a pulse condition. From this study, it was shown that long light will affect the growth and light irradiation of a plant that causes the plant's height reduction and no significant changes in total leaf area when given during the night period [32].

According to Wang et al. [42], plant pigments have specific wavelength absorption patterns known as absorption spectra. Biosynthetic wavelengths for the production of plant pigments are referred to as action spectra. Although different photoperiods for both treatments were applied at the same PFD level, plants showed significant different absorption spectra of photosynthetic pigments, chl *a*, chl *b*, and total chl (Table 1). Perhaps, the applied PFD level (100 μmol $m^{-2}$ $s^{-1}$) had reached certain minimal PFD, which is essential for sufficient synthesis and activity of photosynthetic pigments and electron carrier. It was reported by Ref. [39] that plants with smaller chl contents seemed to use the chl more efficiently than plants with excessive chl. In this case, chl *a*, chl *b*, and total chl in the leaves have statistically different treatments; the chl contents under T2 were the lowest. This indicates that *B. chinensis* grown under T2 might be using chl more efficiently than grown under T1. Plants grown under both treatments appeared to synthesize more chl *a* compared to chl *b* (Table 1); it is due to the wider spectrum for chl *a* absorption and chl *a* is the molecule that makes photosynthesis possible [4].

The photosynthesis rates recorded (Fig. 1) showed that there are significant differences among T1 and T2. Another factor that contributed to the decreased in photosynthesis might be the limited $CO_2$ diffusion into the intercellular spaces of the leaf as a consequence of reduced stomatal conductance [24]. This consequence showed that photosynthesis rate was closely related to the changes in the leaf stomatal conductance. *B. chinensis* grown under T2 showed that stomatal conductance might cause high photosynthesis, because once the stomata remain open, photosynthesis activities will always occur to produce energy that will make food for plant growth. This indicates that long exposure to light under T2 (1 h light, 15 min dark in a day) will increase the activity of photosynthesis, stomatal conductance, and transpiration rate. In addition, higher stomatal frequency could facilitate $CO_2$ uptake and thus maintain a high photosynthetic activity [5]. Light is the energy source for photosynthetic organisms, and light intensity plays an important role in plant growth. Low light conditions inhibit plant growth and productivity by affecting gas exchange [47]. In our study, although gas exchange at T1 showed a low frequency compared with T2, it shows that 100 μmol $m^{-2}$ $s^{-1}$ photon flux density is suitable for their growth when induced health photosynthesis, stomatal conductance, and transpiration. The smaller stomatal frequency could restrain photosynthesis rates by increasing diffusive resistance to $CO_2$ uptake, which might reduce the burden of photosynthetic organs [25]. Stomatal conductance was strongly influenced by light quality, with highest numbers in leaves grown under RB LED light [44]. It can therefore be concluded that light quality has an important permanent effect on gas exchange during leaf development.

WUE showed that there are significant differences between T1 and T2 that consist of frequency of photosynthesis over the transpiration (Fig. 1). The transpiration rate was reduced significantly thus resulting in an increased of WUE as water stress induces stomatal closure. This is in agreement with Arndt et al. [2], which can be attributed to the postponement of the damaging effects caused by water deficit in trees through stomatal control.

## 5 Conclusion

As a conclusion, it was shown that the growth of *B. chinensis* was promoted by RB light-emitting diode, 100 μmol $m^{-2}$ $s^{-1}$ photon flux density under pulse photoperiod (1 h light, 15 min dark in a day) in comparison with the normal photoperiod (12 h light, 12 h darks). In this study, we investigate the photomorphogenesis response of the combination of red (R) and blue (B) light in different photoperiods. Based on this study, it appears that the combined RB LEDs induced relatively higher growth under pulse photoperiod (1 h light, 15 min dark in a day) than normal photoperiod (12 h light, 12 h dark). More importantly, it induces high growth and photomorphogenesis in control environment.

**Acknowledgments** This work was supported by the Fundamental Research Grant Scheme (4 F144) and GUP Tier 2 (002 J8) by the Ministry of Higher Education (Malaysia).

## References

1. Anzelika K, Renata MC, Stase D, Silva Z, Genadij K, Gintautas T, Pavelas D, Arturas Z (2008) In vitro culture of Chrysanthemum plantlets using light-emitting diodes. Cent Eur J Biol 3:161–167
2. Arndt SK, Clifford SC, Wanek W, Jones HG, Popp M (2001) Physiological and morphological adaptations of the fruit tree *Ziziphus rotundifolia* in response to progressive drought stress. Tree Physiol 21:1–11
3. Brown CS, Schuerger AC, Sager JC (1995) Growth and photomorphogenesis of pepper plants under red light-emitting diodes with supplemental blue or far-red lighting. J Am Soc Hort Sci 120:808–813
4. Calatayud A, Barreno E (2004) Response to ozone in two lettuce varieties on chlorophyll a fluorescence, photosynthetic pigments and lipid peroxidation. Plant Physiol Biochem 42:549–555
5. Chartzoulakis K, Bosabalidis A, Patakas A, Vemmos S (2000) Effects of water stress on water relations, gas exchange and leaf structure of olive tree. J Acta Hortic Sin 537:241–247
6. Cosgrove D (1981) Rapid suppression of growth by blue light. Plant Physiol 67:584–590
7. Dewir YH, Chakrabarty D, Hahn EJ, Paek KY (2007) Flowering of Euphorbia milii plantlets in vitro as affected by paclobutrazol, light emitting diodes (LEDs) and sucrose. Acta Hort 764:169–173

8. DiCristina K, Germino M (2006) Correlation of neighbourhood relationships, carbon assimilation, and water status of sage brush seedlings establishing after fire. West N Am Nat 66(4):441–449
9. Dougher T, Bugbee B (2004) Long-term blue light effects on the histology of lettuce and soybean leaves and stems. J Am Soc Hort Sci 129:467–472
10. Duong TN, Hong LTA, Watanabe H, Goi M, Tanaka M (2003) Efficiency of a novel culture system by using light-emitting diode (LED) on in vitro and subsequent growth of micropropagated banana plantlets. Acta Hort 616:121–127
11. Hahn E-J, Kozai T, Paek K-Y (2000) Blue and Red light-emitting diodes with or without sucrose and ventilation affect in vitro growth of *Rehmannia glutinosa*. Plantlets 43(4):247–250
12. Felker FC, Doehlert DC, Eskins K (1995) Effects of red and blue light on the composition and morphology of maize kernels grown in vitro. Plant Cell Tissue Org Cult 42:147–152
13. Hahn EJ, Bae CH, Lee YB (1998) Growth and leaf-surface characteristics of chrysanthemum plantlets between micropropagation and microponic system. J Kor Soc Hort Sci 39:838–842
14. Hahn EJ, Kozai T, Paek KY (2000) Blue and red light-emitting diodes with or without sucrose and ventilation affects in vitro growth of *Rehmannia glutinose* plantlets. Plant Biol 43:247–250
15. Hanyu H, Shoji K (2002) Acceleration of growth in spinach by short-term exposure to red and blue light at the beginning and at the end of the daily dark period. Acta Hortic 580:145–150
16. Smith H (1994) Sensing the light environment: the functions of the phytochrome family. In: Kendrick RE, Kronenberg GHM (eds) Photomorphogenesis in plants, 2nd edn. Kluwer, Dordrecht, pp 377–416
17. Jao RC, Fang W (2004) Effects of frequency and duty ratio on the growth of potato plantlets in vitro using light-emitting diodes. Hort Sci 39:375–379
18. Jao RC, Lai CC, Fang W, Chang SF (2005) Effects of red light on the growth of Zantedeschia plantlets in vitro and tuber formation using light-emitting diodes. Hort Sci 40:436–438
19. Johkan M, Shoji K, Goto F, Hashida S, Yoshihara T (2010) Blue light-emitting diode light irradiation of seedlings improves seedling quality and growth after transplanting in red leaf lettuce. Hort Sci 45:1809–1814
20. Kasajima S, Inoue N, Mahmud R, Kato M (2008) Developmental responses of wheat cv. Norin 61 to fluence rate of green light. Plant Prod Sci 11:76–81
21. Kim HH, Goins GD, Wheeler RM, Sager JC (2004) Green-light supplementation for enhanced lettuce growth under red- and blue-light-emitting diodes. Hort Sci 39:1617–1622
22. Kim SJ, Hahn EJ, Heo JW, Paek KY (2004) Effects of LEDs on net photosynthetic rate, growth and leaf stomata of *Chrysanthemum* plantlets in vitro. Sci Hort 101:143–151
23. Lin K-H, Huang M-Y, Huangc W-D, Hsuc M-H, Yangd Z-W, Yang C-M (2013) The effects of red, blue, and white light-emitting diodes on the growth, development, and edible quality of hydroponically grown lettuce (*Lactuca sativa* L. var. capitata). Sci Horticul 150:86–91
24. Lawlor DW (2002) Limitation to photosynthesis in water stressed leaves: stomata versus metabolism and the role of ATP. Ann Bot 89:871–885
25. Lawson T, Caemmerer S, Baroli I (2011) Photosynthesis and stomatal behaviour. J Prog Bot 72:265–304
26. Lee AE, Tewari RK, Hahn EJ, Paek KY (2007) Photon flux density and light quality induce changes in growth, stomatal development, photosynthesis and transpiration of *Withania somnifera* (L.) Dunal. plantlets. Plant Cell Tissue Organ Cult 90:141–151
27. Lian ML, Murthy HN, Paek KY (2002) Effect of light emitting diodes (LEDs) on the in vitro induction and growth of bulblets of Lilium oriental hybrid'Pesaro'. Sci Hort 94:365–370
28. Liu Mengxi X, Zhigang YY, Yijie F (2011) Effects of different spectral lights on Oncidium PLBs induction, proliferation, and plant regeneration. Plant Cell Tiss Organ Cult 106:1–10
29. MacKinney G (1941) Absorption of light by chlorophyll solutions. J Biol Chem 140:315–322
30. McCree KJ (1972) The action spectrum, absorptance and quantum yield of photosynthesis in crop plants. Agric For Meteorol 9:191–216
31. Miyashita Y, Kitaya Y, Kozai T, Kimura T (1995) Effects of red and far-red light on the growth and morphology of potato plantlets in vitro: using light emitting diode as a light source for micropropagation. Acta Hort 393:189–194

32. Maffei M, Scannerini S (1999) Photomorphogenic and chemical responses to blue light in *Mentha piperita*. J Essent Oil Res 11:730–738
33. Moshe R, Dalia E (2007) On the effect of light on shoot regeneration in petunia. Plant Cell Tissue Org Cult 89:49–54
34. Nhut DT, Takamura T, Watanabe H, Murakami A, Murakami K, Tanaka M (2002) Sugar-free micropropagation of *Eucalyptus citriodora* using light-emitting diode (LEDs) and film-rockwool culture system. Environ Contr Biol 40:147–155
35. Nhut DT, Takamura T, Watanabe H, Okamoto K, Tanaka M (2003) Responses of strawberry plantlets cultured in vitro under superbright red and blue light-emitting diodes (LEDs). Plant Cell Tissue Org Cult 73:43–52
36. Nhut DT, Takamura T, Watanabe H, Okamoto K, Tanaka M (2005) Artificial light source using light-emitting diodes (LEDs) in the efficient micropropagation of *Spathiphyllum* plantlets. Acta Hort 692:137–142
37. Avercheva OV, Berkovich YA, Erokhin AN, Zhigalova TV, Pogosyan SI, Smolyanina SO (2009) Growth and photosynthesis of Chinese cabbage plants grown under light-emitting diode-based light source. Fiziologiya Rastenii 56(1):17–26
38. Puspa RP, Ikuo K, Ryosuke M (2008) Effect of red-and blue-light emitting diodes on growth and morphogenesis of grapes. Plant Cell Tissue Org Cult 92:147–153
39. Saebo A, Krekling T, Applegren M (1995) Light quality affects photosynthesis and leaf anatomy of birch plantlets in vitro. Plant Cell Tissue Org Cult 41:177–185
40. Shin KS, Mrthy HN, Heo JW, Hahn EJ, Paek KY (2008) The effect of light quality on the growth and development of in vitro cultured Doritaenopsis plants. Acta Physiol Plant 30:339–343
41. Tanaka M, Takamura T, Watanabe H, Endo M, Yanagi T, Okamoto K (1998) In vitro growth of Cymbidium plantlets cultured under super bright and blue light-emitting diodes (LEDs). J Hort Sci Biotech 73:39–44
42. Wang H, Gu M, Cui J, Shi K, Zhou Y, Yu J (2009) Effects of light quality on $CO_2$ assimilation, chlorophyll-fluorescence quenching, expression of Calvin cycle genes and carbohydrate accumulation in *Cucumis sativus*. J Photochem Photobiol B Biol 96:30–37
43. Wongnok A, Piluek C, Techasilpitak T, Tantivivat S (2008) Effects of light emitting diodes on micropropagation of Phalaenopsis orchids. Acta Hort 788:149–156
44. van Ieperen W, Savvides A, Fanourakis D (2012) Red and blue light effects during growth on hydraulic and stomatal conductance in leaves of young cucumber plants. Acta Hort 956
45. Yanagi T, Okamoto K, Takita S (1996) Effects of blue and blue/red lights of two different PPF levels on growth and morphogenesis of lettuce plants. Acta Hortic 440:117–122
46. Yorio NC, Goins GD, Kagie HR (2001) Improving spinach, radish, and lettuce growth under red light-emitting diodes (LEDs) with blue light supplementation. Hort Sci 36:380–383
47. Zavala JA, Ravetta DA (2001) Allocation of photoassimilates to biomass, resin and carbohydrates in *Grindelia chiloensis* as affected by light intensity. J Field Crop Res 69:143–149
48. Zhang YS, Huang X, Chen YF (2009) Experimental course of plant physiology. Higher Education Press, Beijing (in Chinese)

# Quiet Standing of Time-Delay Stability with Derivative Feedback Control

**Fitri Yakub, Ahmad Zahran Md. Khudzari, and Yasuchika Mori**

## 1 Introduction

The investigation of time-delay systems has received substantial interest in the last five decades. Conceptually, time delays arise in the property of a physical system by which the action response is delayed to an applied force or in its effect and often appears in many practical systems and mathematical formulations such as chemical processes, electrical and control systems, and economical systems [1]. Figure 1 shows an example in control system applications including the delay existence on account of the time it takes regularly to perform and operate in order to obtain the desired information, to develop decisions of control, and to make and execute the decision making.

Time delay always happens like in traffic flow, a model that refers to the drivers' delayed reactions, in which the reaction delays rely upon the physiological and cognitive states of the driver and vary under stimuli actions and physical conditions. Because drivers require a minimum amount of time in external events to become conscientious, these delays are very important and critical characteristics of human driver's behavior which combine sensing, selection, perception, and response to make a decision.

F. Yakub (✉)
Graduate School of System Design, Tokyo Metropolitan University, Tokyo, Japan

Malaysia-Japan International Institute of Technology, Universiti Teknologi Malaysia, Kuala Lumpur, Malaysia
e-mail: fitri@ic.utm.my

A.Z. Md. Khudzari
IJN-UTM Cardiovascular Engineering Centre, Faculty of Biosciences and Medical Engineering UTM, Johor Bahru, Malaysia
e-mail: zahran@biomedical.utm.my

Y. Mori
Graduate School of System Design, Tokyo Metropolitan University, Tokyo, Japan
e-mail: ymori@tmu.ac.jp

K. Ab. Hamid et al. (eds.), *The Malaysia-Japan Model on Technology Partnership: International Proceedings 2013 of Malaysia-Japan Academic Scholar Conference*,
DOI 10.1007/978-4-431-54439-5_6, 

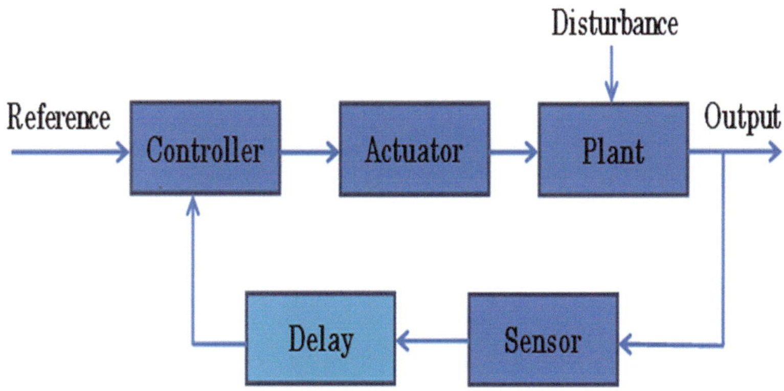


**Fig. 1** Delay in feedback system

Delay also occurs when analyzing vehicle dynamics stability through antilock braking system, collision-free traffic flow controllers, and active front steer systems [2]. Think about driving a car and while turning the steering wheel, the tires do not respond to the short time period. It means there is a delay between the steering wheel and the tires. This delay may invite collisions and result in traffic jams and irregular waves, effects that lead to the presence of casualties on highways and productivity deficit or losses because of increased travel times [3].

In most cases, the presence of delays may be destructive to the operation of the dynamical system since it is frequently a source of system instability and also complicates system analysis and synthesis. For example, a stable feedback system without any delay may behave unstable due to some delay effects [4]. However, somehow there are beneficial aspects of the time delay.

Main advantages of time delay are in feedback systems that generate the reference signal for the minimum required knowledge of the systems [5]. In fact, the reference signal is not necessarily known for feedback time delay of the system under control. The introduction of a delay in the system may stabilize or destabilize the system [6]. Potentially stabilizing and controlling the effect of delay systems is an inspiration and motivation for exploiting the existence of delays in dynamical systems over the last five decades [7, 8].

Another example is in supply chain management applications; appropriate and suitable tuning and modification of the spindle speed in metal machining process may help in adjusting the delay in purpose to avoid the response chattering that may deteriorate the process. Adding some delays with intention will allow supply chain managers to make a decision to observe consumer demand and trends for better stocking and purchasing decisions [9].

The main objective and scope of this paper are to study the effects of time delay on system stability in various limitations and advantages that arise that mainly focused on delay differential equation (DDE) model for a linear time-invariant (LTI) system. Delay effect of derivative feedback for a single delay system is mentioned in Sect. 2. Section 3 is dedicated to case study in biology area which is mainly focused on quiet-standing analysis. Finally, this paper is concluded in Sect. 4.

## 2 Delay as a Derivative Feedback

The most important and popular design method of an LTI system is an eigenvalue assignment. One of the eigenvalue assignment approaches is to use state-derivative feedback based on Ackermann's formula in which the gain matrix is calculated. The concept has been extended to state-derivative feedback that is useful for various practical systems including control of vibration or vibration suppression. By considering the van der Pol equation, which is one of the classical equations has known as a prototype for systems with exhibiting limit cycle oscillations. It has the following equation:

$$\ddot{x}(t) + \varepsilon\left(x^2(t) - 1\right)\dot{x}(t) + x(t) = g(x, t; \tau) \tag{1}$$

with $x \in \mathrm{R}$ and $\varepsilon > 0$ for energy dissipation and generation at high and low amplitudes, respectively. The limit cycle oscillation for the dynamics of the van der Pol equation disappears under the effect of state-derivative feedback with time delay when $\varepsilon$ is zero. Hence, the right-hand side of (1) can be considered as

$$g(x, t; \tau) = \varepsilon k_1 \dot{x}(t - \tau) + \varepsilon k_2 \ddot{x}(t - \tau) \tag{2}$$

where $k_1$ and $k_2$ are the feedback gains and $\tau$ is a positive delay. Next, with the damping van der Pol in (1) and feedback gain in (2), the system description in (1) becomes

$$\begin{aligned} &\ddot{x}(t) + x(t) = \varepsilon\left(x^2(t) - 1\right)\dot{x}(t) + k_1 \\ &\dot{x}(t - \tau) + k_2\ddot{x}(t - \tau) \end{aligned} \tag{3}$$

Let us linearize the system in (3) about the zero equilibrium that may give the analysis equation for infinitesimal perturbation as a sample path technique:

$$\ddot{x}(t) + x(t) = \varepsilon\dot{x}(t) + k_1\dot{x}(t - \tau) + k_2\ddot{x}(t - \tau) \tag{4}$$

By defining $x_1 = x$ and $x_2 = \dot{x}_1$ in the state-space form with the chosen feedback gain to stabilize the system is given in the following state form:

$$\dot{x}(t) = \begin{bmatrix} 0 & 1 \\ -1 & \varepsilon \end{bmatrix} x(t) + \begin{bmatrix} 0 & 0 \\ k_1 & k_2 \end{bmatrix} \dot{x}(t - \tau) \tag{5}$$

In the case when $\varepsilon = 0.1$ and the state-derivative feedback term is zero which is $k_1 = k_2 = 0$, the system in (5) becomes unstable given its eigenvalues are $0.05 \pm 0.999j$. For the case arbitrarily selected, the real part of the desired rightmost eigenvalue is $-1$, when $\tau = 0.085$ s and $T = 5$ s; then, the required gain is found to be $k = [0.69–8.84]$. Figure 2 shows the responses for states $x_1$ and $x_2$ are

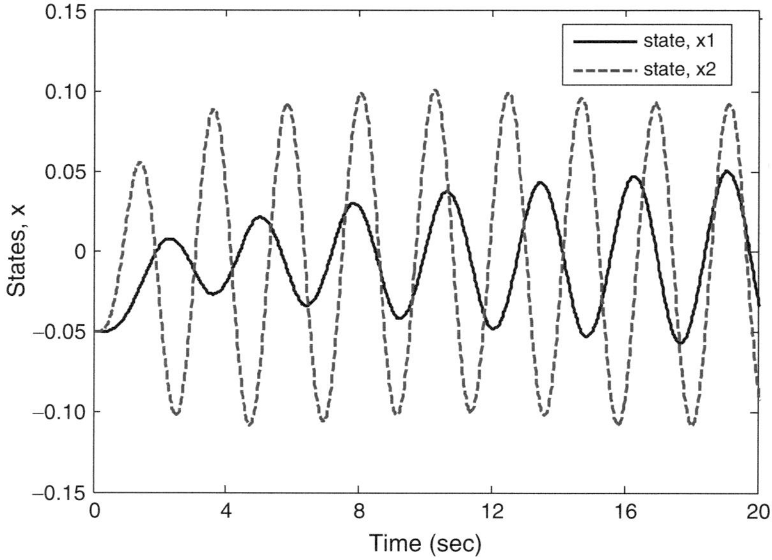


**Fig. 2** Comparison of system states without feedback

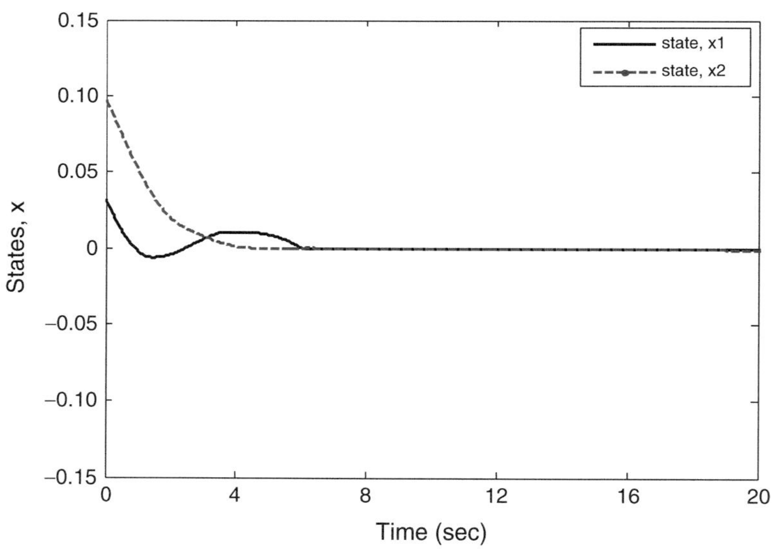


**Fig. 3** Comparison of system states with feedback

unstable for systems without feedback controller, while Fig. 3 shows when the state-derivative feedback controller is applied to systems; it can stabilize the system for both states. In this case, all the other eigenvalues are to the left side plane and the rightmost eigenvalues have exactly their real parts at $-1$.

Now, let us consider the second-order linear system of the following form:

$$\ddot{x}(t) - 0.5\dot{x}(t) + x(t) = u(t) \tag{6}$$

where $x \in \mathrm{R}$ and $u \in \mathrm{R}$, making the system unstable for the given input, $u(t) = 0$ because the eigenvalues have positive real part due to the effect of negative damping, and the result amplitude of the solutions increases exponentially. The derivative feedback controller of (6) is given by

$$u(t) = -k\dot{x}(t) \tag{7}$$

with a value of gain $k > 0.1$; it changes the *open-loop poles and moves it from unstable in* the *right-half* of the s-*plane* toward the left-half plane for stable poles. Here, as the alternative way, we may use the time-delay feedback control law as follows:

$$u(t) = x(t - \tau) - x(t) \tag{8}$$

with a given gain *k*; Equation (8) can be explained and related to a confine difference control law as the form

$$u(t) = \frac{-k[x(t) - x(t - \tau)]}{\tau} \tag{9}$$

For an example where the small values of the delay $\tau$ are chosen such as $\tau < 0.1$, (8) can approximate the derivative feedback control without time delay in (7) for $k = \tau$. Based on linear system in (6), it is clear that the system can be stable and balanced by changing the two poles from the right-half plane into the left-half plane for any value of $\tau \in (10.1, 1.72)$. From the above example, it demonstrated and explained that the system for stability of closed loop can be achieved with taking into account some delays to approximate and resemble the signal feedback derivatives by designing the suitable and appropriate controllers [10].

## 3 Quiet-Standing Case Study

Human bipedal stance is very important involving the upper extremities for execution of tasks. A large body mass keeps the body in straight posture with a small base of support and its center of mass situated high above that give inherently unstable position. Human bipedal stance for quiet stance situation of equilibrium control is represented by the mechanism shown in Fig. 4. In control mechanism point of view, the main focus on quiet stance is to maintain and control the vertical arrangement and center of mass equilibrium of the human body. It consists and involves

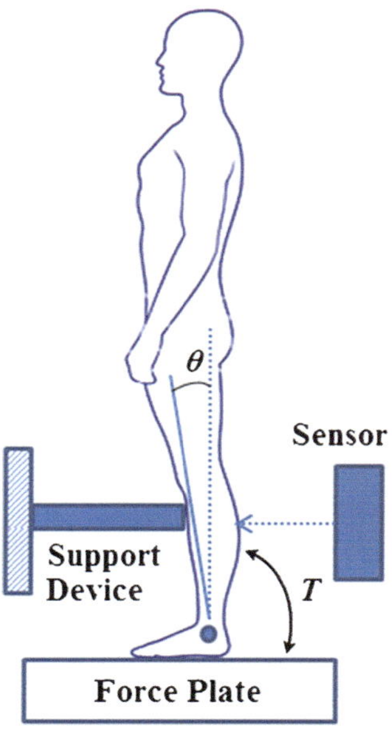


**Fig. 4** Quiet standing (may refer to reference in [9] for further knowledge)

analyzing muscle activity at the ankle joint which represents the human ankles during standing.

Analysis of postural control during quiet-standing balance is a critical factor and proposes awareness on how humans manage and organize their stability in vertical circumstance and ability to control and maintain the human body and considers how humans walk and avoid falling in daily routine activities. The motor command which is the ankle joint torque during quiet stance can be generated either passively or actively in order to maintain, stabilize, and control the body. A passive torque component generated mechanically corresponds to the tension or stiffness produced by muscle tonus and the tension or stiffness of the joint tissues, i.e., tendons and ligaments [11]. However, the passive torque is not sufficient enough to prevent the body from falling by generating the required torques. On the other hand, active torque component is produced by the central nervous system that controls and stimulates contractions of muscle depending on the human body dynamics and kinematics as a feedback information. This torque is influenced and activated by the external disturbance generating impulsive body sway [12]. Thus, the active torque as additional component is needed to ensure a behavior stability system.

The entire set of dynamic equilibrium equations for quiet stance can be easily derived to set up the relationship between the ground reaction forces and the postural sway angle. Figure 5 shows a free body diagram of the inverted pendulum which represents the entire human body as a single segment approximation as shown in Fig. 4 excluding the feet rotating about point $A$ as an ankle joint. $F_v$ and $F_h$ are vertical and horizontal forces acting at ankle joint, respectively, $T$ is moment torque acting at ankle joint produced by body muscles, $m$ and $I$ are the mass and

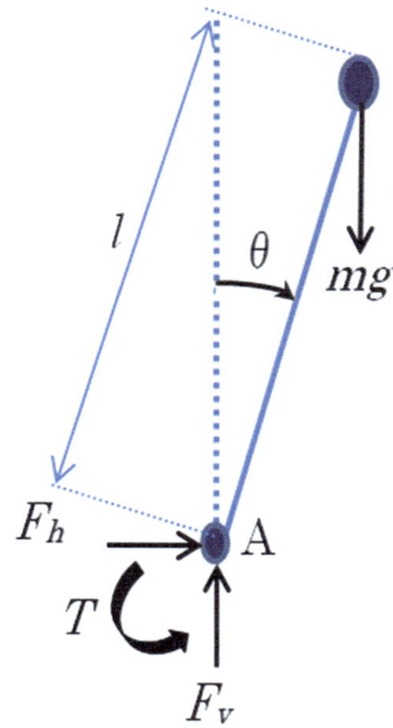


**Fig. 5** Free body diagram of quiet standing

moment of the inertia of body above ankle, respectively, $l$ is the distance between human body center of mass and the ankle, $\theta$ is absolute sway angle with respect to fixed vertical reference at point $A$, and $\epsilon$ is the torque disturbance, which is sufficiently small compared with other torque contributions [13]. Thus, for model of human inverted pendulum, the dynamic equation can be approximated as follows:

$$I\ddot{\theta} = mgl\sin\theta + T + \varepsilon \tag{10}$$

The ankle joint torque, $T$, is modeled as

$$T = k\theta + b\dot{\theta} + f_P(\theta_\tau) + f_D\left(\dot{\theta}_\tau\right) \tag{11}$$

where $k$ is the passive stiffness parameter and $b$ is the passive viscosity parameter. Both parameters represent the passive feedback torques for the human ankle joint that related and connected to the constitutive mechanical impedance without time delay. $\tau$ is the neural transmission delay, $f_p$ and $f_d$ are terms that represent a delay as a function which determines the active and agile neural human feedback that is affected by slope angle, $\theta$, and angular velocity, $\dot{\theta}$.

The DDE of inverted pendulum in Eq. (10) and ankle joint torque equation in (11) are mathematically combined and integrated by employing the technique of forward Euler where $x(t) = [\theta(t), \dot{\theta}(t)]$, $\sigma$ is corresponding amplitude from the sway angle effect, and $\tau$ is the feedback time delay. Moreover, the second-order of motion equation for quiet stance can be reformulated in ordinary DDE as follows:

$$\dot{x}(t) = f(x(t), x(t-\tau)) + \sigma\varepsilon(t) \tag{12}$$

For the quiet stance task, an isometric torque-exertion task can be used by using the model of the neuromusculoskeletal (NMS) process since during quiet standing, it requires a little muscular activity and the changes of the muscle length are very little and small. Therefore, it can easily and concisely be modeled as a second-order low-pass system in critically damped [14]. The transfer function $G(s)$ for NMS system is written as

$$G(s) = \frac{K_G \omega_n^2}{(s+\omega_n)^2} = \frac{K_G \left(\frac{1}{T_s}\right)^2}{\left(s+\frac{1}{T_s}\right)^2} \tag{13}$$

where $T_s$ is the twitch contraction time, which refers to the time distance from the initiation of contraction to the peak of the twitch which is the moment at the time when the created force comes at its peak from the moment during a stimulus or impulse reaches the body muscle, and $K_G$ and $\omega_n$ are the second-order system representing the gain and the natural frequency.

From Eq. (13), the system dynamic characteristics can be uniquely determined and characterized for the body muscle by the twitch contraction time which is equivalent to the inverse of the natural frequency ($T_s = 1/\omega_n$). Take into consideration that both $T_s$ and $\omega_n$ would perform the system dynamics that encourage a phase delay instead of function of constant time delay, using a function of frequency domain, and capture the NMS system dynamics characteristics [15]. Afterward the time delay persuaded through the second-order NMS system is represented by mechanical and chemical dynamics due to the sliding filament action and variation of calcium concentration in the muscle fiber, respectively; the $T_s$ is believed to depend not only on the foot condition and ankle joint, i.e., pressure on ligament, joint angle, and foot stiffness, but also on the corresponding motor task for muscle fiber properties. Therefore, for the quiet-standing postural conditions, it was required under equivalent condition to identify the NMS system.

Figure 6 shows a block diagram consisting of the proportional-derivative (PD) controller combined as neural controller that affects animate nerve organs with given gain parameters $K_P$ and $K_D$ of the closed-loop quiet-standing system. Here, the PID controller is not necessarily desirable; thus, PD controller is chosen for its simplicity. In the case of the larger feedback delay than its critical value, it is known that a PD controller cannot stabilize an unstable equilibrium system. The PD controller critical feedback delay, $\tau_c$, for an inverted pendulum can be given as follows:

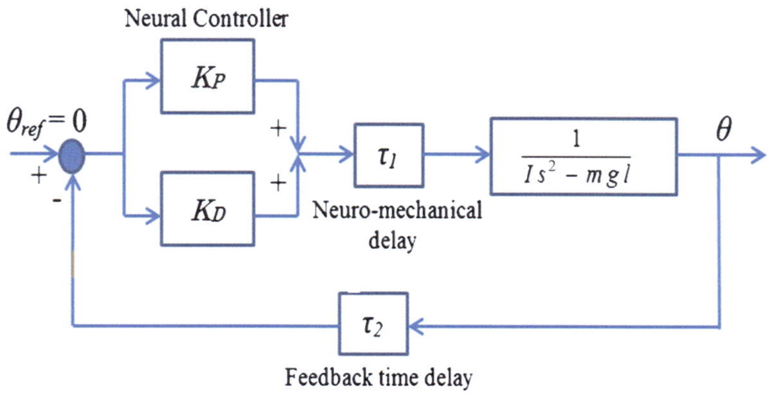


**Fig. 6** Closed-loop control diagram of quiet standing

$$\tau_c = \frac{T_p}{\pi\sqrt{2}} \tag{14}$$

where $T_p$ is the small oscillation period for the inverted pendulum structure hanging at its downward position.

Therefore, with a delay of 200 ms based on reference in [16], the main challenge is how to adjust and eliminate the large neural feedback transmission delay which can cause instability and danger to the system. The use of standard PD controller might face a rigid and tight as give and take which leaves small and narrow boundary or margins for the parameters controller formation. In particular for stability conditions, $K_P$ must be chosen not too large in order to avoid the instability delay effect but must be bigger enough for stability conditions in order to supplement and add the inadequate ankle stiffness. The PD controller works and performs after a length of time $t$ becomes effective as an active correction mechanism comes forth from the neural controller as illustrated in Fig. 6. On the other hand, for the quiet-standing damping factor of sway arrangements, it is required that $K_D$ be on larger value, but keep in mind that the upper bound on this gain parameter is stringent and rigid setting by the feedback delay.

The characteristic equation of quiet standing from the simplifications of the standard block diagram in Fig. 6 is given as

$$f(s;\tau K_P, K_D) = Q_1(s, K_P, K_D) + e^{-\tau s} Q_2(s, K_P, K_D) = 0 \tag{15}$$

where $\tau$ acts as the delayed sensory feedback for the human postural model and $Q_1$ and $Q_2$ are polynomial parameters with a single delay. The aim is to find the combinations of gains $K_P$ and $K_D$ such that for a given delay $\tau$, the NMS quiet-standing model in (13) is stable.

Based on (10) and (11), $\theta_\tau$ and $\dot{\theta}_\tau$ can be approximated by first-order Taylor's series yielding to

$$(I - K_D\tau)\ddot{\theta} + (b + K_D - K_P\tau)\dot{\theta} + (k + K_P - mgl)\theta = 0 \tag{16}$$

Here, we can see that when both inertia and damping remain positive, for an inverted pendulum, the delay tends and aims to diminish the significant damping and inertia for system stability condition. This is because of the eigenvalues can solve the following equation:

$$\lambda^2 + \frac{b + K_D - K_P\tau}{I - K_D\tau}\lambda + \frac{k + K_P - mgl}{I - K_D\tau} = 0 \tag{17}$$

Based on reference in [17], it is explained and demonstrated that in upright posture control situation, in order to gain the asymptotic stability, the feedback controller gains need to be followed and satisfied based on three conditions from which the $K_P$ and $K_D$ gains are given with the following equations:

$$\begin{aligned} &K_P > mgl - k \\ &K_D < I/\tau \\ &K_D > \tau K_P - b \end{aligned} \tag{18}$$

Figure 7 shows the proportional and derivative (*P-D*) plane parameters which illustrates and analyzes for the feedback parameters; a triangle would limits and margin the set of acceptable values. It clearly indicates that when $\tau$ decreases, the triangle area of the *P-D* plane will increase and move to fulfill the entire first portion of the critical value to the right given by *mgl–k*. On the other hand, when $\tau$ increases and reaches a critical value, the triangle area will decrease and vanish, which is given by the following equation:

$$\tau = \frac{b + \sqrt{b^2 + 4I(mgl - k)}}{2(mgl - k)} \tag{19}$$

where $\tau$ is the inverted pendulum system representing the upright posture for physical parameters including *m*, *I*, *l*, *k*, and *b*. Here, we can see that when the condition of $D > (\tau P–b)$ is defective through a Hopf bifurcation, a failure of stability for the position upright stance would happen. This loss of stability is a typical rare occurrence that encourages the dynamical system to be in either a stable or unstable oscillatory manner via equilibrium state instability. This might tend for unstable core type to an unstable oscillation about the equilibrium in upright posture. Even when the condition of $D = (\tau P–b)$, from the linearized equation in (15), the eigenvalues for the real parts would disappear and toward the system instability for upright equilibrium loses.

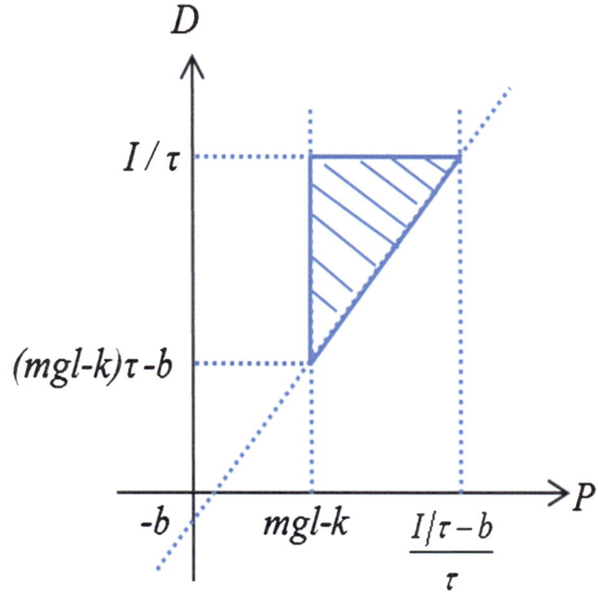


**Fig. 7** Proportional and derivative plane parameters, region of stability (*shaded triangle*)

## 4 Conclusion

In this paper, the effect of delays to linear time-invariant delay differential equations modeled in dynamical systems was expressed and explained. This paper focused on eigenvalue locations and parametric approaches for stability condition rather than Lyapunov techniques. An example with delay as derivative feedback on quiet-standing biological case study was analyzed and discussed. This paper is limited to the study of delays for stability effects based on linear system. Authors believe that delays of controllers for nonlinear system area have grown fast and deserve further research and study.

## References

1. Sontag ED (2004) Some new directions in control theory inspired by systems biology. Syst Biol 1:9–18
2. Bose A, Ioannou PA (2003) "Analysis of traffic flow with mixed manual and semi-automated vehicles", IEEE Trans. Intell Transport Syst 4(4):173–188
3. Treiber M, Kesting A, Helbing D (2006) Delays, inaccuracies and anticipation in microscopic traffic models. Phys A 360(1):71–88
4. Baker CTH, Bocharov GA, Rihan FA (1999) A report on the use of delay differential equations in numerical modelling in the biosciences. Numerical analysis report, No. 343, Manchester, UK. http://citeseer.ist.psu.edu/old/523220.html
5. Logemann H, Townley S (1996) The effect of small delays in the feedback loop on the stability of neutral systems. Syst Contr Lett 27:267–274

6. Frank TD, Friedrich R, Beek PJ (2005) Identifying and comparing states of time-delayed systems: phase diagrams and applications to human motor control systems. Phys Lett A 338:74–80
7. Niculescu S-I, Michiels W (2004) Stabilizing a chain of integrators using multiple delays. IEEE Trans Automat Contr 49:802–807
8. Riddalls CE, Bennett S (2002) The stability of supply chains. Int J Prod Res 40:459–475
9. Sipahi F et al (2011) Stability and stabilization of systems with time delay. IEEE Contr Syst Mag 2011:38–49
10. Linh V, Morgansen KA (2010) Stability of time-delay feedback switched linear systems. IEEE Trans Automat Contr 55(10):2385–2390
11. Morasso PG, Schieppati M (1999) Can muscle stiffness alone stabilize upright standing? J Neurophysiol 83:1622–1626
12. Winter DA, Patla AE, Rietdyk S, Ishac MG (2001) Ankle muscle stiffness in the control of balance during quiet standing. J Neurophysiol 85:2630–2633
13. Schieppati M, Nardone EA (1997) Medium-latency stretch reflexes of foot and leg muscles analyzed by cooling the lower limb in standing humans. J Physiol 503:691–698
14. Loram ID, Maganaris CN, Lakie M (2005) Active, non-spring-like muscle movements in human postural sway: how might paradoxical changes in muscle length be produced? J Physiol 564:281–293
15. Fuglevand AJ, Winter DA, Patla AE (1993) Models of recruitment and rate coding organization in motor-unit pools. J Neurophysiol 70:2470–2488
16. Loram I, Lakie M (2002) Direct measurement of human ankle stiffness during quiet standing: the intrinsic mechanical stiffness is insufficient for stability. J Physiol 545:1041–1053
17. Bottaro A et al (2008) Bounded stability of the quiet standing posture: an intermittent control model. Hum Mov Sci 27:473–495

# Social Determinants of Psychiatric Morbidity Among Malaysian Children

**Ali Adli, YT Ooi, YL Tan Isaac, WY Ho Katrina, HM Chai, Seet Ruijia, KW Teh, and KH Wong**

## 1 Introduction

Social determinants of health interact with each other to elicit a diversification in health and ultimately mortality. Simply defined, it is the health impact of the social environment on people living in a particular community [11], also termed as the causes of risk factors that expose an individual to a disorder or disease, i.e. the causes of causes [3]. Children psychiatric morbidity has become the topic of great interest under the perspective of health disparities. The world is experiencing tremendous growth in the children psychiatric epidemiology over the past 20 years [5].The definitions of caseness differ from encompassing a combination of outcomes such as the child's self-concept, reading ability and life skills [1] or using antisocial behaviours as indicators [7]. Looking into different countries, the overall prevalence of mental disorder in children is 17.7 % in Australia [19] and 18.1 % in Canada [12].

In Malaysia, psychiatric morbidity is one of the top five burden of diseases in our society with the overall prevalence of 48.1 % of adult population having some form of psychiatric morbidity [3]. The psychiatric morbidity in children between the ages of 5 and 15 is noted to be 20.3 % from the same report. Despite the worrying accentuation of this problem in Malaysia, the mental health services available for children have lagged behind when compared to those for adults [9]. Besides, many parents are still unaware of this problem and remain reluctant to attend their children to the services [2, 8].

A. Adli (✉) • Y. Ooi • Y.T. Isaac • W.H. Katrina • H. Chai • S. Ruijia
Department of Paediatrics, Faculty of Medicine, Universiti Kebangsaan, Kebangsaan, Malaysia
e-mail: adli.ali@ppukm.ukm.edu.my

K. Teh • K. Wong
Faculty of Medicine, Universiti Malaysia Sabah, Sabah, Malaysia

K. Ab. Hamid et al. (eds.), *The Malaysia-Japan Model on Technology Partnership: International Proceedings 2013 of Malaysia-Japan Academic Scholar Conference*,
DOI 10.1007/978-4-431-54439-5_7, 

The severity of psychiatric morbidity in children has urged the researchers in the related field to conduct continuous studies. In terms of association, gender, age and racial difference [3, 16–19] are the common socio-demographic factors associated with children psychiatric morbidity. Other etiological causes or risk factors include inherent factors influencing the mental growth and progression, biological factors and tobacco and alcohol exposure during pregnancy [4, 10, 13, 15].

Nonetheless, social determinants of psychiatric morbidity among children provide alternative ways in explaining the phenomenon. Those social determinants include family environment [6, 16], economic hardship [9] and education level of caregiver [16, 22]. These social determinants are able to pose an inequality, particularly to the disadvantaged social groups when compared with advantaged social groups.

The most studied social determinant is the economic status of an individual, in this case, the socio-economic status of the caregivers. There is a strong association between socio-economic status (SES) and psychiatric distress as SES influenced the social support and social capital which are also social determinants for health [9]. In Malaysia, children in a family with lower income are more likely to have psychiatric morbidity [3, 8, 18]. Family environment includes the marital status and increasing number of siblings.

Notwithstanding the fact that numerous studies on social determinants of psychiatric morbidity among children have been published, there are limited resources in Malaysia in which the results are country specific. By addressing the social determinants of children psychiatric morbidity in Malaysia, appropriate interventions can be designed accordingly to tackle and eliminate the health disparities that exist within children. Therefore, in this study, we aimed to determine the social determinants of psychiatric morbidity among children in Malaysia. In addition, we assessed the prevalence of psychiatric morbidity among children in Malaysia.

## 2 Method

This cross-sectional study was carried out from April 2012 to Jun 2012. The study population includes 240 school children and adolescents aged between 5 and 16 years who were studying in primary and secondary levels either in government or private schools in various states of Malaysia. Ethical clearance for the study was obtained from UKMMC, and permission to conduct the study at the school was obtained from the Ministry of Education. Centres which were involved were from the states of Penang, Selangor, Federal Territory of Kuala Lumpur and Sabah. These states were selected as each state has different backgrounds and different social settings. Participants for this study were chosen based on universal sampling. Questionnaires were completed by the caretakers of the children. The questionnaire includes socio-demographic data, social determinants and the Reported Questionnaire for Children (RQC). The RQC is a validated 10-item questionnaire used as a tool to assess children's mental health [20]. A score of 1 and above indicates psychiatric morbidity. The score is modified by the researchers of the Malaysian

Health and Morbidity Survey 2006 to include two additional questions to assess hyperactivity and inability to concentrate. It was found to have a sensitivity score of 91 % and specificity of 82 %; therefore, indicating that one or more positive scores on the questionnaire tend to signify some kind of developmental disability, significant degree of emotional and behavioural disorder and psychotic disorder [3].

The questionnaire was further modified and expanded based on a questionnaire which was used in a small community study undertaken in Malaysia on the psychiatric morbidity assessments among secondary school students [18]. Before the commencement of the study, pretest through face validation with 10 children identified from a primary school was carried out to validate the reliability of the modified questionnaire. The questionnaires were distributed to the parents of students in the schools. All quantitative data were recorded in Microsoft Excel. Before analysis, data were checked to ensure consistency. The dependent variable was children psychiatry morbidity, and the independent variables were the various social determinant factors. To obtain the association between prevalence of children psychiatric morbidity and the respective social determinant factors, chi-square tests and $t$-test were used. An alpha level of 0.05 was used for statistical tests indicating presence of association and 0.01 showing very significant level of association with the confidence level of 95 %. All computation of data was made using Statistical Package for the Social Sciences (SPSS) for Windows Version 17.0.

# 3 Results

## 3.1 Sample Description

A total number of 240 Malaysian children's and their parents were involved in this study. Referring to Table 1, the children were aged between 5 and 15 years with a mean of 10 years 8 months (10.73 ± 2.30) from a sample of 86 males and 146 females. Of this number, Bumiputeras summed up to 94 (39.1 %) with breakdown of 20.4 % Malays, 9.6 % Kadazan, 5.8 % Bajau and 3.3 % Dusun. 90 (37.5 %) were Chinese, 24 were (10.0 %) Indians, 9.6 % were people with multiethnic and 0.8 % were from other races. The caretakers comprised of their parents, grandparents or appointed legal guardians. The average age of their caretakers is 41.38 ± 5.53 year. For the marital status of caretakers, 220 (91.7 %) were married, 7(2.9 %) single parents and 5 (2.1 %) staying together. Among the caretakers, 102 (42.5 %) underwent tertiary education as their highest level of education, followed by 74 (30.8 %) who attended upper secondary school and 35 (14.6 %) who attended lower secondary school. 67 (27.9 %) of them had an average family income of RM5,000 and above, 40 (16.7 %) earn between RM1,000 and RM1,999 and 35 (14.6 %) earn between RM2,000 and RM2,999. Majority of the respondents' caretakers are from the private sector (33.3 %). The unemployment rate among the caretakers is 4.2 %. Majority of the respondents are taught

**Table 1** Socio-demographic details of the respondents

| Characteristics | *n* | % | Mean |
|---|---|---|---|
| *Individual factors* | | | |
| Gender | | | |
| Male | 86 | 35.8 | |
| Female | 146 | 60.8 | |
| Age (years) | | | 10.73 ± 2.30 |
| Respondent's ethnicity | | | |
| Malay | 49 | 20.4 | |
| Chinese | 90 | 37.5 | |
| Indian | 24 | 10.0 | |
| Kadazan | 23 | 9.6 | |
| Dusun | 8 | 3.3 | |
| Bajau | 14 | 5.8 | |
| Others | 23 | 9.6 | |
| Multiethnic | 2 | 0.8 | |
| *Family factors* | | | |
| Age of caretakers | | | 41.38 ± 5.529 |
| Caretakers' marital status | | | |
| Single parent | 7 | 2.9 | |
| Married | 220 | 91.7 | |
| Widowed | 3 | 1.3 | |
| Divorced | 4 | 1.7 | |
| Staying together | 5 | 2.1 | |
| Number of children in household | | | 3.22 ± 1.431 |
| Caretakers' highest level of education | | | |
| Primary school | 9 | 3.8 | |
| Lower secondary school | 35 | 14.6 | |
| Upper secondary school | 74 | 30.8 | |
| Vocational school | 8 | 3.3 | |
| Tertiary education | 102 | 42.5 | |
| Others | 7 | 2.9 | |
| Caretakers' occupation | | | |
| Government | 76 | 31.7 | |
| Private | 80 | 33.3 | |
| Self-employed | 47 | 19.6 | |
| Retired | 1 | 0.4 | |
| Unemployed | 10 | 4.2 | |
| Others | 24 | 10.0 | |
| Caretakers' average family income per month | | | |
| Less than RM400 | 9 | 3.8 | |
| RM400–RM699 | 7 | 2.9 | |
| RM700–RM999 | 16 | 6.7 | |
| RM1,000–RM1,999 | 40 | 16.7 | |
| RM2,000–RM2,999 | 35 | 14.6 | |
| RM3,000–RM3,999 | 33 | 13.8 | |
| RM4,000–RM4,999 | 28 | 11.7 | |
| RM5,000 and above | 67 | 27.9 | |

(continued)

**Table 1** (continued)

| Characteristics | *n* | % | Mean |
|---|---|---|---|
| Number of hours sent to day care | | | 3.78 ± 11.117 |
| Cultural continuity | | | |
| Celebration of traditional festivals at home | 178 | 74.2 | |
| Teaching of mother tongue | 194 | 80.8 | |
| Health behaviour | | | |
| Recent illness of respondent in the past 2 weeks | 28 | 11.7 | |
| Ability to seek medical treatment | 168 | 70.0 | |
| Presence of chronic illness | 5 | 2.1 | |
| *Physical environment* | | | |
| Area of residence | | | |
| Squatters' area | 22 | 9.2 | |
| Rural | 16 | 6.7 | |
| Suburban area | 38 | 15.8 | |
| Urban | 162 | 67.5 | |
| Clean water supply | 229 | 95.3 | |
| Air pollution | 49 | 20.4 | |
| Sound pollution | 55 | 22.9 | |
| *School environment* | | | |
| Type of school | | | |
| National school | 131 | 54.6 | |
| Vernacular school | 96 | 40.0 | |
| Private school | 6 | 2.5 | |
| Education level | | | |
| Preschool | | | |
| Primary school | | | |
| Secondary school | | | |
| Bullying in school | 21 | 8.8 | |
| *Community infrastructure* | | | |
| Safety | 212 | 88.3 | |
| Playground nearby | 159 | 66.3 | |
| Distance to nearest healthcare service (km) | | | 4.25 ± 5.527 |

their mother tongue (80.0 %) and 74.2 % celebrate traditional festivals at home. According to the types of school, 131 children (54.6 %) attended national school, 96 children (40 %) attended vernacular school and 6 (2.5 %) attended private school. Only 21 children (8.8 %) reported being bullied in school.

## 3.2 *Prevalence*

Referring to Table 2, there were 42.5 % of children out of a total sample population with RQC score of 1 and above 102, meaning that they indicated for childhood psychiatric morbidity. Among them, females have a higher prevalence of 25.8 %

**Table 2** Prevalence of psychiatric morbidity

| Characteristics | *n* | % | Mean |
|---|---|---|---|
| *Children psychiatric morbidity* | | | |
| Average Responding Questionnaire for Children Score | | | 0.84 ± 1.384 |
| Prevalence of psychiatric morbidity | | | |
| Score of 1 and above | 102 | 42.5 | |
| Score of 0 | 138 | 57.5 | |
| Gender | | | |
| Male | | 16.7 | |
| Female | | 25.8 | |
| Respondent's ethnicity | | | |
| Malay | | 8.1 | |
| Chinese | | 6.9 | |
| Indian | | 10.9 | |
| Kadazan | | 3.6 | |
| Dusun | | 1.6 | |
| Bajau | | 3.2 | |
| Others | | 8.1 | |
| Multiethnic | | 0.0 | |
| *Distribution of positive responses of each item of RQC* | | | |
| Does the child appear backward or slow to learn? | 45 | 18.8 | |
| Does the child suffer from frequent headaches? | 16 | 6.7 | |
| Does the child nearly never play with other children? | 35 | 14.6 | |
| Does the child get scared or nervous for no good reason? | 28 | 11.7 | |
| Is the child's speech in any way abnormal? | 17 | 7.1 | |
| Does the child sleep badly? | 11 | 4.6 | |
| Does the child wet or soil? | 23 | 9.6 | |
| Does the child ever has a fit or fall for no reason? | 7 | 2.9 | |
| Does the child steal things from home? | 2 | 0.8 | |
| Does the child run away from home frequently? | 2 | 0.8 | |
| Is the child unable to concentrate? | 27 | 11.3 | |
| Is the child extremely active? | 53 | 22.1 | |

but not significantly different from males at 16.7 % after adjustment. The Indian ethnic group had the highest prevalence at 10.9 % followed by the Malays and other ethnic groups at 8.1 %. When combined, the other Bumiputeras collectively have 8.4 % of the sample size with Kadazan at 3.6 %, Dusun at 1.6 % and Bajau at 3.2 %. The Chinese community has a prevalence rate of 6.9 %. The results of the factors assessed by the modified Reporting Questionnaire for Children were hyperactivity 22.1 %, slow learners 18.8 % and poor social behaviour 14.6 %.

In their association with social determinant factors, Tables 3 and 4 show that the respondents' ethnicity [$\chi^2(1, N = 240) = 28.385, p < 0.01$)], Indians were found to be afflicted the most (83.3 % of cases), followed by Bajau (64.3 %) and Dusun (50 %). When prevalence was observed based on gender, females were found to be afflicted more than male with a percentage of 25.8 % but not significantly different

**Table 3** Factors influencing childhood psychiatry disorders (1)

| Factors influencing childhood psychiatry disorders | Screened psychiatric morbidity Case (%) | Control (%) | $\chi^2$ |
|---|---|---|---|
| *Individual factors* | | | |
| Gender | | | |
| Male | 40 (46.5) | 46 (53.5) | 0.823 |
| Female | 59 (40.4) | 87 (59.6) | |
| Age group (years) | | | |
| 5–9 years | 24 (38.7) | 38 (61.3) | 1.441 |
| 10–12 years | 58 (43.0) | 77 (57.0) | |
| 13–15 years | 17 (51.5) | 16 (48.5) | |
| Respondent's ethnicity ($n = 233$)* | | | |
| Malay | 23 (46.9) | 26 (53.1) | 28.385 |
| Chinese | 28 (31.1) | 62 (68.9) | |
| Indian | 20 (83.3) | 4 (16.7) | |
| Kadazan | 10 (43.5) | 13 (56.5) | |
| Dusun | 4 (50.0) | 4 (50.0) | |
| Bajau | 9 (64.3) | 5 (35.7) | |
| Other | 6 (26.1) | 17 (73.9) | |
| Multiethnic | 0 (0.0) | 2 (100.0) | |
| Social capital | | | |
| Number of social activities taken part in the past 12 months | | | |
| Trust towards others* | 35 (36.1) | 62 (63.9) | 5.128 |
| Trust towards parents | 98 (42.8) | 131 (57.2) | 1.728 |
| Dependency on others | 84 (41.8) | 117 (58.2) | 1.334 |
| Willingness to share feelings | 79 (42.9) | 105 (57.1) | 1.047 |
| Food security | | | |
| Adequate food | 97 (43.5) | 126 (56.5) | 1.286 |
| No adequate food | 3 (30.0) | 7 (70.0) | |
| *Family factors* | | | |
| Caretakers' marital status* | | | |
| Single parent | 5 (71.4) | 2 (28.6) | 11.541 |
| Married | 88 (40.0) | 132 (60.0) | |
| Widowed | 1 (33.3) | 2 (66.7) | |
| Divorced | 3 (75.0) | 1 (25.0) | |
| Staying together | 5 (100.0) | 0 (.0) | |
| Caretakers' highest level of education | | | |
| Primary school | 5 (55.6) | 4 (44.4) | 7.937 |
| Lower secondary school | 17 (48.6) | 18 (51.4) | |
| Upper secondary school | 37 (50.0) | 37 (50.0) | |
| Vocational school | 3 (37.5) | 5 (62.5) | |
| Tertiary education | 33 (32.4) | 69 (67.6) | |
| Others | 3 (42.9) | 4 (57.1) | |
| Caretakers' income class | | | |
| Low | 30 (76.9) | 9 (23.1) | 11.326 |
| Middle | 17 (53.1) | 15 (46.9) | |

(continued)

**Table 3** (continued)

| Factors influencing childhood psychiatry disorders | Screened psychiatric morbidity | | $\chi^2$ |
|---|---|---|---|
| | Case (%) | Control (%) | |
| High | 7 (33.3) | 14 (66.7) | |
| Caretakers' average family income per month* | | | |
| Less than RM400 | 5 (55.6) | 4 (44.4) | 20.893 |
| RM400–RM699 | 2 (28.6) | 5 (71.4) | |
| RM700–RM999 | 11 (68.8) | 5 (31.3) | |
| RM1,000–RM1,999 | 23 (57.5) | 17 (42.5) | |
| RM2,000–RM2,999 | 9 (25.7) | 26 (74.3) | |
| RM3,000–RM3,999 | 18 (54.5) | 15 (45.5) | |
| RM4,000–RM4,999 | 12 (42.9) | 16 (57.1) | |
| RM5,000 and above | 19 (28.4) | 48 (71.6) | |
| Caretakers' occupation | | | |
| Government | 30 (39.5) | 46 (60.5) | 3.085 |
| Private | 37 (46.3) | 43 (53.8) | |
| Self-employed | 18 (38.3) | 29 (61.7) | |
| Retired | 0 (0.0) | 1 (100.0) | |
| Unemployed | 6 (60.0) | 4 (40.0) | |
| Others | 10 (41.7) | 14 (58.3) | |
| Difficulty in managing expenditure for the past 12 months* | 38 (66.7) | 19 (33.3) | 20.597 |
| | 63 (35.0) | 117 (65.0) | |
| Health behaviour | | | |
| Recent illness of respondent in past 2 weeks* | 17 (60.7) | 11 (39.3) | 7.268 |
| | 83 (39.5) | 127 (60.5) | |
| Ability to seek medical treatment | 68 (40.5) | 100 (59.5) | 1.798 |
| | 29 (45.3) | 35 (54.7) | |
| Presence of chronic illness in respondents | 4 (80.0) | 1 (20.0) | 6.209 |
| | 94 (40.9) | 136 (59.1) | |
| Area of residence | | | |
| Squatters' area | 9 (40.9) | 14 (59.1) | 4.530 |
| Rural | 9 (59.3) | 7 (43.8) | |
| Suburban area | 11 (28.9) | 27 (71.1) | |
| Urban | 73 (45.1) | 89 (54.9) | |
| | 71 (38.8) | 112 (61.2) | |
| *School environment* | | | |
| Type of school | | | |
| National school | 77 (58.8) | 54 (41.2) | 0.414 |
| Vernacular school | 53 (55.2) | 43 (44.8) | |
| Private school | 3 (50.0) | 3 (50.0) | |
| Education level | | | |
| Primary school | 82 (41.6) | 115 (58.4) | 1.128 |
| Secondary school | 17 (51.5) | 16 (48.5) | |

*$p$ value $< 0.05$

**Table 4** Factors influencing childhood psychiatry disorders (2)

| Characteristics | Group | *N* | Mean | *t-test* | *p* |
|---|---|---|---|---|---|
| Age (years) | Case | 99 | 11.03 ± 2.297 | 1.706 | 0.089 |
| | Control | 133 | 10.51 ± 2.288 | | |
| Age of caretakers | Case | 100 | 41.30 ± 5.771 | −0.186 | 0.853 |
| | Control | 133 | 41.44 ± 5.362 | | |
| Number of children in household | Case | 100 | 3.44 ± 1.566 | 2.007 | 0.046 |
| | Control | 138 | 3.07 ± 1.308 | | |
| Distance to nearest healthcare service | Case | 98 | 4.67 ± 7.051 | 0.993 | 0.322 |
| | Control | 131 | 3.93 ± 4.032 | | |

from the rest of the population. The significance of association of caregivers' marital status to caseness was also proven [$\chi^2(1, N = 240) = 11.541, p < 0.05$)]. Remarkably, of those caregivers staying together, all of their children were found to suffer from some form of psychological morbidity. 75 % of children growing under divorced parents and 71.4 % raised by single parents were also reported with mental health problems. The income class is classified accordance to the National Statistical department criteria. It is found that the income class is significantly associated with caseness with prevalence increases as the income class descends. [$\chi^2(1, N = 240) = 11.386, p < 0.01$].The financial status of caregivers has a significant role in contributing to the prevalence of childhood psychiatric morbidity [$\chi^2(1, N = 240) = 20.893, p < 0.05$].An average monthly income of RM700 to RM999 was seen to be positively associated with children with psychiatric morbidity. On the aspect of expenditures management, 66.7 % of caregivers who experienced difficulty in managing their expenses were associated with their children's psychiatric morbidity. In terms of health behaviour [$\chi^2(1, N = 240) = 7.268, p < 0.05$], the highest prevalence of 60.7 % was observed among respondents reported suffering from illness in the past 2 weeks prior to the survey. It was also found that the number of children per household was significantly associated with children suffering from psychiatric morbidity [$t$-test $= 2.007, p < 0.05$]. On the other hand, it was observed that some demographic data and social determinants play no significant role on the prevalence. These included the age group of respondents, area of residence, type of school and education level, social exclusion, caretakers' age, education level, occupation and health-seeking behaviour.

In order to find out the inequality of social determinants among the sample population, we analysed the factors found to have significant association with childhood psychiatric morbidity with the main social determinants, i.e. income class of caretakers, clearly seen in Table 5. Income class is found to be significantly associated with ethnicity [$\chi^2(1, N = 240) = 27.545, p = 0.01$)], with Indians more prone to be in the lower income class. There is no significant difference between the number of children and income class.

**Table 5** Social determinants associated with income class of caretakers

| Factors associated with income class of caretakers | Income class of caretakers | | | $\chi^2$ |
|---|---|---|---|---|
| | Low (%) | Middle (%) | High (%) | |
| Respondent's ethnicity ($n = 233$)* | | | | |
| Malay | 15 (31.9) | 16 (34.0) | 16 (34.0) | 27.545 |
| Chinese | 21 (23.9) | 42 (47.7) | 25 (28.4) | |
| Indian | 12 (50.0) | 7 (29.2) | 5 (20.8) | |
| Kadazan | 7 (30.4) | 14 (60.9) | 2 (8.7) | |
| Dusun | 0 (0.00) | 7 (87.5) | 1 (12.5) | |
| Bajau | 6 (42.9) | 3 (21.4) | 5 (35.7) | |
| Other | 9 (39.1) | 5 (21.7) | 9 (39.1) | |
| Multiethnic | 1 (50.0) | 0 (0.00) | 1 (50.0) | |

*$p < 0.01$

# 4 Discussion

In our present research to identify the social determinants of psychiatric morbidity among Malaysian children aged 5–15, the prevalence rate is found to be 42.5 % which is twofold higher when compared to the prevalence (20.3 %) found by the Third National Health and Morbidity Survey in 2006. The large difference in prevalence is mostly likely due to two factors, elapsed time and sampling area. The difference between the two previous national health and morbidity survey of ten years has also seen an increase in prevalence, likewise in our study as corrective measures planned has yet to bear fruit. In addition to this, the recent increase in childhood suicidal cases in Malaysia may have led to sensitization among parents regarding their children's abnormal behaviours. There is also due to the limitation on the distribution of respondents which are not adequate to represent the entire country as in the previous study.

On the individual level, recent illnesses play a role in increasing the caseness. On family aspects, ethnicity, marital status, income class and family financial status turned out to be significantly associated with childhood psychiatric morbidity. Based on our chi-square analysis, ethnicity showed obvious significant effects on childhood psychiatry (both $p < 0.01$).

Among the ethnic groups, the Indian community is more susceptible to psychiatric morbidity. This finding is similar to the previous national study, whereby Indians stood at the top with 26.7 % of prevalence among all races [14]. This is most probably due to the stratification of socio-economical status in Malaysia, whereby the Indian community, which is the minority race when compared with the Malays and Chinese, generally tend to have a lower socio-economical status, educational level and family income in the Malaysian society. The significant association of income class and ethnicity in the current study also supports this statement. The lower SES and income create stress within a household and thus resulting in a non-conducive environment for the children. Besides, studies proved that ethnicity has influence on the parents' ability to tolerate children's behaviour. It has been shown that British parents have higher tolerance level towards hyperactivity than most Americans [21]. Thus, this might be the explanation why there is significant difference among Indians and other races, in terms of the difference level of tolerance to the child's behaviour. In view of the caretakers' marital status, the highest number of psychologically morbid children is found to be those staying with cohabiting caretakers.

The study also revealed that low financial status is significantly associated with childhood psychiatric morbidity. Their mental health conditions are found to be negatively affected with decreasing financial capability. Financially restricted family is usually associated with lowly educated parents, whereby the lack of knowledge and awareness in terms of children's health, together with the financial constraints, might impede the child with psychological problem from getting proper treatment.

This type of study has its own limitations. Information was gathered by self-reporting, including those information about their children's health and subject to recall bias. Therefore, in the case of the medically attended conditions, diagnostic precisions are limited by the respondent's memory of the diagnostic terms mentioned by the doctors hence influenced by his perceptions. Furthermore, if an illness has not been medically attended to, a proper and accurate diagnosis is substantially lacking. The household survey requires the use of numerators to gather information. Other problems include the possible refusal of respondents to be interviewed and the inability to trace them as identified by the sample. In addition, there were difficulties in getting larger sample size during the course of the research. Based on verbal accounts from school authorities and written notes from parents, there were some level of stigmatization and fear of labelled or associated with mental illness. Further studies are recommended to correct the weaknesses of this study so that generalizability and validity are ensured.

To address the mental health inequality among Malaysian children, we proposed a multilevelled approach to the issue influenced by two main social determinants, socio-economical status and ethnicity as well as the perception of the disease. We proposed four main solutions which are to elevate socio-economical status of high-risk population, improve access and use of psychiatric management, improve screening programmes and community intervention programmes as well as improve awareness, subsequently reducing social stigmatization and fear. The Ministry of Health should also elevate community mental health services from a subset of primary care and allocate equal funding to both child and adult services. Child mental health services require unique specificities and needs. Hence, it should be viewed as a separate field of healthcare, whereby additional funding and policy development should be considered. A mental health programme should be modelled from other programmes abroad alongside integrating a multidisciplinary team, working together with parents, teachers and counsellors to intervene in states of higher prevalence of psychiatric morbidity. Workshops on child development, parenting, discipline techniques and mental health should be implemented into the programme. In aiding the government, we propose that medical schools adopt schools around their vicinity to carry out the programme with the Ministry of Health.

We suggest a partnership with NGOs and mainstream media to plan and produce an impactful health campaign on a national scale to break the stigma and fear of childhood mental disorders.

## 5 Conclusion

Determining the social determinants can help us to reveal the cause or causes of health disparities so that appropriate solutions can be designed to solve this major world problem. We think that the most effective solution to eradicate health disparities is by doing many-sided interventions which take into account the multiple levels of the socio-economical model, with specific attention to the social, economic, cultures, policies, health and environmental conditions. This solution can be applied throughout the world.

## References

1. Amato PR, Ochiltree G (1987) Child and adolescent competence in intact, one-parent and step-families: an Australian study. J Divorce 110:75–96
2. Bussing R, Zima BT, Gary FA, Garvan CW (2003) Barriers to detection, help-seeking, and service use for children with ADHD symptoms. J Behav Health Serv Res 30(2):176–189
3. CSDH (2008) Closing the gap in a generation: health equity through action on the social determinants of health. Final Report of the Commission on Social Determinants of Health. World Health Organization, Geneva
4. Cornelius MD, Day NL (2009) Developmental consequences of prenatal tobacco exposure. Curr Opin Neurol 22(2):121–125
5. Costello EJ, Mustillo S, Erkanli A, Keeler G, Angold A (2003) Prevalence and development of psychiatric disorders in childhood and adolescence. Arch Gen Psychiatry 60(8):837–844
6. Lipman EL, Offord DR, Boyle MH, Racine YA (1993) Follow-up of psychiatric and educational morbidity among adopted children. J Am Acad Child Adoles Psychiatry 32(5):1007–1012
7. Gibson HB (1969) Early delinquency in relation to broken homes. J Child Psychol Psychiatry 10:195–204
8. Kramer T, Evans N, Garralda ME (2000) Ethnic diversity among child and adolescent psychiatric (CAP) clinic attenders. Child Psychol Psychiatry Rev 5:169–175
9. Myer L (2008) Social determinants of psychological distress in a nationally-representative sample of South African adults. Soc Sci Med 66:e1828–e1840
10. Luthar SS (1993) Annotation: methodological and conceptual issues in research on childhood resilience. J Child Psychol Psychiatry 34:441–453
11. Marmot M (2005) Social determinants of health inequalities. Lancet 365:1099–1104
12. Offord DR, Boyle MH, Szatmari P et al (1987) Ontario child health study. II. Six-month prevalence of disorder and rates of service utilization. Arch General Psychiatry 44:832–836
13. Piven J, O'Leary D (1997). Neuroimaging in autism. In: Peterson BS, Lewis M (eds) Child and adolescent psychiatric clinics of North America: neuroimaging, vol 6, Chap 2. W. B. Saunders, Philadelphia, pp 305–324
14. Institute of Public Health, Ministry of Health Malaysia (2008) Psychiatric morbidity, a report of the third national health and morbidity survey, 2006. Institute of Public Health, Ministry of Health Malaysia, Malaysia

15. Ramli H, Ohn K, Krishnaswamy S, Kasmini K, Hassan S (1987) A prevalence survey of psychiatric morbidity in a rural Malaysian village: a preliminary report. Singapore Med J 28:530–533
16. Adlina S, Suthahar A, Ramli M, EdariahSoe-Soe-Aye AB, Mohd Ariff F, Narimah AHH, Nuraliza AS, Karuthan C (2007) Pilot study on depression among secondary school students in Selangor. Med J Malaysia 62(3):218–222
17. Toh CL, Ding LM, Peng R et al (1997) Psychiatric morbidity in children and adolescents. Public Health Institute, Ministry of Health, Malaysia
18. Normar Rashid Y, Zain AM, Jan N (2000) Psychiatric morbidity assessment among students. Malaysian J Psychiatry 8(2):31–38
19. Zubrick SR, Nguyen H, Silburn S et al (1995) Linkages between child health and child adversity. Commonwealth Department of Health, Housing and Community Services, Canberra
20. Giel R, De Arango MV, Climent CE, Harding TW, Ibrahim HHA, Ladrigoignacio L et al (1981) Childhood mental disorders in primary health care: results of observations in four developing countries. Pediatrics 68(5):677–683
21. Mann C, Lubar J, Zimmerman A, Miller C, Muenchen R (1992) Quantitative analysis of EEG in boys with attention deficit hyperactivity disorder: controlled study with clinical implications. Paediatric Neurol 8:30–36
22. Saltzman WR, Lester P, Beardslee WR, Layne CM, Woodward K, Nash WP (2001) Mechanisms of risk and resilience in military families: theoretical and empirical basis of a family-focused resilience enhancement program. Clin Child Fam Psychol Rev 14(3):213–230

# Modelling Community Response for Ecological Significance Based on Laboratory Simulations of Variable Copper Exposure

**Khun C. Tan and Carolyn Oldham**

## 1 Introduction

Aquatic ecosystems receiving contaminated urban runoff are being ecologically degraded globally [1]. The deleterious effects have been attributed to multiple and highly variable stressors in the runoff. Current approaches for managing urban runoff are mainly concerned with meeting guideline limits and focus less on the potential ecological changes to aquatic ecosystems. Similarly, although ambient biological monitoring can provide information on the changes in community structure, it cannot establish causality, and hence, it cannot be used to identify the stressor(s) to be targeted for management actions [2]. In order to protect the aquatic ecosystems, a risk assessment of ecological changes by runoff is necessary.

Central to the ecological risk assessment is the relationship between stressor and ecological effect that is essential for establishing causality and for predicting the effect concentration for a desired level of protection [3]. The relationship between exposure and effect is often evaluated by statistical or regression analysis. Until recently, a statistical approach to hypothesis testing of critical effect concentrations, notably the no-observable and lower-observable effect concentrations, had been used. The Organisation for Economic Co-operation and Development has declared this approach scientifically inappropriate due to unconstrained Type II errors and a high dependency on the test concentration range used in the determination [4, 5]. On the other hand, a regression approach that relies on a mathematical model assumes a continuous relationship. The accuracy of the approach is highly dependent on how well the model describes the exposure–effect relationship.

K.C. Tan (✉)
Universiti Tunku Abdul Rahman, Jalan Universiti, Bandar Barat, 31900 Kampar, Perak, Malaysia
e-mail: tckhun@utar.edu.my

C. Oldham
The University of Western Australia, Crawley, Australia

K. Ab. Hamid et al. (eds.), *The Malaysia-Japan Model on Technology Partnership: International Proceedings 2013 of Malaysia-Japan Academic Scholar Conference*,
DOI 10.1007/978-4-431-54439-5_8, © Springer Japan 2015

The logistic model traditionally used has been found inadequate [6], especially for positive responses [7], and other regression functions have since been developed. Brain and Cousens [8] modified the function by including an extra term that will provide for a peak in the curve. Additionally, as the relationships described by regression models are empirical, they are deficient as a learning tool for understanding the ecological relevance of that relationship. It seems the use of such models is solely for predicting effect concentrations for risk assessment. An alternative approach is the conceptual modelling of ecological processes that has a sound mechanistic basis to enable a better understanding of the links between stressors and effects. An example is the model of scope for growth based on an energy budget [9].

Direct determination of runoff stressor-effect relationship using field data is not feasible due to the complexity of runoff phenomena. It is necessary to use controlled laboratory simulations in combination to evaluate the relationship. In a study by the authors, a combination of field and laboratory study was used to investigate the ecological effects of runoff discharge from two urban catchments on the receiving Swan River estuary in Western Australia [10]. The field study was conducted to assess the physico-chemical quality of the catchment runoff discharge and the associated changes in the periphyton community in the receiving estuarine water. Copper was identified to be a significant stressor in the urban runoff. This paper describes the modelling of the community response to the laboratory simulation of variable exposure to copper in the runoff.

## 2 Materials and Methods

### *2.1 Sampling*

An assemblage of periphyton organisms, indigenous to the Swan River estuary, Western Australia, was colonized on glass substrata, under controlled laboratory conditions, in a series of seven copper solutions of nominal concentrations between 0.05 and 0.3 mg/L Cu for 2 weeks. The colonized communities were subsequently exposed for 24 h, to a series of seven copper solutions of higher nominal concentration between 0.16 and 4 mg/L Cu. Both the long-term and short-term treatments, plus a control, respectively, were done in triplicate. Conditions of the long-term exposure were made less toxic by the addition of ethylenediaminetetraacetate [11]. The community response in photosynthetic efficiency was determined as PSII quantum yield using a fluorometer (WALZ DIVING-PAM). The community response was taken after the long-term exposure of 2 weeks and after the subsequent short-term exposure of 24 h.

## 2.2 Modelling of Community Responses

The model of the periphyton community to the variable copper exposure is described by the following equation:

$$Y = k(1 + fx)/(1 + \exp(b + gx)) + d \quad (1)$$

where $Y$ is the community photosynthetic efficiency; $x$ is the exposure copper concentration, in mg Cu/L; $k$ is an inertia parameter defining the initial photosynthetic performance of the community; $f$ is an enhancing factor; $b$ is a decay parameter; $g$ is a suppressing factor; and $d$ is a correction factor.

The resident organisms within a community respond differentially to the enhancing and suppressing effects of copper. The model integrates the differential responses to both the long-term and short-term exposures into a positive and a negative component. The positive component, denoted by the numerator in the model, has two parameters: $k$ and $f$. The parameter $k$ denotes the community inertia to persist in its initial status of photosynthesis performance that is independent of the copper concentration of the short-term exposure but is related to the previous long-term exposure. Therefore, between communities, $k$ accounts for the changes in the community caused by the long-term exposure. The parameter $f$ defines the enhancing effect of the short-term exposure on the community, and a first-order effect was assumed in the first instance.

The negative response component, denoted by the denominator in the model, describes the decay in photosynthetic activity of the community. An exponential decay was assumed in the first instance. The parameter $b$ defines the initial decay that is independent of the copper concentration of the short-term exposure. As with $k$, $b$ is related to the previous long-term exposure condition. The parameter $g$ is the suppressing factor of ambient copper, which is the short-term exposure concentration. The parameter $d$ is a correction factor, introduced to account for background signal, including any residual signal in the test sample.

The model parameters were estimated by regression analysis using the "nls" function of the R computer program (Version 1.7.1 for Windows by Ihaka R and Gentleman R 1996, accessed date 16 Jun 2003). The response data of the control community and the communities of long-term treatment at 0.1, 0.2, and 0.3 mg/L Cu were used for this purpose. For computing purposes, missing data were replaced by the mean values, and for logarithm computation, zero response measurement was replaced by the detection limits (defined as 2× standard deviation; standard deviation was determined by using all low values near zero value).

## 2.3 Model Diagnosis

Goodness of fit was evaluated from residual analysis of the predicted and observed data points. The goodness of fit was evaluated against the traditional logistic model by comparing values of the Akaike information criteria (AIC), including the second derivative [12], and the values of the Bayesian information criteria (BIC).

The model was validated using response data of communities not used for the model calibration process, i.e. the communities long-term treated in 0.05, 0.15, and 0.25 mg/L Cu. The model was used for evaluating ecological risks at various effect sizes, notably the no-effect concentration (NEC) and the 50 % affected concentration ($EC_{50}$). Two additional effect concentrations were proposed. One is the pivotal concentration (PC), which is defined by the exposure concentration at which a decline in yield is observed. This is attributed to the dominance of suppressing effect of copper. Therefore, PC would correspond to the response peak at maximum yield ($Y_{max}$). The second new effect concentration is $EC_{0.5Y}$, defined as the exposure concentration corresponding to the community performances at 50 % efficiency, the tipping point from ecological fitness. The difference between $EC_{0.5Y}$ and $EC_{50}$ is that $EC_{0.5Y}$ defines a performance yield, whereas $EC_{50}$ defines a relative performance to a control performance.

The effect concentrations were determined by point estimation using the Bootstrap method. Resampling with $n_R = 99$ was applied to each series of independently observed response measurements of $n = 72$ from three replicates of long-term $\times$ 3 replicates of short-term $\times$ (7 test solutions + 1 control). Case resampling was non-parametric, random with replacement, stratified, and balanced [13]. This method was preferred as it does not set explicit requirements on the residuals and has a more general applicability than resampling of residuals [13, 14]. The 95 % confidence interval was estimated by bootstrap resampling with $n_R = 999$, performed on the previously obtained 99 bootstrapped response curves [15]. For both resampling, the value of $n_R$ was selected such that no further improvements were attainable with further increases [16].

The no-effect concentration (NEC), defined as the limiting concentration at which the yield declined below the yield at $x = 0$, was determined by solving for the roots of the curve function, taking the value of the larger root. The $EC_{50}$ value was estimated at 50 % of initial yield. The peak of the response curve corresponding to (PC, $Y_{max}$) was determined by double differentiation of the curve. Estimates of NEC and $EC_{Ymax}$ with values less than zero were taken to have zero value. This occurred mainly for curves with $f < 0$. The yield $Y_0$, at $x = 0$, can be measured or estimated from (2)

$$Y_0 = \frac{k}{1 + \exp(b)} + d \tag{2}$$

Another relevant endpoint is the tolerance width of the community and is given by the difference between NEC and $EC_{50}$. This endpoint expresses the rate of declining performance with additional copper stress that can provide an indication of the exposure concentration allowance the community can withstand before its performance collapses to below par (i.e. <50 % yield).

New relationships were created by parametric correlation and regression analysis of the variables derived using the model together with observed quantities.

# 3 Results

The predicted and observed data were linearly correlated with $R^2 > 0.99$, and the standardized residuals were distributed randomly within ±2 standard deviations, not indicating any distinct trend (Fig. 1). Using the proposed model equation, the curves fitted are noticeably better than those fitted using the logistic model, especially for long-term treatment at 0.2 mg/L Cu where positive responses were observed (Fig. 2). The AIC including the second derivatives and BIC for positive response curves by the proposed model are more negative in value than by the traditional logistic model, indicating improved goodness of fit (Table 1) in the correct order.

The model parameters $b$, $d$, $f$, $g$, and $k$, estimated by direct regression on the observed data and by bootstrap regression, are mostly comparable (Table 2). The parameter values range within an order: $f$ and $k$ range in the order of less than zero, while $b$ and $g$ range in the order of one. The value of $b$ is negative. The value of $f$ is maximum at long-term treatment at 0.1 mg/L Cu. Apart from $b$ and $g$, which are linearly correlated ($R^2 = 0.87$), all other parameters are nonlinearly correlated, as expected from their relationships through the model equation. Both $d$ and $k$ form quadratic relationship with long-term exposure concentration, respectively ($R^2 = 0.95$). The other parameters varied sinuously along with the long-term copper concentration (cubic fit, $R^2 = 1$).

The magnitudes of the effect concentrations were in the order of PC < NEC < $EC_{0.5Y}$ < $EC_{50}$ (Table 3). The values of these effect points, within 95 % confidence interval, are quite distinct between communities, overlapping progressively towards higher-effect end at and above $EC_{0.5Y}$ and higher long-term exposure concentration. The tolerance width varied between 0.26 and 0.80 mg/L Cu, maximum at 0.1 mg/L Cu and minimum at 0.2 mg/L Cu.

In general, $EC_{0.5Y}$ correlates significantly well with all other effect variables except $Y_{50}$, while $EC_{50}$ does not except with $EC_{0.5Y}$. Both $EC_{0.5Y}$ and $EC_{50}$ form a log relationship with long-term exposure concentration ($R^2 > 0.89$). When data of all communities (i.e. modelled and test) are included, the $R^2$ values alter nominally (<0.1). Correlation of $Y_0$ with $EC_{0.5Y}$ is linear ($R^2 > 0.97$), while with long-term copper concentration, it is a quadratic relationship ($R^2 > 0.96$).

The inertia parameter $k$ was more significantly correlated with the low-effect variables of $Y_0$, $Y_{max}$, and $EC_{0.5Y}$, whereas the decay model parameters $b$ and $g$ are significantly correlated more with the higher-effect variable of $Y_{50}$. Both parameters $b$ and $g$ correlate linearly with tolerance width ($R^2 > 0.90$). The parameter $b$ also

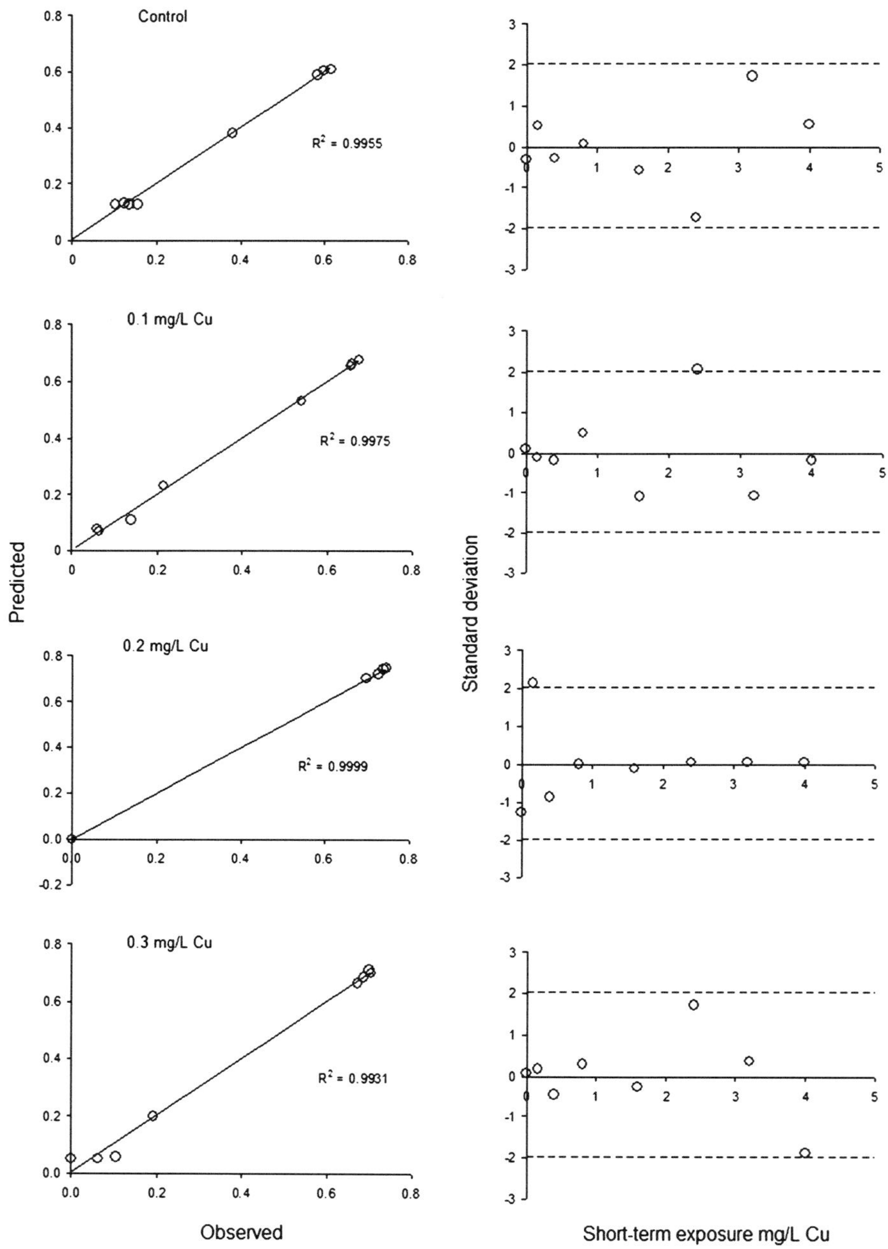


**Fig. 1** Correlations between predicted and observed responses (in PSII quantum yield) of the periphyton communities (2-w colonized at 0.1, 0.2, and 0.3 mg/L Cu) to short-term (24-h) exposure at higher concentrations (range 0.16–4 mg/L Cu), plus the control community

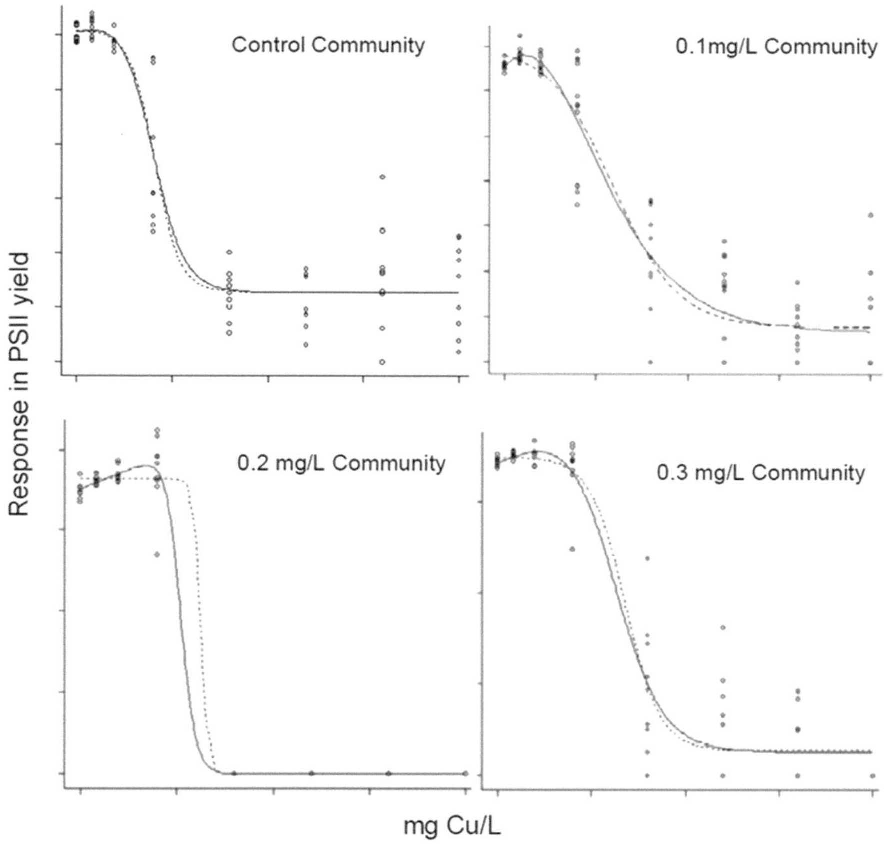


**Fig. 2** Curve fitting of the community response to variable copper exposure using the proposed model (*solid line*) in comparison to using the traditional logistic model (*dashed line*); the communities were exposed to copper long term (2 w at 0.1, 0.2, and 0.3 mg/L Cu), followed by short term at higher concentrations (24 h, 0.16–4 mg/L Cu), plus control

**Table 1** Comparison of goodness of fit of the proposed model against the traditional logistic model by Akaike information criteria (AIC) and Bayesian information criteria (BIC)

| | Model: Y ~ $k$ (1 + $fx$)/(1 + exp ($b$ + $gx$)) + $d$ | | | | Proposed model – logistic model[a] | | |
|---|---|---|---|---|---|---|---|
| Long-term exposure (mg/L Cu) | df | 1st-order AIC | 2nd-order AIC | BIC | 1st-order AIC | 2nd-order AIC | BIC |
| Control | 67 | −171 | −168 | −160 | 2 | 2 | 4 |
| 0.1 | 67 | −137 | −134 | −125 | 0 | 0 | 2 |
| 0.2 | 67 | −280 | −278 | −269 | −8 | −8 | −5 |
| 0.3 | 67 | −156 | −153 | −145 | 1 | 1 | 1 |
| Overall mean | 67 | −194 | −191 | −183 | −1 | −1 | −1 |

[a]More negative for better fit

**Table 2** Statistics of model parameters estimated with and without bootstrap resampling ($n_R = 99$)

| Long-term exposure (mg/L Cu) | Parameter statistics | $b$ | $d$ | $f$ | $g$ | $k$ |
|---|---|---|---|---|---|---|
| Control community | Bootstrap (SE) | 4.8 (1.7) | 0.13 (0.01) | 0.14 (0.27) | 6.1 (1.6) | 0.48 (0.02) |
| | Bias | −0.2 | 0 | 0.08 | 0.4 | 0.01 |
| | Observed (sig.) | −4.8 (1) | 0.13 (0) | 0.14 (1) | 6.1 (1) | 0.48 (0) |
| 0.1 | Bootstrap (SE) | −2.3(1.1) | 0.07 (0.02) | 0.54 (0.4) | 2.7 (0.6) | 0.71 (0.25) |
| | Bias | −1.4 | 0 | −0.46 | 0.6 | 0.12 |
| | Observed (sig.) | −0.9 (1) | 0.07 (0.001) | 1 (1) | 2.1 (0) | 0.83 (0.1) |
| 0.2 | Bootstrap (SE) | −9.9 (4.4) | 0 (0) | 0.13 (0.03) | 8.4 (3) | 0.7 (0.01) |
| | Bias | −1.6 | 0 | 0.01 | 1.2 | 0 |
| | Observed (sig.) | −13.4 (1) | 0 (1) | 0.14 (0.01) | 12.9 (1) | 0.7 (0) |
| 0.3 | Bootstrap (SE) | −4.8 (1.8) | 0.05 (0.02) | 0.18 (0.16) | 4 (1.2) | 0.64 (0.04) |
| | Bias | −0.4 | 0 | 0.02 | 0.3 | 0.01 |
| | Observed (sig.) | −4.8 (0.01) | 0.05 (0.001) | 0.18 (1) | 4 (0) | 0.64 (0) |

correlates with PC in a quadratic relationship ($R^2 > 0.93$). The enhancement parameter $f$, however, is not correlated (at sig. level 0.1) parametrically or non-parametrically with any of the effect variables. Its threshold is <0.2. The predicted values of $k$, $Y_0$, and $EC_{0.5Y}$ for the test communities (i.e. those communities long-term exposed to 0.05, 0.15, and 0.25 mg/L Cu) using the model are comparable to the values derived from the response curve plots of the test communities, respectively (Table 4).

## 4 Discussion

The model performance was good as indicated by the closeness between observed and predicted values for the test communities. The goodness of fit of the proposed model was generally comparable to the traditional logistic model but was better where responses were positive. This is demonstrated by the response curve of the long-term treatment at 0.2 mg/L. The improvement is attributed to the inclusion of the $f$ term [8]. Standard regression logistic models such as Gaussian and Weibull, and other related model functions which are based on a monotonic, sigmoidal relationship, are known to be inadequate for describing positive responses [6, 17, 18]. The positive response to an initial low concentration of contaminant is referred to as hormesis, which is often attributed to some stimulating effects.

The model provided a means of predicting the exposure risks to copper. Traditionally, $EC_{50}$ is the primary measure of toxic effect, selected purely because it is the most reliable estimate statistically with the smallest confidence interval of 95 %

**Table 3** Effect endpoints as mean (95 % confidence interval) on Bootstrap resampling ($n_R$ = 99 and 999)

| Long-term exposure (mg/L Cu) | Pivotal point | | Threshold point | | | 50 % yield | 50 percentile affected | | Tolerance width (mg/L Cu) |
|---|---|---|---|---|---|---|---|---|---|
| | PC (mg/L Cu) | $Y_{max}$ (yield) | NEC (mg/L Cu) | Modelled $Y_0$ (yield) | Observed $Y_0$ (yield) | $EC_{0.5Y}$ (mg/L Cu) | $EC_{50}$ (mg/L Cu) | $Y_{50}$ (yield) | |
| Control ($n$ = 72) | 0.123 (0.103–0.138) | 0.612 (0.611–0.613) | 0.22 (0.188–0.247) | 0.605 (0.603–0.606) | 0.601 (0.589–0.612) | 0.64 (0.63–0.65) | 0.82 (0.81–0.83) | 0.3663 (0.3657–0.3667) | 0.6 |
| 0.1 ($n$ = 72) | 0.217 (0.211–0.224) | 0.67 (0.665–0.674) | 0.415 (0.402–0.426) | 0.655 (0.650–0.658) | 0.655 (0.644–0.666) | 0.91 (0.89–0.93) | 1.21 (1.19–1.23) | 0.3605 (0.3586–0.3621) | 0.8 |
| 0.2 ($n$ = 72) | 0.712 (0.689–0.740) | 0.76 (0.757–0.763) | 0.954 (0.932–0.977) | 0.702 (0.701–0.703) | 0.699 (0.684–0.715) | 1.13 (1.11–1.15) | 1.21 (1.20–1.23) | 0.3515 (0.3510–0.3520) | 0.26 |
| 0.3 ($n$ = 71) | 0.426 (0.412–0.445) | 0.426 (0.412–0.445) | 0.705 (0.687–0.723) | 0.687 (0.686–0.688) | 0.688 (0.680–0.696) | 1.14 (1.13–1.16) | 1.33 (1.32–1.34) | 0.3688 (0.3683–0.3693) | 0.62 |

**Table 4** Comparison of model predictions and observed values for test communities (long-term treatment at 0.05, 0.15, and 0.25 mg/L Cu)

| Long-term exposure in mg/L Cu ($x$) | 0.05 | 0.15 | 0.25 |
|---|---|---|---|
| Derived $EC_{0.5Y}$ | 1.02 | 1.12 | 1.23 |
| Predicted using the derived relationship: $EC_{0.5Y} = 0.0671\ \ln(x) + 1.1801$ | 1 | 1.05 | 1.09 |
| Deviation as % of predicted | −4 | −6 | −12 |
| Observed $Y_0$ | 0.644 | 0.692 | 0.713 |
| Predicted using the derived relationship: $Y_0 = -1.8597x^2 + 0.8519x + 0.6012$ | 0.637 | 0.683 | 0.696 |
| Deviation as % of predicted | −1.1 | −1.3 | −2.5 |
| $k$ derived from observed data of test community | 0.58 | 0.69 | 0.71 |
| Predicted using the derived relationship: $k = -7.2158x^2 + 2.6348x + 0.4902$ | 0.6 | 0.72 | 0.7 |
| Deviation as % of predicted | 5 | 4 | −2 |

and not because of any ecological significance [19]. The effect concentration $EC_{0.5Y}$ was indicated to be a more meaningful effect concentration on the basis that it is just as reliable statistically plus it has ecological relevance in relating copper effects to the long-term exposure concentration of copper and also to the inertia parameter $k$ and most of the other effect variables.

The significance of NEC as the conventional effect concentration for regulatory purposes may not be acceptable in the case of copper as an essential nutrient element of the periphyton community [7]. Our previous study has demonstrated that the community response to subsequent copper exposure was dependent on the pre-exposure conditions. With increased baseline contamination to start off with, the initial community performance is very likely to be adverse and not at optimum. However, this consideration has no bearing on the endpoint PC and therefore should be considered as an alternative endpoint to NEC. PC is lower in magnitude than NEC and hence more protective. Furthermore, PC has ecological relevance in that it defines the location where the turning point in the community performance with further increased exposure is. The implications of such turning point for managing essential trace elements, such as copper, are significant [20]. The tolerance curve for essential trace elements takes on an inverted U-shape with the lower turning point at the deficient end and the upper turning point at the excessive end of the concentration gradient. Hence, the ambient level should be managed such that it is within the deficient and toxic ends. The effect concentration PC may provide a viable alternative to risk assessment for protecting ecological health, particularly for chemical contaminants which are also micronutrients, such as copper.

The model provided an understanding of how the long-term and short-term exposures to copper separately affected the community photosynthesis performance. The short-term response is likely to be dominated by individual reactions to the copper stress, whereas the long-term response is likely to be dominated by the coherent forces within community, acting to adapt to the copper stress. The latter response type was not apparent in the short term because of the time scale for feedback mechanism. Generally speaking, the time scale of the short-term response would be within a generation, whereas that of the long-term response would be a few generations. The long-term response therefore involved a series of adapting changes over several generations. The tolerance width was widest for the maximum value of *f* and narrowest for the minimum value, from which it is inferred that the enhancement factor *f* is associated with the uncoordinated responses of individuals reacting to the stress from short-term copper exposure. The high variability and consequently insignificant correlation of *f* with any of the effect variables can be attributed to the expected chaotic, reactive mode of responses to short-term, acute stresses. On the other hand, the inertia parameter *k* varied in a more defined manner, forming a definitive relationship with the copper concentration of the long-term exposure. Its close relation to the lower effect response of $Y_0$ or $EC_{0.5Y}$ indicated its association with the community resistance to change as a whole. In contrast, the decay parameters *b* and *g* are more closely related to higher-effect responses and also with tolerance width, indicating their close association with the community's sensitivity to the suppressing effects of copper toxicity. The parameters *b* and *g* are highly correlated to each other and may be better rewritten as a quotient of *g*/*b* (see Eq. (3) below). There is a close association between *b* and PC.

Optimum performance was attained when the positive component (comprising *f* and *k*) of the model was maximal and the negative component (comprising *b* and *g*) was minimal. This occurred under conditions of long-term treatment at 0.1 mg/L Cu. The opposite effect was at 0.2 mg/L Cu treatment. Therefore, it is inferred that for the ecosystem being studied, an ambient long-term exposure at 0.1 mg/L Cu would be optimum as it resulted in the periphyton community being the most efficient in photosynthetic performance as well as the most tolerant with the widest tolerance range to short-term acute copper exposure.

The model structure can be simplified by having only the short-term exposure copper concentration variable (*x*) and linking to the long-term exposure concentration through Eq. (2). If the long-term exposure concentration is of interest, the model can be rewritten as Eq. (3), with the community short-term response $Y_{\mathrm{ST}}$ as a function of the community long-term response $Y_{\mathrm{LT}}$:

$$Y_{ST} = Y_{LT} + \frac{f}{\exp(gx/b)} \tag{3}$$

The reformatted Eq. (3) is modular, comprising the community responses of the various exposure conditions. Such a modular format has a desirable attribute of extending the model application to include other preconditioning effects through $Y_{\mathrm{LT}}$ linkage [21].

The response–exposure curve can be better presented using a 3D plot with the long-term exposure concentration as the 3rd axis. It should however be noted that the long-term exposure concentrations are apparent values because of ethylenediaminetetraacetate in the media. Similarly, the model parameters were determined under a set of specified conditions, which may limit the model interpretation outside these conditions. Further studies are needed to extend the capabilities of the model for other concentration ranges or exposure conditions.

# 5 Conclusion

The proposed model $Y = k(1 + fx)/(1 + \exp(b + gx)) + d$ adequately described the periphyton community response to the variable long-term and subsequent short-term exposure to copper. There is improved goodness of fit for positive responses of hormesis by the proposed model compared to the conventional logistic model. The model facilitated the estimation of effect endpoints for ecological risk assessment and for evaluation of changes in community from the variable copper exposure. The model has enabled identification of the enhancing and suppressing effects of the variable copper exposure on the communities, hence making it a useful learning tool for understanding the effects of variable copper exposure on the ecologically important periphyton community. Through the model, new endpoints of ecological relevance were identified as PC and $EC_{0.5Y}$, and new relationships were derived for interpretation of ecological relevance.

**Acknowledgements** Thanks to Louis Evans, Curtin University of Technology, and Carolyn Oldham, the University of Western Australia, for organizing the student scholarship and project funding, and sincere thanks to Brian Jones, Department of Fisheries, Western Australia, Australia, and Peter Chapman, EVS Environmental Consultants, Canada, for their valuable discussion.

## References

1. Tsihrintzis VA, Hamid R (1997) Modelling and management of urban stormwater runoff quality: a review. Water Resour Manag 11:137–164
2. Scherman PA, Muller WJ, Palmer CG (2003) Links between ecotoxicology, biomonitoring and water chemistry in the integration of water quality into environmental flow assessments. River Res Appl 19:483–493
3. Linhurst RA, Bourdeau P, Tardiff RG (1995) A conceptual approach for ecological research, monitoring, and assessment. In: Linhurst RA, Bourdeau P, Tardiff RG (eds) Methods to assess the effects of chemicals on ecosystems. Wiley, Chichester
4. Koepp H (1997) NOEC or what? SETAC Eur News 8:3–4
5. ANZECC (2000) The national water quality management strategy Paper No. 4: Australian and New Zealand guidelines for fresh and marine water quality. Australian and New Zealand Environment and Conservation Council, Canberra
6. Newman MC, Ownby DR, Mezin LCA, Powell DC, Christensen TL, Lerberg SB, Anderson BA (2000) Applying species -sensitivity distributions in ecological risk assessment: assumptions of distribution type and sufficient numbers of species. Environ Toxicol Chem 19:508–515
7. Chapman PM (2002) Ecological risk assessment (ERA) and hormesis. Sci Total Environ 288:131–140
8. Brain P, Cousens R (1989) An equation to describe dose responses where there is simulation of growth at low doses. Weed Res 29:93–96
9. Calow P, Sibly RM (1990) A physiological basis of population processes: ecotoxicological implications. Funct Ecol 4:283–288
10. Tan CK (2006) An ecotoxicological study of urban runoff, Doctoral Thesis. The University of Western Australia
11. Stauber JL, Florence TM (1987) Mechanism of toxicity of ionic copper and copper complexes to algae. Mar Biol 94:511–519

12. Sakamoto Y, Ishiguro M, Kitagawa G (1986) Akaike information criterion statistics. D Reidel Publishing Company, Dordrecht
13. Davison AC, Hinkley DV (1999) Bootstrap methods and their application. Cambridge University Press, Cambridge
14. Efron B, Tibshirani R (1993) An Introduction to the Bootstrap. Chapman and Hall, New York
15. Grist EPM, Leung KMY, Wheeler JR, Crane M (2002) Better bootstrap estimation of hazardous concentration thresholds for aquatic assemblages. Environ Toxicol Chem 21:1515–1524
16. Chernick MR (1999) Bootstrap methods. Wiley, New York
17. van Ewijk PH, Hoekstra JA (1993) Calculation of the EC50 and its confidence interval when subtoxic stimulus is present. Ecotoxicol Environ Saf 25:25–32
18. Scholze M, Boedeker W, Faust M, Backhaus T, Altenburger R, Grimme LH (2001) A general best-fit method for concentration-response curves and the estimation of low-effect concentrations. Environ Toxicol Chem 20:448–457
19. Schindler DW (1996) Ecosystems and ecotoxicology: a personal perspective. In: Newman MC, Jagoe CH (eds) Ecotoxicology: a hierarchical treatment. Lewis Publishers, Boca Raton
20. Chapman PM, Wand F (2000) Issues in ecological risk assessment of metals and metalloids. Hum Ecol Risk Assess 6:965–988
21. Lassiter RR (1986) Design criteria for a predictive ecological effects modelling system. Aquatic Toxicol Environ 921:42–54

# Part II
# Mechanical and Manufacturing Engineering

# Application of Integrated Fuzzy-AHP for Design Concept Evaluation: A Case Study on Mold Design Selection

**Faiz Mohd Turan and Badrul Omar**

## 1 Introduction

In today's industries, product design has become the main focus of competition in a highly competitive environment and fast-growing global market [1]. Benchmarks used to determine competitive advantage of a manufacturing company are customer satisfaction, shorter product development time, higher quality, and lower product cost [2–4]. Today's product designer is being asked to develop high quality products at an ever-increasing pace [5]. To meet this new challenge, new and novel design methodologies that facilitate the acquisition of design knowledge and creative idea for later reuse are much sought after. In the same context, Liu and Boyle [6] describe that current challenges faced by the engineering design industry are the need to attract and retain customers, the need to maintain and increase market share and profitability, and the need to meet the requirements of diverse communities. Thus, a good design process should take into account the aforementioned criteria as early as possible in order to ensure the success of a product [1].

The product development process is a transformation process from customer requirements to a physical structure while considering the various design constraints [7]. Since long, new product development has been considered as an essential element for organizational competitiveness and success [8]. Product development also plays a critical role in the survival and success of manufacturing enterprises, and many researchers have improved their cognation of the need to manage it strategically [9–13]. However, truly effective product development

F.M. Turan (✉)
Faculty of Manufacturing Engineering, Universiti Malaysia Pahang, 26600 Pekan, Pahang, Malaysia
e-mail: faisz74@gmail.com

B. Omar
Faculty of Mechanical and Manufacturing Engineering, Universiti Tun Hussein Onn Malaysia, Parit Raja, Batu Pahat, 86400, Johor, Malaysia

K. Ab. Hamid et al. (eds.), *The Malaysia-Japan Model on Technology Partnership: International Proceedings 2013 of Malaysia-Japan Academic Scholar Conference*, DOI 10.1007/978-4-431-54439-5_9, 

remains difficult [14]. The rapid pace of technology development has led to shorter product life cycles for many product categories, most notably in consumer electronics. The need to stay competitive has shrunk product development time through the use of simultaneous and collaborative design processes.

One important step in designing new products is generating conceptual designs [1]. Design concept evaluation is a complex multi-criteria decision-making (MCDM) process that includes a set of technical activities, which are the refinement of customer requirements into design functions, new concept development, and embodiment engineering of a new product [7]. Lotter [15] and Ullman [16] point out that 75 % of the manufacturing cost is committed early in the design process. Under such circumstances, design concept evaluation in the early phase of product development plays a critical role as it has a significant impact on downstream processes [17, 18].

In order to help the designers to have better-informed decision before making judgment, systematic design evaluation method is needed. Among the various tools developed for design concept evaluation, fuzzy set theory and Analytical Hierarchy Process (Fuzzy-AHP) methods have received the most attention due to their abilities in handling uncertainty and MCDM [1, 19]. An ideal design evaluation method needs to use less number of design criteria, use optimum number of pair-wise comparisons, and have a supportive tool to verify and validate the rank of alternatives obtained.

However, in many practical situations, the human preference model is uncertain and decision makers might be reluctant or unable to assign exact numerical values to the comparison judgments. Consequently, the decision makers will need a process of reconsideration of design alternatives in relation to design criteria, and it may not help them to reduce the number of design criteria exactly. In addition, the final weight of design alternatives may not give significant difference which will have an impact on the designers or decision makers in making a judgment. A sole conventional Fuzzy-AHP is thus insufficient when applied to ambiguous problems.

The proposed design evaluation method will integrate Fuzzy-AHP with another effective method in order to provide another alternative to the designers. The literature search indicates that no work has been done with the above proposed methodology in design concept evaluation. The implementation of the proposed novel method will be divided into three stages, and these are screening, evaluating, and verifying which refer to useless number of design criteria, minimize the number of pair-wise comparisons, and have a supportive tool to verify and validate the rank of alternatives obtained. Following the methodology as outlined above, it perhaps can fulfill the aforementioned requirement of ideal design evaluation.

## 2 Proposed Method

The general framework of the approach has been depicted in Fig. 1.

Prescriptive design process model of Pahl and Beitz [20] as a based, the proposed design concept evaluation will focus on conceptual design and embodiment design stage. The designer or decision maker will initially set up the design

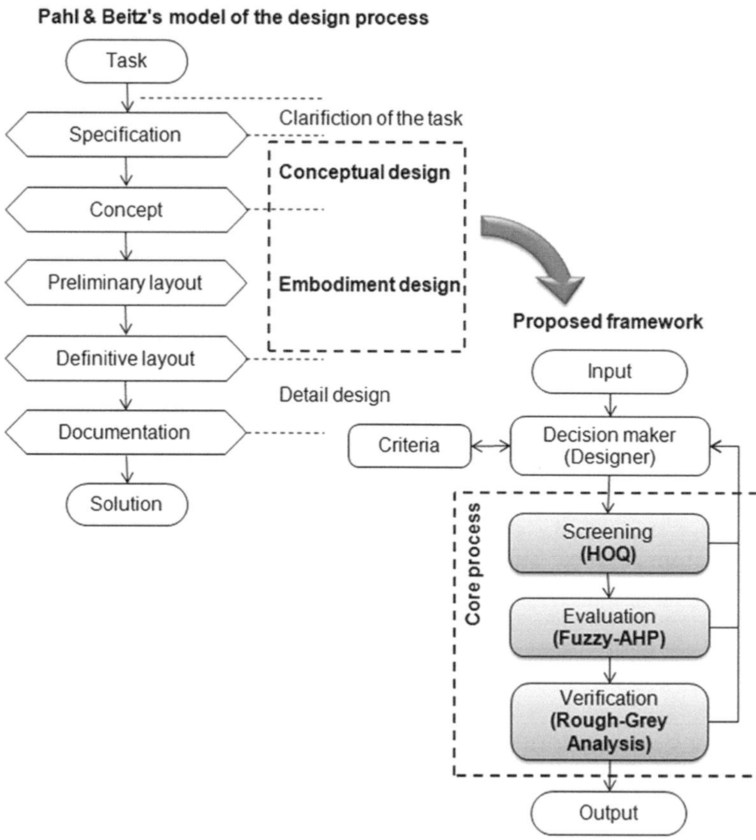


**Fig. 1** General framework of proposed approach

structure according to the recommended procedure from specification to definitive layout process. Then, they can create a general hierarchy and then identify the relevant criteria or sub-criteria, and one can identify the criteria and sub-criteria and then put them into hierarchy. The output can be made using a screening process, followed by an evaluation and verification method from each hierarchy with its relevant criteria.

In this research, HOQ method has been used for screening or pre-evaluating the alternatives suggested by the designer. Then the Fuzzy-AHP method will be used for obtaining the weights of alternatives from the point of view of each decision maker. Finally, the rank of alternatives will be verified and validated using Rough-Grey Analysis method.

### 2.1 Screening Process Using HOQ Method

First of all the customer needs and the engineering characteristics have to be identified.

- Identification of customer needs.
- Establishing engineering characteristics.
- Estimating WHATs/HOWs correlations by design team.
- WHATs/HOWs correlations derived from customers' evaluations.

### 2.2 Evaluation Process Using Fuzzy-AHP Method

The proposed Fuzzy-AHP based methodology provides a framework for prioritization of alternatives at early stages of design process. The methodology can be divided into four steps as described in the following paragraphs.

- Benchmarking and building of model hierarchical.
- Construction of pair-wise comparison matrices (PCM).
- Calculation of eigenvectors of elements by solving fuzzy PCM.
- Calculation of overall prioritization weights for each alternative.

The overall or total prioritization weight (TW) of an alternative was calculated by considering the individual weights of all the relevant criteria. Mathematically, it can be represented as follows [21]:

$$\mathrm{TW}_{Ak} = \sum_{k} \underset{i \in Uij}{W_{Ui}} \times \underset{Uij \in Ak^{*}}{W_{Uij}} \; W_{Ak} \; \forall k \tag{1}$$

where $\underset{i \in Uij}{W_{Ui}}$ is the relative importance of general criterion $U_i$ that is relevant to the secondary criteria $U_{ij}$. $\underset{Uij \in Ak}{W_{Uij}}$ is the Relative importance of secondary criteria $U_{ij}$ that are relevant to the alternatives $A_k$. $W_{Ak}$ is the Relative importance of an alternative $A_k$ with regard to its next higher level secondary criterion. $A_K$ is the alternatives, $k = 1, 2, 3$.

### 2.3 Verification Process Using Rough-Grey Analysis

Rough-Grey Analysis approach is very suitable for solving the group decision-making problem under uncertain environment. The selection procedures are summarized as follows [22–24]:

- Establishment of grey decision table.
- Normalization of grey decision table.

- Determination of the suitable alternatives.
- Making the ideal alternative for reference.
- Selecting the most suitable alternative.

The Grey relational grade (GRG) between each comparative sequence $\otimes x_i$ and the reference sequence $\otimes x_0$ can be derived from the average of Grey relational coefficient (GRC), which is denoted as

$$\Gamma_{0i} = \sum_{k-1}^{n} \frac{1}{n} \gamma(\otimes x_0(k), \otimes x_i(k)) \quad (2)$$

where $\Gamma_{0i}$ represents the degree of relation between each comparative sequence and the reference sequence. The alternative corresponding to the maximum value of GRG can be considered as the most suitable alternative.

### 2.4 *Case Example*

In this case study, the criteria and alternatives formulation will involve initial criteria selection from technical documents and survey results from questionnaire. The application is to select the best mold design for video camera's top cover among three developed concept designs, which have been designed by design engineers. These alternatives are depicted in Fig. 2.

From the point of view of design engineers, all three alternatives can be potentially implemented. There are five decision makers whose views are deemed important and they should be taken into account for making a decision. They are from production, maintenance, engineering, quality control, and production control department.

Table 1 lists the initial criteria for this case study. The table describes the criteria in more detail. There are a total of 23 criteria being selected at this initial stage.

## 3 Results and Discussions

### 3.1 *Screening Using HOQ*

Experts in the multidisciplinary team identified the relationships between each pair of customer's requirements and technical requirements. The output from HOQ matrix which is the HOQ summary is shown in Table 2. The summary of HOQ includes the relative weight or relative importance of each quality characteristic. The weight obtained will be ranked and filtered, and used as a reference for the next process. Table 3 is the modified HOQ summary after selecting the top seven criteria.

**MOLD DESIGN 1 (Alternative 1)**

| | |
|---|---|
| *Type of mold* | 3-plate |
| *Number of cavities* | 4 |
| *Mold insert material* | P5 |
| *Types of gates* | pin point gate |
| *Types of ejection system* | stripper plate |
| *Cooling systems* | spiral cooling |

**MOLD DESIGN 2 (Alternative 2)**

| | |
|---|---|
| *Type of mold* | 3-plate |
| *Number of cavities* | 4 |
| *Mold insert material* | P5 |
| *Types of gates* | pin point gate |
| *Types of ejection system* | ejector pin |
| *Cooling systems* | spiral cooling |

**MOLD DESIGN 3 (Alternative 3)**

| | |
|---|---|
| *Type of mold* | 3-plate |
| *Number of cavities* | 2 |
| *Mold insert material* | P5 |
| *Types of gates* | pin point gate |
| *Types of ejection system* | stripper plate |
| *Cooling systems* | spiral cooling |

**Fig. 2** Design alternatives

## 3.2 Evaluation Using Fuzzy-AHP

Figure 3 depicts the hierarchical structure of alternatives and other general criteria and secondary criteria to prioritize alternatives for selecting the best material in order to optimize the cost and performance of the product. The criteria ($U_i$) represent a combination of strategic index and key factors in design selection based on screening results obtained from previous process. At the next level, three alternatives that significantly influence the criteria were considered.

Table 4 presents the results of prioritization weights calculations for the alternatives with respect to the criteria. In this study, the consistency ratio (CR) values for all of the pair-wise comparison matrices have been found to be less than 0.1 which is consistent and acceptable. It also shows the largest Eigen value ($\lambda_{max}$), consistency index (CI) and CR validating the pair-wise comparison. The final results of overall prioritization weight for each alternative are presented in Table 5.

The Fuzzy-AHP analysis for the proposed approach suggests that Design 1 with weight of 0.3455 should be given the highest priority. Among the three alternatives selected in this study, the second most important alternative is Design 2 with a weight of 0.3365 followed by Design 3 (0.3179).

**Table 1** Initial criteria

| No | Selected initial criteria |
|---|---|
| 1 | Cavity design—Number of cavities |
| 2 | Cavity design—Cavity layout (equal runner, symmetrical, diaphragm) |
| 3 | Mold inserts—Mold inserts dimension |
| 4 | Mold inserts—Mold inserts material |
| 5 | Mold inserts—Mold inserts finishing |
| 6 | Mold structure—2 plate |
| 7 | Mold structure—3 plate |
| 8 | Mold structure—Operation mode (manual, semi-auto, auto) |
| 9 | Undercut release mechanism—Manual |
| 10 | Undercut release mechanism—Collapsible core |
| 11 | Undercut release mechanism—Side core |
| 12 | Undercut release mechanism—Split mold |
| 13 | Undercut release mechanism—Unscrewing (electric motor, hydraulic motor, hydraulic rack and pinion) |
| 14 | Feeding—Gating (side gate, submarine gate, pin point gate, disc gate) |
| 15 | Feeding—Runner (conventional, insulated runner, hot runner) |
| 16 | Temperature control—Cooling channel design (copper pin, baffer-type, babbler-type, spiral cooling, no core cooling) |
| 17 | Ejection—Air |
| 18 | Ejection—Stripper plate |
| 19 | Ejection—Ejection pin |
| 20 | Ejection—Ejection sleeve |
| 21 | Manufacturing—Cycle time |
| 22 | Manufacturing—Machine type |
| 23 | Manufacturing—Mold base requirements |

## 3.3 *Verification Using Rough-Grey Analysis*

There is a grey information system $T = (U, A, V, f_\otimes)$ for selection of alternatives. The grey decision table is expressed by $T = (U, A \cup D, f_\otimes)$. $U = \{S_i, i = 1, 2, 3\}$ are three potential alternatives for seven attributes $A = \{a_j, j = 1, 2, \ldots, 7\}$. The seven attributes include qualitative attributes and quantitative attributes. $a_2$, $a_3$, $a_4$, and $a_5$ are benefit attributes; the larger values are better. $a_2$, $a_6$, and $a_7$ are cost attributes; the smaller values are better. The selection structure is shown in Fig. 4.

The grey decision table is formed as shown in Table 6. Next step is to normalize the grey decision table. As a result, the grey normalized decision table is shown in Table 7.

The grey relational attribute (GRA) is a numerical measure of the relationship between comparative values and objective values, and the numeric values are among 0 and 1. By the rule that the design corresponding to the maximum value of GRG is the most suitable design, the grade is as shown in Table 8.

This is verification process using Rough-Grey analysis method. From the GRG results, the most suitable design is Design 1 ($\Gamma_{01} = 1.000$), followed by Design 2 ($\Gamma_{02} = 0.667$), and Design 3 ($\Gamma_{03} = 0.333$). All of the results are consistent with the results of Fuzzy-AHP.

**Table 2** HOQ summary

| No | Quality characteristics ("Functional Requirements" or "Hows") | Requirement weight | Relative weight (relative importance) (%) |
|---|---|---|---|
| 1 | Cavity design—Number of cavities | 297.83 | 6.56 |
| 2 | Cavity design—Cavity layout (equal runner, symmetrical, diaphragm) | 247.83 | 5.46 |
| 3 | Mold inserts—Mold inserts dimension | 146.74 | 3.23 |
| 4 | Mold inserts—Mold inserts material | 101.09 | 2.23 |
| 5 | Mold inserts—Mold inserts finishing | 86.96 | 1.92 |
| 6 | Mold structure—2 plate | 226.09 | 4.98 |
| 7 | Mold structure—3 plate | 239.13 | 5.27 |
| 8 | Mold structure—Operation mode (manual, semi-auto, auto) | 279.35 | 6.16 |
| 9 | Undercut release mechanism—Manual | 194.57 | 4.29 |
| 10 | Undercut release mechanism—Collapsible core | 233.70 | 5.15 |
| 11 | Undercut release mechanism—Side core | 233.70 | 5.15 |
| 12 | Undercut release mechanism—Split mold | 240.22 | 5.29 |
| 13 | Undercut release mechanism—Unscrewing (electric motor, hydraulic motor, hydraulic rack and pinion) | 194.57 | 4.29 |
| 14 | Feeding—Gating (side gate, submarine gate, pin point gate, disc gate) | 255.43 | 5.63 |
| 15 | Feeding—Runner (conventional, insulated runner, hot runner) | 267.39 | 5.89 |
| 16 | Temperature control—Cooling channel design (copper pin, baffer-type, babbler-type, spiral cooling, no core cooling) | 229.35 | 5.06 |
| 17 | Ejection—Air | 116.30 | 2.56 |
| 18 | Ejection—Stripper plate | 146.74 | 3.23 |
| 19 | Ejection—Ejection pin | 146.74 | 3.23 |
| 20 | Ejection—Ejection sleeve | 129.35 | 2.85 |
| 21 | Manufacturing—Cycle time | 160.87 | 3.55 |
| 22 | Manufacturing—Machine type | 157.61 | 3.47 |
| 23 | Manufacturing—Mold base requirements | 205.43 | 4.53 |

# 4 Conclusion

The results of the example presented in this work show that the idea of using the integration of Fuzzy-AHP with HoQ and Rough-Grey Analysis provides designers with another alternative to the existing methods, for the performance of design evaluation in the early stages of product development. The proposed framework has successfully helped the designers to reduce product development time and cost.

This research not only benefits the area of design evaluation in product development but it can be applied to any other area associated with a decision-making process. The efficacy of the proposed method could be extended by applying it in different conditions or to products of different complexity.

**Table 3** Modified HOQ summary

| No | Quality characteristics ("Functional Requirements" or "Hows") | Requirement weight | Relative weight (relative importance) (%) |
|---|---|---|---|
| 1 | Cavity design—Number of cavities | 297.83 | 16.30 |
| 2 | Mold structure—Operation mode (manual, semi-auto, auto) | 279.35 | 15.29 |
| 3 | Feeding—Runner (conventional, insulated runner, hot runner) | 267.39 | 14.63 |
| 4 | Feeding—Gating (side gate, submarine gate, pin point gate, disc gate) | 255.43 | 13.98 |
| 5 | Cavity design—Cavity layout (equal runner, symmetrical, diaphragm) | 247.83 | 13.56 |
| 6 | Undercut release mechanism—Split mold | 240.22 | 13.15 |
| 7 | Mold structure—3 plate | 239.13 | 13.09 |

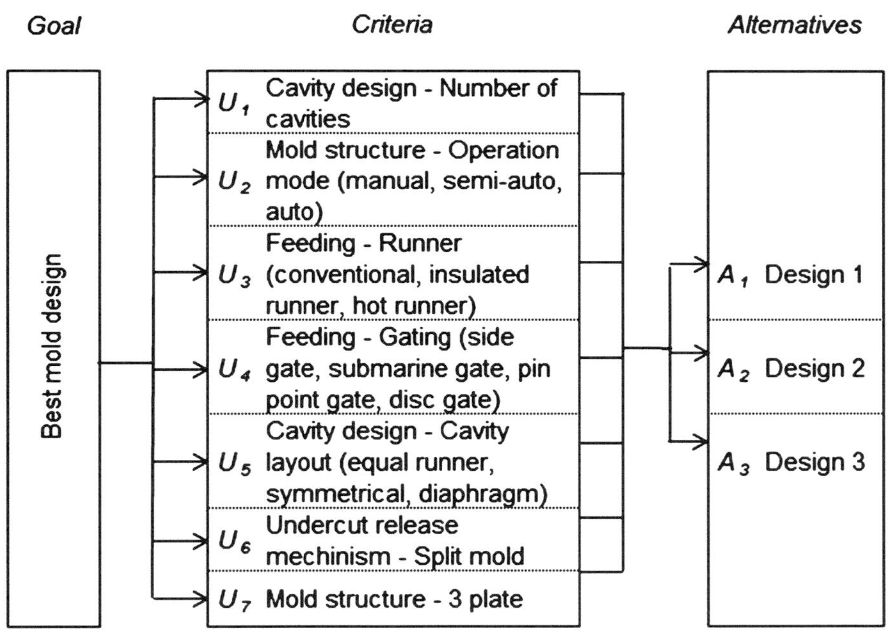


**Fig. 3** Analytic hierarchy structure

**Table 4** Summary of relative importance

| Criteria | | $\lambda_{max}$ | C.I. | C.R. |
|---|---|---|---|---|
| $U_1$ | Cavity design—Number of cavities | 3.213 | 0.106 | 0.085 |
| $U_2$ | Mold structure—Operation mode (manual, semi-auto, auto) | 3.045 | 0.022 | 0.018 |
| $U_3$ | Feeding—Runner (conventional, insulated runner, hot runner) | 3.000 | 0.000 | 0.000 |
| $U_4$ | Feeding—Gating (side gate, submarine gate, pin point gate, disc gate) | 3.000 | 0.000 | 0.000 |
| $U_5$ | Cavity design—Cavity layout (equal runner, symmetrical, diaphragm) | 3.045 | 0.022 | 0.018 |
| $U_6$ | Undercut release mechanism—Split mold | 3.000 | 0.000 | 0.000 |
| $U_7$ | Mold structure—3 plate | 3.000 | 0.000 | 0.000 |

**Table 5** Overall prioritization weight for alternatives

| Goal | Criteria | Criterion relative importance $W_{Ui}$ | Alternative | Alternative relative importance $W_{Ak}$ | Total alternative weight $TW_{Ak}$ | Ranking |
|---|---|---|---|---|---|---|
| Best mold design | $U_1$ | 0.163 | $A_1$ | 0.3735 | $A_1 = 0.3455$ | 1 |
| | | | $A_2$ | 0.3735 | $A_2 = 0.3365$ | 2 |
| | | | $A_3$ | 0.2530 | $A_3 = 0.3179$ | 3 |
| | $U_2$ | 0.153 | $A_1$ | 0.3529 | | |
| | | | $A_2$ | 0.2941 | | |
| | | | $A_3$ | 0.3529 | | |
| | $U_3$ | 0.146 | $A_1$ | 0.3333 | | |
| | | | $A_2$ | 0.3333 | | |
| | | | $A_3$ | 0.3333 | | |
| | $U_4$ | 0.140 | $A_1$ | 0.3333 | | |
| | | | $A_2$ | 0.3333 | | |
| | | | $A_3$ | 0.3333 | | |
| | $U_5$ | 0.136 | $A_1$ | 0.3529 | | |
| | | | $A_2$ | 0.3529 | | |
| | | | $A_3$ | 0.2941 | | |
| | $U_6$ | 0.131 | $A_1$ | 0.3333 | | |
| | | | $A_2$ | 0.3333 | | |
| | | | $A_3$ | 0.3333 | | |
| | $U_7$ | 0.131 | $A_1$ | 0.3333 | | |
| | | | $A_2$ | 0.3333 | | |
| | | | $A_3$ | 0.3333 | | |
| Sum | | | | | 1.0000 | |

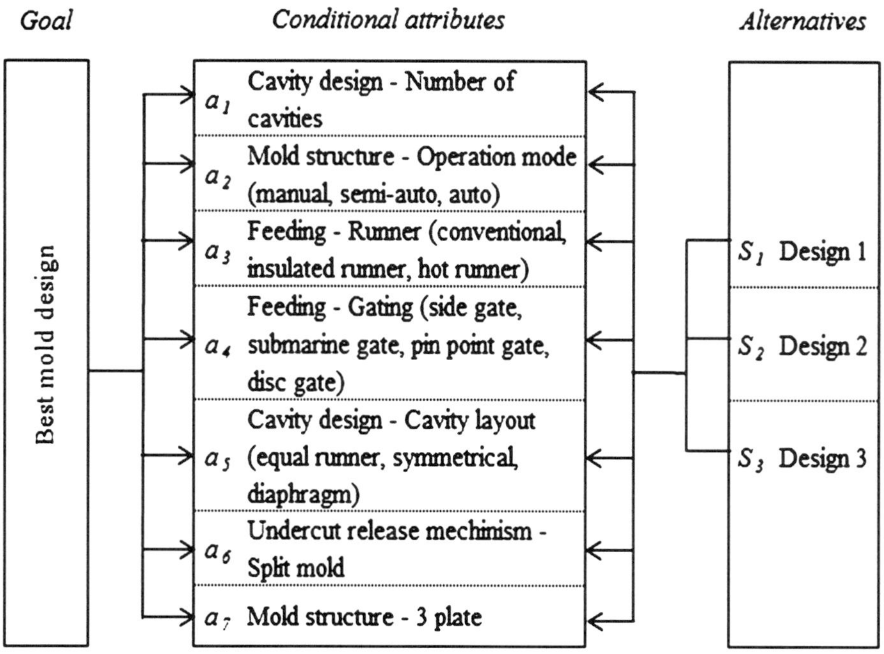


**Fig. 4** Selection structure

**Table 6** Grey decision table

| Alternatives | $S_1$ | $S_2$ | $S_3$ |
|---|---|---|---|
| $a_1$ | [5.7,6.7] | [5.7,6.7] | [3.7,4.7] |
| $a_2$ | [4.3,5.3] | [3.5,4.5] | [4.3,5.3] |
| $a_3$ | [3.7,4.7] | [3.7,4.7] | [3.7,4.7] |
| $a_4$ | [2.9,3.9] | [2.9,3.9] | [2.9,3.9] |
| $a_5$ | [4.3,5.3] | [4.3,5.3] | [3.5,4.5] |
| $a_6$ | [4.3,5.3] | [4.3,5.3] | [4.3,5.3] |
| $a_7$ | [6.3,7.3] | [6.3,7.3] | [6.3,7.3] |
| Decision | 2 | 2 | 1 |

**Table 7** Grey normalized

| Alternatives | $S_1$ | $S_2$ | $S_3$ |
|---|---|---|---|
| $a_1^*$ | [0.867,1] | [0.813,0.947] | [0.467,0.6] |
| $a_2^*$ | [0.867,1] | [0.867,1] | [0.493,0.627] |
| $a_3^*$ | [0.789,0.965] | [0.789,0.965] | [0.789,0.965] |
| $a_4^*$ | [0.804,1] | [0.804,1] | [0.804,1] |
| $a_5^*$ | [0.804,1] | [0.804,1] | [0.804,1] |
| $a_6^*$ | [0.787,1] | [0.787,1] | [0.787,1] |
| $a_7^*$ | [0.787,1] | [0.787,1] | [0.787,1] |
| Decision | 2 | 2 | 1 |

**Table 8** Grey relational grade

| GRG | | $\Gamma_{01}$ | $\Gamma_{02}$ | $\Gamma_{03}$ |
|---|---|---|---|---|
| Conditional attributes | $a_1$ | 0.333 | 0.333 | 0.000 |
| | $a_2$ | 0.333 | 0.000 | 0.333 |
| | $a_3$ | 0.000 | 0.000 | 0.000 |
| | $a_4$ | 0.000 | 0.000 | 0.000 |
| | $a_5$ | 0.333 | 0.333 | 0.000 |
| | $a_6$ | 0.000 | 0.000 | 0.000 |
| | $a_7$ | 0.000 | 0.000 | 0.000 |
| Total | | 1.000 | 0.667 | 0.333 |
| Ranking | | 1 | 2 | 3 |

## References


1. Faiz MT, Badrul O (2013) The integration of HOQ and Fuzzy-AHP for design concept evaluation. Appl Mech Mater 315:25–29
2. Hsu W, Woon IMY (1998) Current research in the conceptual design of mechanical products. Comput Aided Des 30:377–389
3. Subrahmanian E, Rachuri S, Fenves SJ, Foufou S, Sriram RD (2005) Product lifecycle management support: a challenge in supporting product design and manufacturing in a networked economy. Int J Product Lifecycle Manage 1:4–25
4. Shai O, Reich Y, Rubin D (2007) Creative conceptual design: extending the scope by infused design. Comput Aided Des 41:117–135
5. Ye X, Liu H, Chen L, Chen Z, Pan X, Zhang S (2008) Reverse innovative design—an integrated product design methodology. Comput Aided Des 40:812–827
6. Liu S, Boyle IM (2009) Engineering design: perspectives, challenges, and recent advances. J Eng Des 20:7–19
7. Li W, Li Y, Wang J, Liu X (2010) The process model to Aid innovation of products conceptual design. Expert Syst Appl 37:3574–3587
8. Edwards T, Battisti G, McClendon WP Jr, Denyer D, Neely A (2005) Pathways to value how UK firms create more value using innovation strategically, 1st edn. Advanced Institure of Management Research (AIM), London
9. Brown S, Eisenhardt K (1995) Product development: past research, present findings, and future directions. Acad Manag Rev 20:343–378
10. Griffin A, Hauser J (1996) Integrating R&D and marketing: a review and analysis of the literature. J Prod Innov Manag 13:191–215
11. Krishnan V, Ulrich K (2001) Product development decisions: a review of the literature. Manag Sci 47:1–21
12. Chesbrough H, Teece D (2002) Organizing for innovation: when is virtual virtuous? Harv Bus Rev 80:127–134
13. Ayag Z, Odzemir RG (2008) A hybrid approach to concept selection through fuzzy analytic network process. Comput Ind Eng 56:368–379
14. Lee F, Santiago M (2008) Creativity in new product development: an evolutionary integration, 1st edn. Butterworth-Heinemann, Boston
15. Lotter B (1986) Manufacturing assembly handbook. Butterworth-Heinemann, Boston
16. Ullman DG (2009) The mechanical design process, 4th edn. McGraw-Hill, New York
17. Zhai LY, Khoo LP, Zhong ZW (2009) Design concept evaluation in product development using rough sets and grey relation analysis. Expert Syst Appl 36:7072–7079
18. Geng X, Chu X, Zhang Z (2010) A new integrated design concept evaluation approach based on vague sets. Expert Syst Appl 37:6629–6638

19. Scott MJ (2002) Quantifying certainty in design decisions: examining AHP. In: ASME Design Engineering Technical Conferences (DETC2002), 2002
20. Pahl G, Beitz W, Feldhuson J, Grote KH (2007) Engineering design: a systematic approach, 4th edn. Springer, London
21. Nepal B, Yadav OP, Murat A (2010) A fuzzy-AHP approach to prioritization of CS attributes in target planning for automotive product development. Expert Syst Appl 37:6775–6786
22. Bai C, Sarkis J (2010) Integrating sustainability into supplier selection with grey system and rough set methodologies. Int J Product Econ 124:252–264
23. Bai C, Sarkis J (2011) Evaluating supplier development programs with a grey based rough set methodology. Expert Syst Appl 38:13505–13517
24. Li GD, Yamaguchi D, Nagai M (2008) A grey-based rough decision-making approach to supplier selection. Int J Adv Manuf Technol 36:1032–1040

# Numerical Modelling of Dry and Wet Steam Flows in Converging-Diverging Nozzles

**Hasril Hasini, Norhazwani Abd Malek, and Mohd. Zamri Yusoff**

## 1 Introduction

Multiple phase flows are frequently encountered in many contexts such as environment (formation of rain) and industrial applications (such as the condensation of steam in steam turbine and spray combustion) [1]. Decades ago, the basic modelling of two phase flow has been devoted towards these applications but one of the most complex applications involved the condensing flow of steam due to the fact that the process occurs in supersonic conditions. In steam turbine operation, it is known that the last stage of a conventional steam turbine has to operate in wet steam condition. For nuclear fuelled power plant, all stages of the steam turbine are operating in wet steam condition. This condition is undesirable due to the fact that the water droplets formed within the superheated steam will hit the turbine blade at supersonic condition and result in erosion damage to the blade. Even though much progress has been made in wet steam flow, the understanding on the underlying mechanism of the transformation of superheated steam to wet mixture needs detail and careful investigation. The amount of literatures on condensing flows is very limited to the fact that the investigation is very complex and it involves interaction between different phases. Historically, the majority of the works in this field were mainly aimed at the experimental investigation in predicting the wetness fraction and most importantly the location at which the phenomenon is critical[2–6]. Another possible explanation for the lack of published work on nucleating steam in the open literature is the restriction in terms of publishing confidential data by major steam turbine manufacturer who, in most cases, has vast interest in the work. Progress on numerical work related to this field was accelerated with the

H. Hasini (✉) • N.A. Malek • M.Z. Yusoff
Centre for Advanced Computational Engineering, College of Engineering,
Universiti Tenaga Nasional, Jalan IKRAM-UNITEN, 43000, Kajang, Selangor, Malaysia
e-mail: hasril@uniten.edu.my

K. Ab. Hamid et al. (eds.), *The Malaysia-Japan Model on Technology Partnership: International Proceedings 2013 of Malaysia-Japan Academic Scholar Conference*,
DOI 10.1007/978-4-431-54439-5_10, 

successful use of nucleation and droplet growth equation which is based on classical nucleation theory. The inclusion of additional equations has increased the computation time tremendously since most the models were empirical in nature; thus various constants have to be tweaked around for different sets of geometrical nozzle test.

In recent years, rapid development of computing power has enabled numerical calculation of highly complex three-dimensional flow domain. Parallel to this development, commercial CFD codes have also evolved with better solution accuracy in predicting solutions to various types of flows. It is well acknowledged that most commercial CFD code is general-purpose in nature. In some cases, it has been demonstrated that these codes are able to resolve complex flow problem with great accuracy. In this chapter, the development of the numerical formulation for the prediction of dry and wet steam flows in converging-diverging nozzles is described. Emphasis is given on the attempt to model the flow using commercial CFD code, FLUENT. In addition, for dry superheated flows, test is also made with the solution given by in-house CFD solver for dry and condensing flows.

## 2 Governing Equations

The three-dimensional continuity, *x-, y- and z-* momentum and energy equations describing the fluid flow of a compressible flows expressed in conservation form can be written as:

$$\frac{\partial \overline{W}}{\partial t}+\frac{\partial \overline{F}}{\partial x}+\frac{\partial \overline{G}}{\partial y}+\frac{\partial \overline{H}}{\partial z}=0 \tag{1}$$

where $\overline{W}$ is the conserved variable, $\overline{F}$ is the flux in $x$ direction, $\overline{G}$ and $\overline{H}$ are the flux in $y$ and $z$ directions, respectively. For inviscid flows, the conserved variable, fluxes and energy transport terms could be written as

$$\overline{W}=\begin{pmatrix}\rho\\ \rho u\\ \rho v\\ \rho w\\ \rho e_0\end{pmatrix}\quad \overline{F}=\begin{pmatrix}\rho u\\ \rho u^2+P\\ \rho uv\\ \rho uw\\ \rho u h_0\end{pmatrix}\quad \overline{G}=\begin{pmatrix}\rho v\\ \rho uv\\ \rho v^2+P\\ \rho vw\\ \rho v h_0\end{pmatrix}\quad \overline{H}=\begin{pmatrix}\rho w\\ \rho uw\\ \rho vw\\ \rho w^2+P\\ \rho w h_0\end{pmatrix}$$

The wetness term is included into the governing equations by considering the droplet formation and growth that is resulted from the change in Gibbs free energy within the thermodynamic system. In order to grow, a stable nucleus needs to attain minimum radius, $r^*$ (known as critical radius) which is given by:

$$r^*=\frac{2\sigma_r}{\rho_L R T_G \ln[p/p_s(T_G)]} \tag{2}$$

Detailed formulation and derivation of this equation is given in [7]. Based on (2), droplets formed with smaller radius than $r^*$ have the tendency to evaporate while droplets with radius higher than $r^*$ will grow. A comprehensive analysis for tracking individual droplet's formation and growth was reviewed in [8]. The model adopted by FLUENT is based on compressible Navier–Stokes equation with two additional transport equations. The compressible N-S equation is casted in the following manner:

$$\frac{\partial W}{\partial t}\frac{\partial}{\partial t}\int_V Q\mathrm{d}V + \oint[F - G]\mathrm{d}A = \int_V H\mathrm{d}V \quad (3)$$

The additional transport equations are for condensed phase and control equation for unit volume density droplet as given in (4) and (5), respectively.

$$\frac{\partial \rho\beta}{\partial t} + \nabla\left(\rho\vec{v}\beta\right) = \Gamma \quad (4)$$

$$\frac{\partial \rho\eta}{\partial t} + \nabla\left(\rho\vec{v}\eta\right) = \rho I \quad (5)$$

From (4) and (5), $\beta$ is the liquid mass fraction, $\Gamma$ is the mass condensation rate, $\eta$ is the unit volume density of droplet and $I$ is condensate nucleation rate (FLUENT User Guide). The relations between wet steam ($\rho$) and vapour densities ($\rho_v$) are given by

$$\rho = \rho_v/(1 - \beta) \quad (6)$$

# 3 Model Descriptions

In order to test the applicability of the numerical scheme in the calculation of dry and wet steam flows, the scheme is applied to converging-diverging nozzles with different expansion rate. For dry superheated steam case, the numerical and commercial scheme is applied to Binnie and Green nozzle [9]. Figure 1a shows the three-dimensional illustration of the nozzle. The nozzle is 230 mm length, 40 mm height and 40 mm depth. The mesh scheme employed is structured mesh and the total number of cells is 28,125 (125 × 15 × 15). Detail of quadrilateral mesh constructed at the throat region is shown in Fig. 1b. In order to ensure consistencies, similar geometrical mesh is also developed for FLUENT calculation. The condensing flow test, however, is applied to different nozzle [10]. With the appropriate converging and expansion rate, the condensing flow phenomena could be made possible. Figure 2a shows the three-dimensional representation of the nozzle. The approximate dimension of the nozzle is 300 × 80 × 80 mm. Similar to the previous case, quadrilateral mesh scheme was adopted throughout the flow domain. The total number of mesh used is similar to the one used in Binnie and Green nozzle. To test the solution's dependency on the mesh, additional case was also introduced by increasing the number of mesh through mesh refinement.

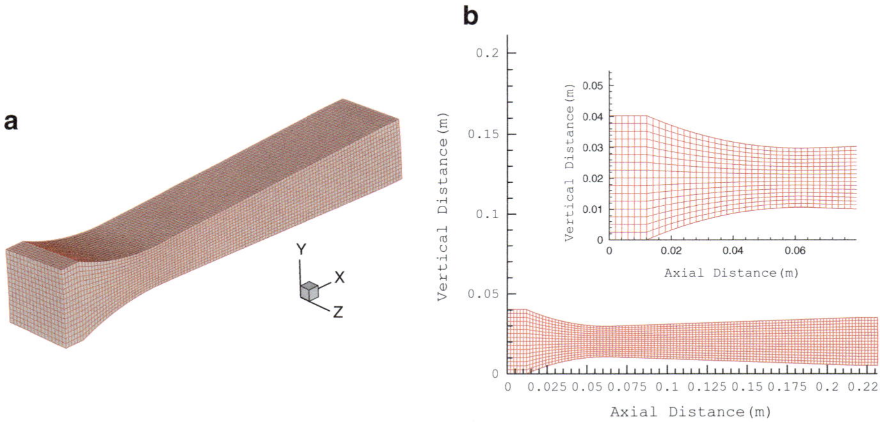


**Fig. 1** Schematic diagram of Binnie and Green nozzle [9]. (**a**) Isometric view (**b**) Front and zoomed throat view

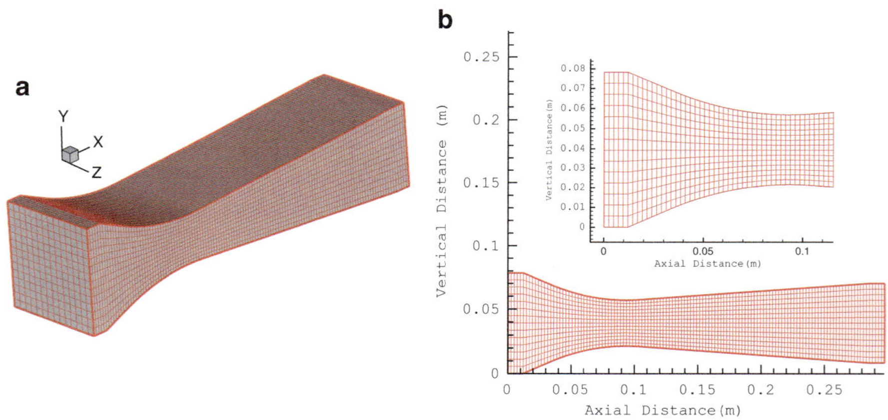


**Fig. 2** Schematic diagram of Binnie and Wood nozzle [10]. (**a**) Isometric view (**b**) Front and zoomed throat view

## 4 Boundary Conditions

For the test in Binnie and Green nozzle, three cases were calculated namely subsonic, supersonic and shock cases. The inlet total pressure for all cases is fixed to 72,700 Pa while the back pressure varies according to the cases. In subsonic case, the back pressure is set to 68,000 Pa giving a pressure ratio of 1.069. In supersonic case, the back pressure is determined by linear interpolation. To obtain shock condition, the back pressure is adjusted until discontinuity in pressure profile along the nozzle length is obtained. In this case, the shock condition could be attained by setting the back pressure to 60,400 Pa. For all cases, the total temperature at the inlet was set to 421.3 K. In Binnie and Wood nozzle, the total inlet pressure and temperature were set to 143,980 and 391.87 K, respectively.

# 5 Result and Discussions

Two tests were conducted on Binnie and Green and Binnie and Wood nozzles for dry and wet steam flow conditions. Using Binnie and Green nozzle, three cases for subsonic, supersonic and shock cases were calculated. The calculation was made using in-house solver for dry and wet steam flow, and extensive comparison was also made with the solutions given by FLUENT. The wet steam calculation is carried out on Binne and Wood nozzle. The prediction was later compared with experimental data of pressure along nozzle axial distance.

Figures 3a and b show the pressure and Mach number along axial nozzle distance, respectively, for subsonic flow. In general, the pressure decreases towards the throat and increases downstream due to the increase in flow area. The trend is the opposite for the Mach number distribution as shown in Fig. 3b. Comparison between the solutions for both in-house numerical code and FLUENT shows reasonably good agreement for inviscid flow. A test was also carried out to investigate the effect of the inclusion of turbulence model to realistically emulate the flow in nozzle. It was found that the pressure and kinetic energy difference between the inlet and throat is lower for test with turbulence as compared to inviscid case. This can be attributed to turbulence loss in flow channel that was not taken into account in the inviscid calculation.

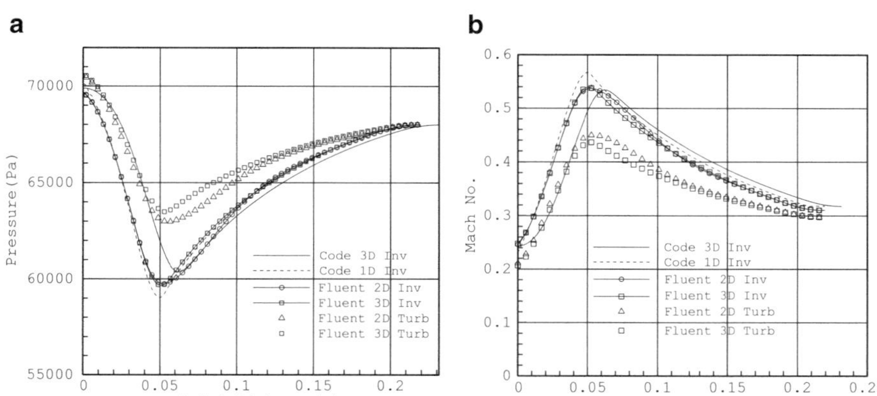


**Fig. 3** Distribution of (**a**) Pressure and (**b**) Mach number along nozzle axial distance for subsonic case

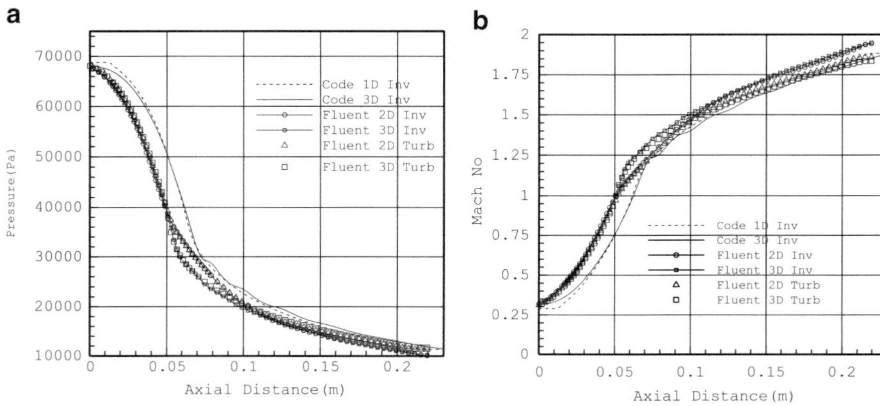


**Fig. 4** Distribution of (**a**) Pressure and (**b**) Mach number along nozzle axial distance for supersonic case

The pressure and Mach number along nozzle axial distance for subsonic case are shown in Figs. 4a and b respectively. General comparison between 1D analytical solution, in-house and commercial CFD codes shows reasonable agreement. Pressure reduces while Mach number increases throughout the nozzle axial distance. From Fig. 4a, there was a slight discrepancies between the solution given by in-house and FLUENT upstream of the throat. The prediction made by in-house code agrees very well with 1D analytical solution while FLUENT underpredicts the pressure from the inlet to the throat region. Downstream of the throat, excellent agreement was predicted for all cases. This trend is also true for Mach number which shows slight discrepancies at regions upstream of the throat.

Figs. 5a and b show the pressure and Mach number distribution along nozzle axial distance for shock case. General overview of pressure and Mach number for all cases shows good trend compared to 1D analytical solution except for turbulence cases which failed to capture the shock occurred slightly downstream of the throat. The shock could be identified through the sudden increase in pressure or sudden drop in velocity (Mach) at about 80 mm from the inlet.

The pressure ratio distribution along nozzle distance for wet steam case is shown in Fig. 6. General trend shows reduction in pressure as the steam flows downstream of the inlet towards the throat. At the onset of condensation, a sudden increase in pressure is anticipated when maximum supercooling is achieved. This consequently will prevent further formation of new nuclei, and the remaining droplet already formed will grow rapidly to restore the system to thermodynamic equilibrium. In this test, the onset of condensation was predicted slightly upstream as compared to the one given by experiment. The magnitude of pressure increase given by FLUENT was greater than experiment. Further downstream of agreement between simulation and experiment was excellent. It is also interesting to note that for test

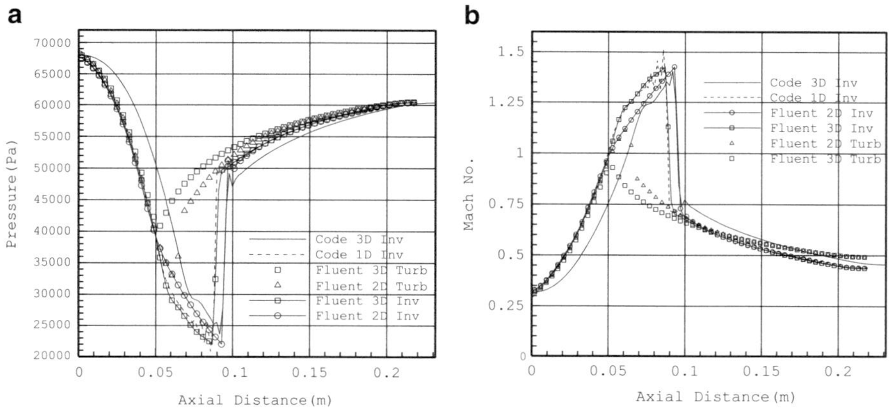


**Fig. 5** Distribution of (**a**) Pressure and (**b**) Mach number along nozzle axial distance for shock case

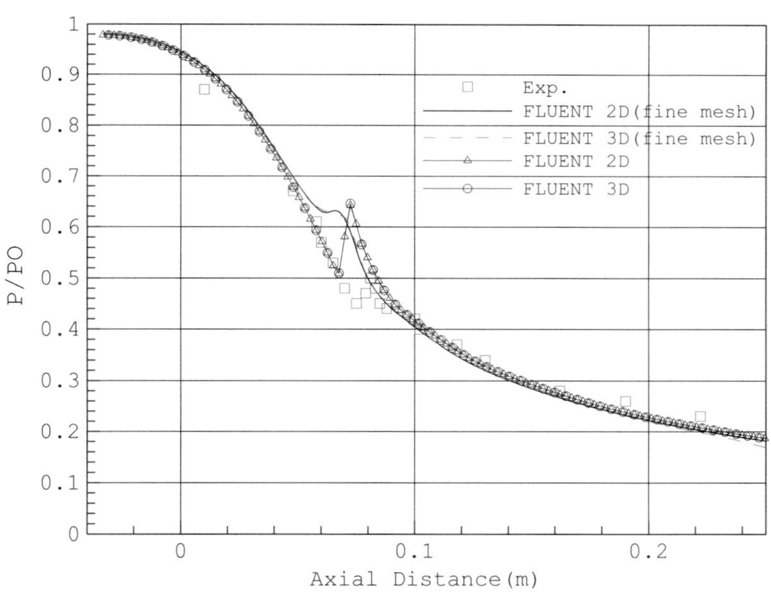


**Fig. 6** Pressure ratio along Binne and Wood nozzle

with higher mesh density, the pressure ratio distribution predicted by FLUENT in particular the magnitude of pressure increase is better. However, the location at which the onset of condensation is predicted is similar to the solution given by the base mesh case.

# 6 Conclusions

Numerical formulations for calculation of dry and wet steam flows in converging-diverging nozzles have been developed. The numerical scheme is applied to Binnie and Green and Binne and Wood nozzles for dry superheated and wet steam respectively. For dry superheated steam case, three conditions were tested namely subsonic, supersonic and shock. The prediction made by the numerical scheme in terms of pressure and Mach number distribution has been good. The calculation is also made using commercial CFD code FLUENT with reasonably good agreement with the one given by in-house code and 1D analytical solution. For wet steam test, the prediction of pressure along nozzle length shows good agreement with experimental data. However, the location at which the onset of condensation occurs was underpredicted. The magnitude of pressure increase given by coarse mesh case during the onset of condensation was however overpredicted. Refining the mesh improves the prediction on this location. From the test, it has been demonstrated that for simple geometry such as converging-diverging nozzle, the dry and wet properties of flowing steam could be qualitatively predicted using commercial CFD code FLUENT. The general trending of properties such as pressure and Mach number could be made accurately. To quantitatively predict the flowing steam properties such as the amount of pressure increase during the onset of condensation, additional improvements to the geometrical and mathematical model need to be considered. This could be made possible using the user-defined function available to users.

# References


1. Avettisian AR, Philippov GA, Zaichik LI (2008) Effects of turbulence and inlet moisture on two-phase spontaneously condensing flows in transonic nozzle. Int J Heat Mass Transfer 51:4195–4203
2. Bakhtar F, Webb RA (1983) Tabular method of estimating nucleating steam flows. Int J Heat Fluid Flow 217–228
3. Bakhtar F, Ebrahimi M, Webb RA (1995) On the performance of a cascade of turbine rotor tip section blading in nucleating steam—Part 1: surface pressure distribution. Proceedings of the Institution of Mechanical Engineers Part C. J Mech Eng Sci 209:115–124
4. Bakhtar F, Ebrahimi M, Bamkole BO (1995) On the performance of a cascade of turbine rotor tip section blading in nucleating steam—Part 2: wake traverses. Proceedings of the Institution of Mechanical Engineers Part C. J Mech Eng Sci 209:169–177
5. Gyarmathy G (2005) Nucleation of steam in high pressure nozzle experiments. In: Proceedings of the 6th European conference on turbomachinery, Lille, France, 458–469
6. Wroblewski W, Dykas S, Gepert A (2009) Steam condensing flow modeling in turbine channels. Int J Multiphase Flow 35:498–506
7. Mahpeykar MR, Lakzian E, Amirirad E (2009) Reduction of thermodynamic losses in a supersonic nucleating steam nozzle by spraying water droplets. Scientia Iranica Trans B Mech Eng 16:253–262

8. Hasini H, Yusoff MZ, Shuaib NH (2010) Preliminary theoretical study of the formation and growth of droplets in nucleating flows. In: AIP proceedings 10th Asian international conference of fluid machinery vol 1225. pp 751–757
9. Binnie AM, Green JR (1943) An electrical detector of condensation in high-velocity steam. Proc R Soc (Lond) A 181:134–154
10. Binnie AM, Wood MW (1938) The pressure distribution in a convergent-divergent steam nozzle. Proc Inst Mech Eng 138:229–266

# How Quality Engineering in Japan Affects Global Product Performance

**Rozzeta Dolah and Zenichi Miyagi**

## 1 Introduction

Quality is an essence in Japan. It is no doubt, compared to others, Japanese products provide highest satisfaction to the users. How does this happen? What is behind their success? How Japan can sustain the excellence of quality ahead of other countries, generally. Thus, this paper discusses the quality engineering (QE) methodology in general which has been the backbone of the Japan's product quality performance. This finding is useful for further implementing, expanding, and disseminating the quality research in other countries than Japan. QE approaches are presented in two ways: firstly on how a company successfully implemented the QE and secondly, a practical case study is shown on what we mean by improving product's performance using QE methodology. The excellence of QE has been practiced and applied by other countries of the world globally, which made Japan as their role model. The term QE is described as a robust parameter design in which a system is insensitive to variation. Arvidsson and Gremyr [1] reviewed on the robust design methodology that contains QE as part of its systematic efforts to achieve insensitivity to noise factors which founded on an awareness of variation and can be applied in all stages of product design. Ilzarbe et al. [2] explained the practical applications of experimental design which has been applied for many years in industry to improve quality. Tanco et al. [3] described how the experimentation

R. Dolah (✉)
Department of Mechanical Engineering, School of Science and Technology, Meiji University, Higashi Mita 1-1-1, Kawasaki, Kanagawa 214-8571, Japan

UTM Razak School of Engineering and Advanced Technology, Universiti Teknologi Malaysia, Jalan Semarak, 54100, Kuala Lumpur, Malaysia
e-mail: rozzetadolah@gmail.com

Z. Miyagi
Department of Mechanical Engineering, School of Science and Technology, Meiji University, Higashi Mita 1-1-1, Kawasaki, Kanagawa 214-8571, Japan

K. Ab. Hamid et al. (eds.), *The Malaysia-Japan Model on Technology Partnership: International Proceedings 2013 of Malaysia-Japan Academic Scholar Conference*, DOI 10.1007/978-4-431-54439-5_11, © Springer Japan 2015

carried out by companies in three European regions. The findings revealed that systematic experimental design is far much better than conventional way of doing experiment to improve the performance of products or processes. Viles et al. [4] emphasized on the importance of planning stage in industrial problems, where there are different factors that strongly affect the results of the study. Statistical design of experiment provides a proper way of planning an experiment in selecting appropriate data. Design of experiments (DOE) such as factorial design, response surface methodology (RSM), and Taguchi methods are widely used compared to traditional one-factor-at-a-time approach. Taguchi methods have been widely applied for optimization in peel test [5–7]. Taguchi method had simplified the classical design of experiment which was found to be too complicated to be applied by engineers in application field [8]. These literature reviews explained the application of QE in respective fields of research. However, none elaborates the linkage between laboratories research and industrial practice in QE. Thus, the gap and findings are explained in this paper and is constructed in the following manner. Firstly, an overview of Taguchi parameter design or robust engineering is given. Secondly, the experience of a company implementing QE is presented. Then, the practical data of a laboratory experiment on orthogonal array L18 is discussed in order to relate the experimental result to the requirement in industry. The implementation of QE is explained based on the strategies used in tackling organization problems. Optimization of peel adhesion strength test is carried out to propose the feasibility of experimental design tools. Finally, through the implementation of QE in organization and method applied in experimental design, a methodology flow is presented for QE methodology.

## 1.1 Taguchi Method

Back in the early 1920s, Sir R.A. Fisher introduced an experimental design method of statistical technique called Design of Experiments (DOE). Started from crop optimization in agricultural experiment, research and development of DOE grew significantly in academic environment. Not many industries applied DOE in production process. The more the research grew, the more complicated it became and the less it reflected the practitioner to apply it. In the 1940s, a Japanese engineer Dr. Genichi Taguchi modified and standardized the technique into a more useful way for practitioner. Here is the significant moment where DOE techniques become extremely useful and friendlier to apply. Transforming the sophisticated method into an applicable and easier to practice method, Dr. Taguchi introduced the simplified DOE to design quality into products and is called Quality Engineering that provides the ability to produce high-quality, low-cost products that fully satisfy customer needs.

Taguchi and Wu [9] introduce their approach using experimental design as a two-step optimization technique: Step 1: Reduce functional variability to increase robustness: A design that can maximize the signal-to-noise ratio which can then optimize the process or product function. It is more difficult to reduce variability than to adjust the mean to target value. Step 2: Adjust sensitivity. Adjust average response to the target value.

This paper focuses on the parameter design, which is an investigation done to identify settings that minimize the variation. Different setting may generate different variations in product or process performance. In classical parameter design developed by R.A. Fisher [10], the experimental design is complex and not easy to use. The main reason for this is a large number of the experiments need to be carried out when the number of parameters increases. For example, a full factorial experimental design for studying four parameters at three levels would have required 81 experimental trials ($3^4$). Adding in two-level noise factor with two repetitions would make number of observations to 324 ($81 \times 2 \times 2$), an unacceptable number due to constraints on experimentation. From the orthogonal arrays by Taguchi, a modified L-9 orthogonal array was chosen. Only 36 observations are implied ($9 \times 2 \times 2$). A loss function is then calculated from the deviation between experimental value and the desired value; or in other words, deviation from the target will create loss to customer. Value of loss function is transformed into a signal-to-noise ratio (SNR), which is a metric for robustness. SNR (unit: dB) is defined as in (1) below:

$$\begin{aligned} \text{SNR}, \eta &= \text{power of signal}/\text{power of noise} \\ &= (\text{sensitivity})^2/(\text{variability})^2 \\ &= \beta^2/\sigma^2 \end{aligned} \tag{1}$$

Inversed SNR is the variance per unit input [11]. As in (2), loss is proportional to the inverse of the SNR. The larger the SNR, the smaller the loss, thus the better the quality is.

$$\text{Loss} \propto \sigma^2 \propto 1/\text{SNR} \tag{2}$$

S/N ratio measures variability around the mean. It represents the ratio of sensitivity to variability. Therefore, higher S/N ratio is better as robust condition due to minimum variability. There are two categories of SNR: non-dynamic SNR when there is no signal factor, and dynamic SNR when signal factor exists. Signal factor is a controllable variable to actualize the intention. Ideally, a system with zero or minimum noise is desired. This means, after optimization, the noise level gap must be as small as possible to produce an ideal function as shown in Fig. 1. In this study, $Y$ is the output that is peel strength and represents a zero-point proportional equation [11] with dynamic SNR. $M$ is the signal factor that is specimen width. Beta, $\beta$, is the measurement sensitivity to different inputs; thus the slope must be steep. Therefore, the ideal function is $Y = \beta M$. Three elements of SNR are to improve the linearity, sensitivity, and variability of a system.

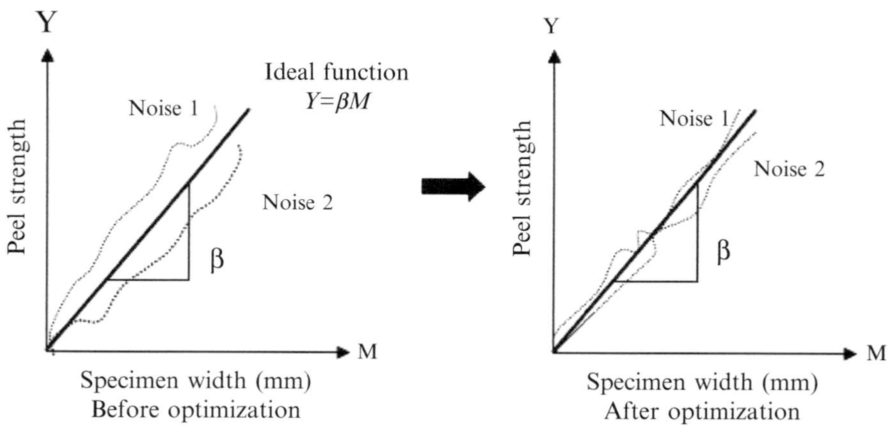


**Fig. 1** Variability improvement after optimization

# 2 Methodology

## 2.1 *QE Implementation in a Company*

At the initial stage of QE introduction, opposition from engineers made QE hard to deploy. This is due to the strong force from QE promoter and lack of understanding among the top management. Fuji Xerox overcomes the problem by training the problem solving teams via outside and in-house seminars. The skillful member will then train the engineers through many practices. Those engineers who have a full understanding of the robustness concept formed an internal study team and made textbooks with case studies. Those documents are established in internal seminar for Robust Design. It is concluded from this positive growth the importance of management's role in supporting engineers to study voluntarily. In addition, current engineers and future trainers are given advice by trainers to ensure QE continuity. A revised internal QE seminar is held to focus on quality characteristic in technology development. Eight Key Factors for Success (KFS) [12] has been identified to ensure quality engineering is successfully implemented. The eight factors are policy making by top management, responsibility to promote QE, promotion activities, QE training consistency, theme clarification of QE status, continuous meetings, consultant and expert support, and result review.

### 2.1.1 Engineering Tool for QE Methodology

Annual QE forum emphasized the engineering tools that are QE and DOE. Fuji Xerox has differentiated the use of QE tool based on process and purpose. DOE is utilized at the research stage to fix the themes of product and process and further verify the feasibility of the research. Taguchi method is used extensively to find the design parameters that result in the product or process robustness. It is an immensely useful tool for product development to establish the technology. Three main steps in technology development are preparing a strategy by setting the objective, selecting technology in the first development step and robust design in the second development step. Criterion in research and technology development process is defined. The objective of research is to find for "Blue Bird" [13], meaning to create breakthrough technologies valuable to customers. The purpose of DOE in troubleshooting the problem is to find the factors that change the mean value of characteristics. The difference between DOE and Taguchi method is critical to understand ensuring the suitability of the tool based on purpose. DOE and Taguchi method have been placed separately in manufacturing process flow.

## 2.2 *QE Methodology in Practical Case Study*

As explained by Fuji Xerox, Taguchi method had been used in research and technology development process to find the design parameters that minimize the variation. It is a QE tool to make the process or product insensitive to changes in the noise factor or variation. A case study, an optimization of T-peel adhesion test using Taguchi method, is done to propose the feasibility of QE in practical experimentation. Standardized test method of T-peel test in measuring peel strength is established by JIS K6854 [14] and ASTM D1876 [15]. The limitation of the standardized method is the test only fits for rigid materials and not for flexible film. Big variation in peel strength measurement due to specimen failure to hold the T-shape during peeling is a significant problem when standardized method is used on flexible film. This problem statement has motivated the researcher to come up with a system that can satisfy the industry requirement, which in this case is flexible packaging film. Thus, a new testing apparatus had been established to overcome this problem for flexible film. The case study is discussed on T-peel test optimization of flexible packaging film using the new apparatus. The objective is to obtain the minimum variation of peel strength. The goal of research and the technology used to deliver the goal have been integrated by applying QE. Three main steps mentioned in the Fuji Xerox's strategy of implementing QE are as follows: [13], objective setting (to satisfy the testing capability), technology selection (new apparatus for flexible film instead of using established method) and to enable the functionality and finally robust design (optimization of T-peel test for minimum variation in flexible film). The study was carried out to identify a factor's level that would minimize the variation in peel strength.

### 2.2.1 Ideal Function, P-Diagram, and Signal Strategy

A dynamic ideal function was identified in this study based on various range of specimen width with $Y = \beta M$. P-diagram in Fig. 2 is constructed to give a whole

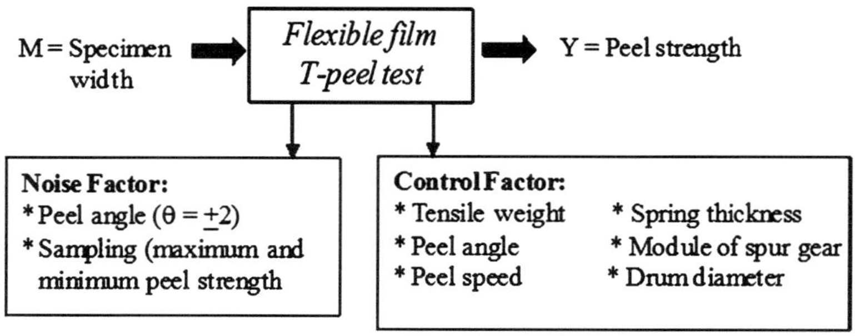


**Fig. 2** P-diagram of practical case study

picture of the parameters studied. The function of Al-CPP T-peel test is to measure peel strength. Thus, the response or output of T-peel test is peel strength, which is measured in Newton (N). The input of T-peel test is known as signal factor. Signal factor, in this study, is specimen width and is a controllable variable to actualize the intention (variation in peel strength) to achieve robust condition regardless of various width condition [9]. Signal-to-Noise ratio (SNR) with dynamic response in (3) is used due to the existence of signal factor with a specimen width of 5, 10, and 15 mm to measure the peel strength linearity.

$$\text{SNR}, \eta = 10\log(1/r)\left[\left(S_\beta - V_e\right)/V_N\right] \tag{3}$$

Where $S_\beta$ = variation caused by the linear effect, $V_e$ and $V_N$ = error variance (error variance/DOF), $r$ = total number of measurements under signal. (note: $r$ is also the effective divisor due to level changes of signal factor. DOF is degree of freedom.)

### 2.2.2 Noise Strategy

Noise factor is an uncontrollable factor during normal production or use, but can be controlled during the experiment. Noise factors are likely to produce variability in the response. For noise factor (outer array), historical data has proven that the peel angle would vary during exchanging the peel angle setting and during peeling process. Peel angle deviation will affect the peel strength; thus peel angle is considered as sources of variability. Noise in peel angle is defined as deviation in $\pm 2^\circ$ due to angle variation during adhesive peeling process caused by natural movement of the specimen. Maximum and minimum value of peel strength at $+2^\circ$ and $-2^\circ$ angle are taken for result. Thus, there are two noise levels that are N1 and N2 under each signal factor level. The intended condition is N1 has higher peel strength than N2 (N1 > N2). N1 consists of peel angle with deviation $+2^\circ$ and maximum peel strength is taken as a result. On the other hand, N2 level consists of peel angle deviation $-2^\circ$ and minimum peel strength is taken as a result in outer array. Peel strength is determined by measuring the strength required to peel the

**Table 1** Parameters and orthogonal array L18

| Control Factor | Unit | Level 1 | Level 2 | Level 3 |
|---|---|---|---|---|
| A: Tensile weight | g | 4 | 8 | |
| B: Peel angle | ° | 60 | 90 | 120 |
| C: Peel speed | mm/s | 6 | 9 | 12 |
| D: Data region | % | 30 | 50 | 70 |
| E: Spring thickness | mm | 0.3 | 0.4 | 0.5 |
| F: Module of spur gear | | 0.5 | 1.0 | 2.0 |
| G: Drum diameter | mm | 20 | 30 | 40 |
| *Signal Factor* | | *Levels* | | |
| M: Specimen width | mm | 5 | 10 | 15 |
| *Noise Factor* | | *Level N1* | *Level N2* | |
| Peel Angle | θ | +2 | −2 | |
| Peel strength sampling | N | Maximum | Minimum | |

interlayer of cast polypropylene (CPP) away from the aluminum layer. Peel side of flexible film packaging affects the optimum condition of peel strength variation [16]. In this paper, peel angle is read from aluminum side of specimen.

### 2.2.3 Control Factor Selection

The objective of this T-peel test is to satisfy the industry requirement of getting the minimum variation for flexible film. Thus, select control factors that may affect variability in the response, and possibly the mean of the response. The controllable factors or inner array is chosen based on testing and design condition which possible to affect the variance. The controllable factor selection is also considered based on previous experiment result, preliminary test, theory and available knowledge, and expert's opinion.

### 2.2.4 Orthogonal Array Selection

The design space is large, and it needs a strategy to explore. After determining the control parameters and its level, they are assigned into an orthogonal array as shown in Table 1. An orthogonal array is used for optimization that is to maximize the signal-to-noise ratio [11]. A balance set of experimentation runs is provided by orthogonal array. Design of experiments using orthogonal array L18 is utilized with 1 two-level factor (tensile weight) and 6 three-level factors (peel angle, peel speed, data region, spring thickness, module of spur gear, and drum diameter) as shown in Table 2 with only 108 experiments (18 runs × 3 signal level × 2 noise level).

**Table 2** Parameters and orthogonal array L18

| | A | B | C | D | E | F | G | 5 mm N1 | 5 mm N2 | 10 mm N1 | 10 mm N2 | 15 mm N1 | 15 mm N2 |
|---|---|---|---|---|---|---|---|---|---|---|---|---|---|
| 1 | 1 | 1 | 1 | 1 | 1 | 1 | 1 | | | | | | |
| 2 | 1 | 1 | 2 | 2 | 2 | 2 | 2 | | | | | | |
| 3 | 1 | 1 | 3 | 3 | 3 | 3 | 3 | | | | | | |
| 4 | 1 | 2 | 1 | 1 | 2 | 2 | 3 | | | | | | |
| 5 | 1 | 2 | 2 | 2 | 3 | 3 | 1 | | | | | | |
| 6 | 1 | 2 | 3 | 3 | 1 | 1 | 2 | | | | | | |
| 7 | 1 | 3 | 1 | 2 | 1 | 3 | 2 | | | | | | |
| 8 | 1 | 3 | 2 | 3 | 2 | 1 | 3 | | | | | | |
| 9 | 1 | 3 | 3 | 1 | 3 | 2 | 1 | | | | | | |
| 10 | 2 | 1 | 1 | 3 | 3 | 2 | 2 | | | | | | |
| 11 | 2 | 1 | 2 | 1 | 1 | 3 | 3 | | | | | | |
| 12 | 2 | 1 | 3 | 2 | 2 | 1 | 1 | | | | | | |
| 13 | 2 | 2 | 1 | 2 | 3 | 1 | 3 | | | | | | |
| 14 | 2 | 2 | 2 | 3 | 1 | 2 | 1 | | | | | | |
| 15 | 2 | 2 | 3 | 1 | 2 | 3 | 2 | | | | | | |
| 16 | 2 | 3 | 1 | 3 | 2 | 3 | 1 | | | | | | |
| 17 | 2 | 3 | 2 | 1 | 3 | 1 | 2 | | | | | | |
| 18 | 2 | 3 | 3 | 2 | 1 | 2 | 3 | | | | | | |

## 3 Results and Discussions

There are two main plots for QE analysis that are SNR and sensitivity response plot. SNR plot shown in Fig. 3 is obtained by computing the marginal average of SNR [8] at each level of a parameter. Calculation example for run 1 is shown below by using result of run 1 in Table 3:

$$\mathrm{SNR}, \eta = 10\log(1/r)\ [(S_b - V_e)/V_N] \tag{4}$$

$$S_\beta = \frac{((4.45+4.53)5 + (8.51+8.97)10 + (13.51+12.94)15)^2}{2(5^2+10^2+15^2)}$$

$$V_e = S_e/f_e = (S_T - S_\beta - S_{N\mathrm{x}\beta})/4 \quad f_e = 4 \tag{5}$$

$$S_T = 4.45^2 + 4.53^2 + 8.51^2 + 8.97^2 + 13.51^2 + 12.94^2$$

$$S_{N\mathrm{x}\beta} = ((4.45)5 + (8.51)10 + (13.51)15)^2 + ((4.53)5 + (8.97)10 + (12.94)15)^2)/(5^2 + 10^2 + 15^2) - S_\beta$$

$$V_N = S_e'/f_e' = (S_T - S_b)/5 \quad f_e' = 5 \tag{6}$$

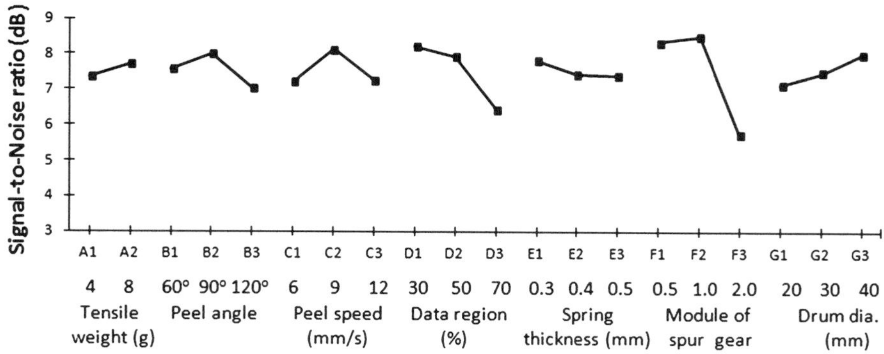


**Fig. 3** Signal-to-Noise ratio response plot

**Table 3** Experiment result for run 1

| | Specimen width (mm) | | | | | |
|---|---|---|---|---|---|---|
| Run | 5 | | 10 | | 15 | |
| 1 | 4.45 | 4.53 | 8.51 | 8.97 | 13.51 | 12.94 |

$$\begin{aligned}\text{SNR}, \eta &= 10\log_{10}\left(1/2\left(5^2+10^2+15^2\right)\right)\left[\left(S_\beta - V_e\right)/V_N\right]\\ &= 11.20\ \text{dB}\end{aligned}$$

$$\begin{aligned}\text{Sensitivity}, \beta &= 10\log(1/r)\left(S_\beta - V_e\right)\\ &= 10\log_{10}\left(1/2\left(5^2+10^2+15^2\right)\right)\left(S_\beta - V_e\right)\\ &= -1.10\ \text{dB}\end{aligned} \tag{7}$$

The SNR for all 18 runs is calculated using SNR calculation above. Then, the mean SNR so-called process average is calculated to find the effect of each control factor. The effect is separated out at different levels because the experimental design is orthogonal. The maximum level of SNR value in optimum condition implies the minimum variation as the signal is bigger than noise. Sensitivity response plot, as in Fig. 4 often called Beta plot derived from (7), shows the sensitivity of response value at each level. It focuses on sensitiveness of response upon level's change.

A confirmation run is done to check the reproducibility of the experiment. SNR in optimum condition is compared with worst condition. DB gain for confirmation SNR differed by 2.86 dB than estimated SNR. The experiment is considered successful, as dB gain difference is less than 3 dB or 30 %. The dB gain difference is caused by the worst condition as confirmation SNR deviates a little bit from the estimated SNR for worst condition. Table 4 summarized the optimum and worst condition and dB gain. Second step in two-step optimization is to adjust the controllable factor to target value. The second step is done when certain target is desired. Based on sensitivity plot in Fig. 4, the best factor to adjust is drum diameter (factor G) because of high sensitivity, and SNR is roughly even. Thus, the variability in peel strength is not influenced by different level of that factor. Factors with even sensitivity and uneven SNR as C, D, and F are particularly useful to improve

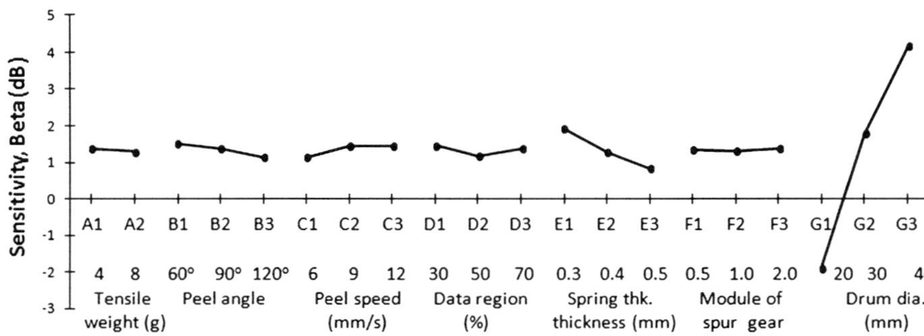


**Fig. 4** Sensitivity response plot

**Table 4** Optimum condition and SNR gain

| Type | Condition | Estimated SNR (dB) | Confirmation SNR (dB) |
|---|---|---|---|
| Optimum | A1 B1 C3 D1 E2 F2 G3 | 14.91 | 14.82 |
| Worst | A2 B2 C1 D3 E1 F3 G2 | 4.30 | 7.07 |
| SNR dB gain | | 10.61 | 7.75 |
| Gain difference | | 2.86 dB = 26.95 % | |

variation because the value of peel strength has no change. Meanwhile, QE in Fuji Xerox is started for problem identification that motivates what kind of improvement to be done. Based on three case studies, problems can be from industry requirement, customer dissatisfaction (Sakanobe et al., Robust design of photoreceptor cleaning system in electro-photographic printer, unpublished), technology obsoleteness (Sakai, Optimizing colorimetry accuracy of LEDcolor sensor, unpublished), cost reduction factor, system improvement (Akiyama, Application of orthogonal array to software testing—HAYST method, unpublished), and such. Sakanobe et al. "Robust design of photoreceptor cleaning system in electro-photographic printer, unpublished" emphasizes the relationship between output (Y) and problem statement to generate signal factor that transforms the energy. Optimization is conducted with the ideal function. In laboratory, problem is known from available standards and further optimization is done for the betterment of the newly developed apparatus. Similarly, the output Y (peel strength) is related with the known problem (big variation) to generate the ideal function. Quality characteristic is defined from the measured value of the objective, which referred to response, results, or output, Y [17]. Both flows in laboratory and industry [18] focused on selection of quality characteristic which describes the desired result. Ideal function and P-diagram are identified after problem statement is done. Confirmation run in Fuji Xerox is done on trial manufacture while case study is done with laboratory scale. In Fuji Xerox, quality is monitored after-launch to society upon the in-house quality result is official.

It is proven that QE engineering tool can be applied in any environment, be it industrial application, service, or research field. QE methodology flow is derived from QE implementation findings of Fuji Xerox and practical case study. A result

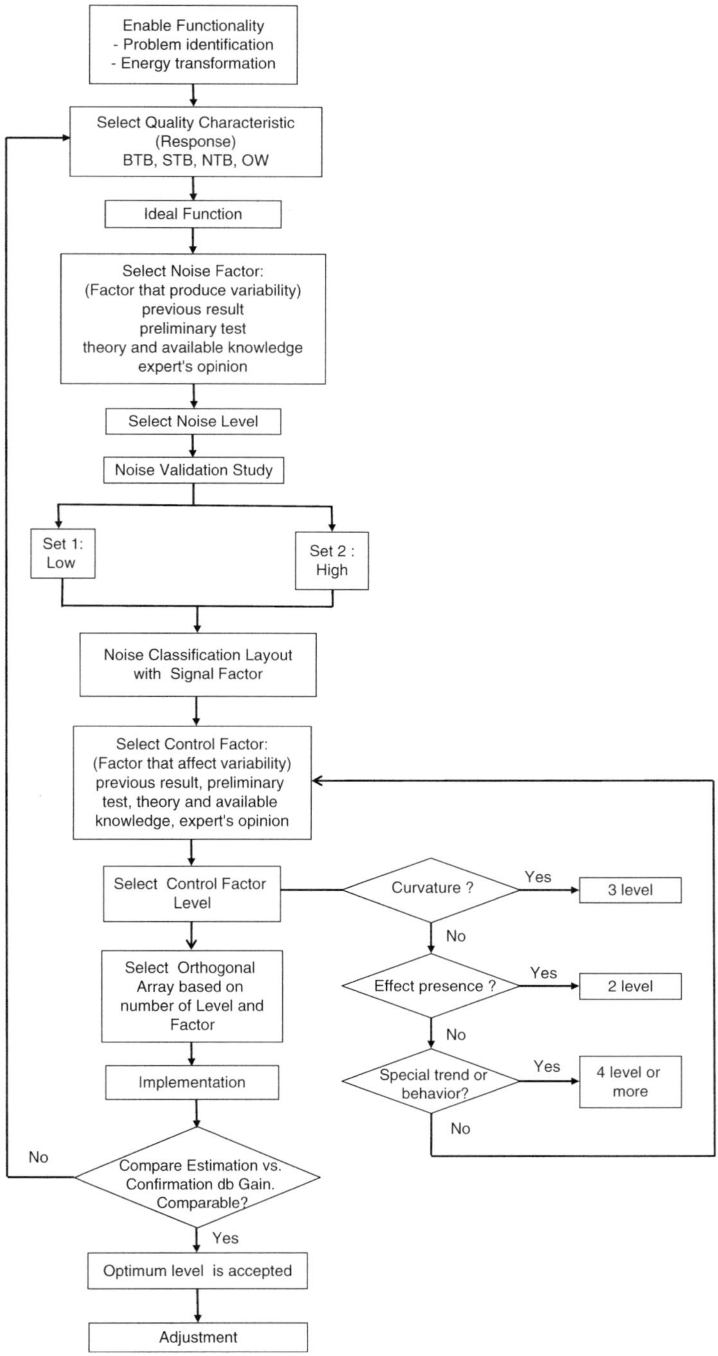


**Fig. 5** Quality engineering methodology framework

from methodology comparison [17] has produced a QE methodology framework as shown in Fig. 5 to obtain a robustness of product or process. This paper demonstrates the experience from orthogonal array L18 experiment in selecting control and noise level by combining the industrial and laboratories using quality engineering.

## 4 Conclusion

This paper discusses the implementation of QE in an organization and QE application in process or product optimization through practical case study. Japan's QE has proven successful and is emphasized during the design stage before manufacturing or production to find the design parameters and ensure the product's robustness, thus increasing product performance to satisfy customer needs. Fuji Xerox hypothesis of KFS has helped promoting QE in research, technology, and product development activities. QE promotion activities accelerate the implementation in an organization. Top-down approach is undeniably a driving force for a successful QE implementation. The case study represents how QE is implemented in one of the product optimization stages. Identifying the experiment's objective is crucial that affects the selection of noise and control factors. The engineering tool employs the engineering and statistic knowledge to obtain product robustness. A brief framework is presented for QE implementation in organization and procedures on QE methodology. In QE methodology, planning before implementation is a key element for performing a successful experimental design. Learning from QE is crucial to be practiced, not only in engineering but also in other fields such as health and services. QE is practical, and meant to improve product to maximum performance, thus making the product to be the best with least variation. Undeniably, it is time to change the old practice of experimental design into QE to produce a world-class quality product.

## References

1. Arvidsson M, Gremyr I (2008) Principles of robust design methodology. Qual Reliab Eng Int 24:23–35
2. Ilzarbe L, Alvarez MJ, Viles E, Tanco M (2008) Practical applications of design of experiments in the field of engineering: a bibliographical review. Qual Reliab Eng Int 24:417–428
3. Tanco M, Viles E, Ilzarbee L, Alvarez MJ (2008) How is experimentation carried out by a companies? A survey of three European region. Qual Reliab Eng Int 24:973–981
4. Viles E, Tanco M, Ilzarbe L, Alvarez MJ (2009) Planning experiments, the first real task in reaching a goal. Qual Eng 21:44–51
5. Matsuda M, Miyagi Z, Hirai S (2010) Reliability evaluation of t-peel test method for laminated flexible films. Annual Meeting of the Japan Society of Mechanical Eng. Kanto Branch, 2010

6. Moon Soo P, Jae Choon C, Sung Hyun K, Dong Joo S, Hyung Mi J, Choon Keun L, Mi-Suk C, Youngkwan L (2011) Optimization of desmear process for high adhesion of insulating film in printed circuit boards (PCBs) via Taguchi method. Int J Adhesion Adhesives 31:466–472
7. Hirai S, Ashizawa H, Miyazawa H, Naito S, Miyagi Z. Performance Evaluation of T-peel Apparatus for Flexible Materials
8. Anthony J (2006) Taguchi or classical design of experiments: a perspective from a practitioner. Sens Rev 26(3):227–230
9. Taguchi G, Wu Y (1985) Introduction to off-line quality control. Central Japan Quality Control Association, Japan
10. Yates F (1970) Experimental design: selected papers by Frank Yates. Griffin, London
11. Taguchi G, Chowdhury S, Wu Y (2005) Taguchi's quality engineering handbook. Wiley, Hoboken
12. Saitoh K, Yoshizawa M, Tatebayashi K, Doi M (2003) A study about how to implement quality engineering in research & development (Part 1). J Qual Eng Soc 11:100–107
13. Saitoh K, Yoshizawa M, Tatebayashi K, Doi M (2003) A study about how to implement quality engineering in research & development (Part 2). J Qual Eng Soc 11:64–69
14. JIS (2008) Japanese Industrial Standard Handbook. In: Adhesives Determination of peel strength of bonded assemblies-Part 3: Adhesive-180 degree peel test for flexible-to-flexible bonded assemblies (T-peel test), ed. Japan
15. ASTM (2007) Annual Book of ASTM standards section 15: general products, chemical specialties, and end use products vol. 15.06
16. Dolah R, Miyagi Z (2013) Effect of Peel Side on Optimum Condition for Measuring Flexible Film Peel Strength in T-Peel Adhesion Test. J Test Eval 42(1):1–2. doi:10.1520/JTE20120342. ISSN 0090-3973
17. Roy RK (2001) Design of experiments using the Taguchi approach. Wiley, Canada
18. Dolah R, Miyagi Z, Tatebayashi K (2012) Quality engineering implementation in an organization: a brief finding on QE methodology. J Manage Prod Eng Rev 3(4):26–34

# Developing a Heuristic Robotics Integrated Production Model for Economic Production Quantity in Mass Customization Manufacturing

**S.E.S. Bariran, K.S.M. Sahari, and A.H. Mahlouji**

## 1 Introduction

Mass customization manufacturing (MCM) has been introduced as an alternative solution to the common obstacles faced in traditional Flexible Manufacturing Systems (FMS) and Reconfigurable Manufacturing Systems (RMS), hindering the product customization. On the other hand, determining the economic production quantity of a mass customization production system is a critical task that also needs to be customized. In other words, unlike the assumptions made under classical EPQ models, the perfectly reliable production process and the fixed set-up costs are no longer applicable with the state-of-the-art robotics enhanced MCM systems today.

As an increasingly upward trend, the MCM is shifting more and more toward the robotics enhanced production with a gradual elimination of the role of man power as the primary production assistant, replacing the human role by alternative robotics agents. This evolutionary shift in production mode is accompanied by certain risk and returns that depending on the process and level of robotics integration has significant effects on the major production outputs such as total production volume/ cost, final product quality, and the EPQ among others. Therefore, the main objective of this chapter is to propose a heuristic model for determining economic production quantity in mass customization manufacturing systems based on robotics associated parameters rather than the conventional inventory and demand estimations used in the EPQ calculation. The results of the study compare the total risks and returns of robotic integrated production based on triple characteristics of maintenance, safety, and training imposed by such configuration. For each production quantity, the absolute value of the difference between risk and return is

S.E.S. Bariran (✉) • K.S.M. Sahari
Department of Mechanical Engineering, Centre for Advanced Mechatronics and Robotics, Universiti Tenaga Nasional, 43000 Kajang, Malaysia
e-mail: ses.bariran@outlook.com

A.H. Mahlouji
College of Graduate Studies, Universiti Tenaga Nasional, 43000 Kajang, Malaysia

K. Ab. Hamid et al. (eds.), *The Malaysia-Japan Model on Technology Partnership: International Proceedings 2013 of Malaysia-Japan Academic Scholar Conference*, DOI 10.1007/978-4-431-54439-5_12, © Springer Japan 2015

calculated as an index to determine the EPQ. Besides the innovative methodology proposed, the results also provide managerial insight into the probable effects of these factors on EPQ in mass customization manufacturing systems.

## 2 EPQ: Classical vs. Customized Modeling

Classical models of EPQ did not take into consideration the effect of several restricting factors that existed in the real production setting. In response to this deficiency, extensive research has been conducted in previous decade to investigate different perspectives of production systems as relevant to the economic production quantity. In the following three subsections, the issue of EPQ modeling will be covered from three different aspects. First, the concept and history of EPQ modeling will be briefly highlighted. Second, the classical scheme for EPQ modeling is mathematically presented, and finally followed by a selective review on the modern era of EPQ modeling.

### *2.1 Theoretical Concept and Overview*

Economic Production Quantity or EPQ model determines the optimal production quantity a manufacturer or supplier should produce to minimize the total production and inventory costs per unit product by balancing the associated production and inventory costs. The EPQ model was developed by E.W. Taft in 1918 as an extension of the Economic Order Quantity or EOQ model. The difference between these two methods underlies in their theoretical assumptions. The EPQ model assumes the company will produce its own quantity or the parts are going to be shipped to the company while they are being produced; therefore, the orders are available or received in an incremental manner while the products are being produced. On the other hand, the EOQ model assumes the order quantity arrives complete and immediately after ordering [1]. In some literature Economic Manufacturing Quantity model (EMQ) is an alternative to Economic Production Quantity model (EPQ). Similar to the EOQ model, EPQ is a single product lot scheduling method.

### *2.2 Classical Mathematical Modeling*

EPQ model can be mathematically derived from the traditional EOQ model. Equations (1) and (2) show the EOQ-based total cost and optimal order quantity, where TC is the total annual ordering cost, $D$ is the annual demand, $Q$ is order quantity, $H$ is annual holding cost, and finally $S$ denotes for setup or ordering cost [1].

$$\mathrm{TC_{EOQ}} = \frac{D}{Q}S + \frac{Q}{2}H \tag{1}$$

$$\mathrm{EOQ} = \sqrt{\frac{2\mathrm{DS}}{H}} \tag{2}$$

As a general approach, in EOQ model, total cost (TC) can be minimized by ordering for the amount of EOQ as given in (2). The same rule can be applied in deriving the EPQ model with the exception that unlike EOQ, in EPQ model inventory arrives in increments and is drawn down right upon arrival [1].

In order to derive the Economic Production Quantity (EPQ) model, first, the EOQ-based total cost should be adjusted as given in (3), where $I_{\mathrm{MAX}}$ shows the maximum inventory volume and is calculated by (4). Then, from (3) and (4), adjusted order quantity for the EPQ model is developed in (5) [1].

$$\text{Adjusted Total Cost}: \mathrm{TC_{EPQ}} = \frac{D}{Q}S + \frac{I_{\mathrm{MAX}}}{2}H. \tag{3}$$

$$\text{Maximum Inventory}: \mathrm{TC_{EPQ}} = Q\left(1 - \frac{d}{p}\right). \tag{4}$$

$$\text{Adjusted Total Cost}: \mathrm{EPQ} = \sqrt{\frac{2\mathrm{DS}}{H\left(1 - \frac{d}{p}\right)}} \tag{5}$$

## 2.3 *Customized EPQ Modeling*

### 2.3.1 EPQ Customized with Maintenance Factors

Pioneered by Giri and Yun [2], the effect of random machine failures and the resulted backlog on the optimal production quantity was numerically investigated [2]. One year later, C-Lin and Gong enhanced the so-called pioneer model by expanding the model assumption to encompass fixed time for corrective maintenance for a single product subjected to exponential decay and a no-resumption inventory policy [3].

As a climax for the similar research, the issue of imperfect maintenance and minimal repair combined with inspection errors in an imperfect production system was the core objective of the study by Lin et al. [4]. However, it was not until 2012 that for the first time in EPQ literature, the effect of stochastic machine breakdown and repair was integrated into the EPQ model by Saberi-Lahiji et al. in a typical production system with multiple identical machines producing not-identical products with specific demand rate [5].

### 2.3.2 EPQ Customized with Quality Factors

The effect of imperfect quality characteristics on economic production quantity was originally coined in 2000 by Salameh and Jaber. The model was developed on the basis that the defective products will be withdrawn from inventory after a 100 % inspection to be sold at a discounted rate later. An extended version of this model was later proposed by Jaber et al. [6]. A learning curve was developed and empirically validated in this study proving that the defect rate per lot will reduce during the production process [6].

The effect of quality on EPQ may also occur due to defects and inspection errors. Despite the practical significance of this issue, it was not taken into consideration until Yoo et al. [7] modeled the effect of both imperfect production quality and inspection errors that may cause either a profit loss or defect sales return if the inspection falsely detects a conforming item as defected or vice versa respectively. The model also included the imposed cost from the rework/salvage of the screened/returned items [7].

### 2.3.3 EPQ Customized with Shop Floor Mixed Factors

In a collaborative study by Pasandideh et al., triple characteristics of space constraints, shortage-caused backlog, and discrete delivery of orders were investigated for multiproduct EPQ modeling [8].

The extended version of the Teng's model for fluctuating demand was proposed by Pal et al. [9], where the case was further developed to encompass the stochastic random demand characteristics to determine buffer inventory and production quantity in an imperfect production system subject to facility maintenance/repair [9].

Another multi-criteria covering model was developed by Taleizadeh et al. combining the effects of random defective items and repair failure in a single machine production system that imposed limited production capacity and shortage as well [10,11].

## 3 EPQ and Robotics Integrated Production

Robotics integrated production model is a unique configuration of the current MCM systems with enhanced robotics agents as alternative for human factors. The primary objective of this production model is to simultaneously maximize the production rate and product quality while minimizing the total production cost.

However, some unique characteristics of the so-called robotics integrated production systems such as maintenance, training, and safety have not been properly addressed in the EPQ modeling. First, all the previous customized EPQ modelings with the maintenance cause are solely restricted to the failure of the primary production machine and has neglected the effect of associated robotics maintenance that has been integrated into production system. Second, after a case of robotics caused maintenance, the production process and the robotic operation should be re-synchronized requiring a training agent that will necessitate extra start-up time and cost. Finally, are the obligatory safety measurements enforced by quality standards such as ISO/TS 16949 as instructed under clauses 6.3 and 6.4 under infrastructure and work environment requirements?

## 4 Solution: Ideal Function Modeling

A RIP ideal function is proposed in this section by incorporating the triple risk-return characteristics of RIP systems known as maintenance, training, and safety. The ideal function is composed of two sub-functions namely Ideal Risk Function (IRiF) and Ideal Return Function (IRef). IRiF is a function to estimate the potential cost-associated risks of the new RIP model and accordingly, IRef has the same functionality in case of potential revenue-associated returns of the RIP model. A combination of these two functions provides a preference index known as Total Risk-Return Index or TRRI which is independent of production parameters and can be simply used in determining the most cost-effective production mode based on production volume and cycle time as illustrated in (6–8).

$$\text{Ideal Risk Function : IRiF} = \sum_{i=0}^{PQ} \text{EPQ} * (\text{MT}_{Ci} + \text{TR}_{Ci} + \text{ST}_{Ci}). \tag{6}$$

$$\text{Ideal Return Function : IReF} = \sum_{j=0}^{PQ} \text{EPQ} * \left(\text{MT}_{Rj} + \text{TR}_{Rj} + \text{ST}_{Rj}\right). \tag{7}$$

$$\text{Total Risk Return Index : TRRI} = \sum_{i,j=0}^{PQ} = \left(\text{IRiF}_i + \text{IReF}_j\right) \tag{8}$$

In above equations MT, TR, and ST are the corresponding coefficient for maintenance, training, and safety measurements. The index of $C_i$ and $R_i$ is used to denote the respective cost and revenue coefficient for production volumes ranging from $i$, $j = 0$ to PQ which show the max production capacity (refer to Table 1).

**Table 1** Comparison of production mode preference for IRiF and IRef

| Production mode preference | | | | | | | | |
|---|---|---|---|---|---|---|---|---|
| | | T-MCM (Machine II) | | | RIP-MCM (Machine I) | | | |
| Weekly production quantity (WPQ) (*$10^3$ U) | Cycle time (CT) (seconds) | IRif | IReF | TRRI | IRif | IReF | TRRI |
| $WPQ_1 \leq 8$ | $CT \geq 75$ | 0 | 1 | 1 | 1 | 0 | 0 |
| $WPQ_2 \leq 13$ | CT = 45–75 | 0 | 0 | 0 | 1 | 1 | 1 |
| $WPQ_3 \leq 20$ | $30 < CT < 45$ | 0 | 0 | 0 | 1 | 1 | 1 |

# 5 Real Case Study: Plastic Injection Molding

The model analysis is based on a case study in a plastic injection molding company capable of producing a variety of automotive polymeric appearance parts such as bumper and grille via injection molding process in mass volumes and different product models on a continuous working schedule of 6 working days weekly and 3 shifts of 8-h per day. Two Krauss-Maffei injection machines with the same calming capacity of 1,300 tons were selected for the case study. Both machines are used for mass production of the same product (auto front bumper) and may in case be used for the production of same family products such as grille by changing the mold and machine setups. Machine "A" has recently been equipped with a robotic arm to proceed with the postinjection operation such as vacuum and trimming that in case of machine "B" is done by a human operator. Company's management has decided to gradually equip all other injection machines with the same robotic equipment to increase the production volume by reducing cycle times and to achieve a lower quality defect rate that is currently 10 % as reported by the company's quality engineering team [12].

# 6 Results and Analysis

The ideal function was applied in the case study to predict the respective risk-return values. The analysis was based on three different products that were monitored for a continuous 1-week period. The products belong to the same product family but with different production cycle times as illustrated in Table 1 under CT column. For each cycle time (CT) category, the equivalent production quantity is calculated under weekly production quantity (WPQ) column in the same table, assuming the maximum production capacity. Using ideal risk function and ideal return function, total risk return index in calculated for each product category. TTRI results for RIP-MCM and T-MCM are graphically presented in Figs. 1 and 2, respectively. Production mode preference results are then summarized in Table 1 by respective zero-one preference denotations. Results indicate that for $WPQ_2$ and $WPQ_3$, RIP-MCM has a better TRRI index while for $WPQ_1$, T-MCM is preferred. On the

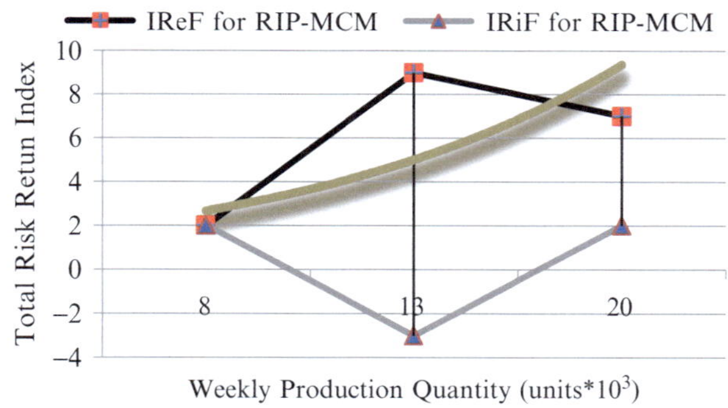


**Fig. 1** Ideal Function Analysis for RIP-MCM

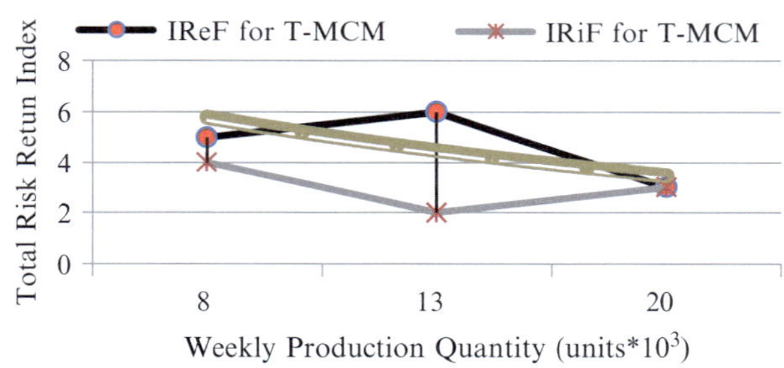


**Fig. 2** Ideal Function Analysis for Traditional-MCM

other hand, the corresponding exponential trend lines show that RIP-MCM is capable of maintaining a stable increase for risk-return index for all production volumes ranging from 8 to 20 K units. For each production mode, EPQ is simply determined as the volume for which there is a maximum positive return and minimum negative risk. Therefore, in this case study, $WPQ_2$ is selected as the EPQ for both production models due to its maximum simultaneous risk-return values.

## 7 Conclusion

In this study, a heuristic model for economic production quantity (EPQ) was developed for robotics integrated mass customization manufacturing systems. Based on the results of a real case study in plastic injection modeling industry, the proposed TRRI index was proved to be both a simple and effective tool for determining EPQ by calculating the absolute value of the so-called TRRI index. The findings also reveal that, depending on the mass production quantity and product standard cycle times, production mode should be switched between traditional MCM and RIP to optimize the risk-return ideal functions. Among three different family products with different cycle times that were investigated, robotics integrated production (RIP) model showed better performance as compared to traditional model for products with cycle time ranging from 30 to 75 s. It is also worth emphasizing that the decision for an optimal mass customization production scenario is critically dependent on total production capacity and product cycle time, as two major control factors of MCM nondeterministic environment especially in case of robotics integrated production systems. As part of future research, it is also suggested that the model be applied in different MCM settings rather than the one in the case study to test for the validity and reliability of the results with respect to the sensitivity of the RIP model to the production volume and product cycle time.

## References

1. Sanders RDRNR (2005) Operations management. Wiley, New York
2. Giri B, Yun W (2005) Optimal lot sizing for an unreliable production system under partial backlogging and at most two failures in a production cycle. Int J Prod Econ 95:229–243
3. Lin GC, Gong D-C (2006) On a production-inventory system of deteriorating items subject to random machine breakdowns with a fixed repair time. Math Comput Model 43:920–932
4. Lin Y-H, Chen J-M, Chen Y-C (2011) The impact of inspection errors, imperfect maintenance and minimal repairs on an imperfect production system. Math Comput Model 53:1680–1691
5. Sabri-Laghaie K, Mansouri M, Motaghedi-Larijani A, Jalali-Naini G (2012) Combining a maintenance center M/M/c/m queue into the economic production quantity model with stochastic machine breakdowns and repair. Computers & Industrial Engineering 63:864–874
6. Jaber M, Goyal S, Imran M (2008) Economic production quantity model for items with imperfect quality subject to learning effects. Int J Prod Econ 115:143–150
7. Yoo SH, Kim D, Park M-S (2009) Economic production quantity model with imperfect-quality items, two-way imperfect inspection and sales return. Int J Prod Econ 121:255–265
8. Pasandideh SHR, Niaki STA, Yeganeh JA (2010) A parameter-tuned genetic algorithm for multi-product economic production quantity model with space constraint, discrete delivery orders and shortages. Adv Eng Softw 41:306–314
9. Pal B, Sana SS, Chaudhuri K (2013) A mathematical model on EPQ for stochastic demand in an imperfect production system. J Manuf Syst 32:260–270
10. Taleizadeh A, Najafi A, Akhavan Niaki S (2010) Economic production quantity model with scrapped items and limited production capacity. Scientia Iranica Trans E: Ind Eng 17:58–69
11. Taleizadeh AA, Wee H-M, Jalali-Naini SG (2013) Economic production quantity model with repair failure and limited capacity. Appl Math Model 37:2765–2774
12. SPK Co (2011) Sahand Plastic KavoshCo. Producer of Palstic Industrial Parts. SPK Co, Tehran, Iran

# Systematic Way for Problem Identification of Car Braking System by Using TRIZ

N.F. Mesbah and S.M. Yusof

## 1 Introduction

The passenger car is one of the most popular types of transportation around the world for people to travel from one place to another in the safest manner. The upgrading of vehicle safety has become one of the most important targets in the vehicle industry. The brake system plays an essential role to enhance the active vehicle safety of a car. It must ensure the safe control of a vehicle during its normal operation and must bring the vehicle to a smooth stop within the shortest possible stopping distance under emergency conditions [1]. The basic goals of braking systems are to decelerate a vehicle during stopping, to maintain vehicle speed during downhill operation and to hold a vehicle stationary on a grade.

In recent years, braking systems have undergone tremendous changes in terms of performance, technology, design and safety [2]. In trying to improve braking system, the brake engineer has two challenges:

1. Create enough deceleration to stop the car as quickly as the driver wishes, without exceeding the driver's comfort level with regard to pedal effort or pedal travel.
2. Manage the resulting heat energy so as not to damage the brake system or the rest of the vehicle.

There are many ways to create or develop new ideas such as through brain-storming, benchmarking, and the Pugh method, which may not clearly point to the ways for solving problem or highlight the right solutions [3].

These methods usually have the ability of identifying or uncovering the problem and its root cause, but lack the capability to actually solve those problems. Amongst several creativity tools, TRIZ provides systematic problem-solving approach for

N.F. Mesbah (✉) • S.M. Yusof
Razak School of Engineering and Advanced Technology, Universiti Teknologi Malaysia (UTM), Kuala Lumpur, Malaysia
e-mail: nurulfatiha.mpputm@gmail.com; shari@fkm.utm.my

K. Ab. Hamid et al. (eds.), *The Malaysia-Japan Model on Technology Partnership: International Proceedings 2013 of Malaysia-Japan Academic Scholar Conference*, DOI 10.1007/978-4-431-54439-5_13, 

innovation. Usually, it is used for patent circumvention, system and process improvement, cost reduction and new development of products. TRIZ also offers the delivery of systematic innovation, acceleration of problem solving in creative ways, confidence that all possibilities of new solutions have been covered and overcoming the innovation killers such as psychological inertia, limited knowledge and contradiction [4]. In addition, TRIZ provides effective tools which enable engineers to predict future technology and produce new ideas from it.

In any problem-solving method, problem identification is the first part that plays an important role. This paper will discuss on how to identify the problem by using the TRIZ method to improve the braking performance of the hydraulic brake in a selected passenger car.

# 2 Methodology

## *2.1 TRIZ Method*

TRIZ was developed by a Russian mechanical engineer, Genrich Altshuller, in 1946 [5]. He was a former Soviet inventor. The word TRIZ is the Russian acronym of "Theory of Inventive Problem Solving". Research on TRIZ began with the hypothesis that there are universal principles of invention that are the basis for inventive innovations. If these principles could be identified and extracted, they could be applied to make the process of invention more direct. A short word of this is "Somebody somewhere has already solved this problem". Creativity is now finding that solution and adapting it to this particular problem. Altshuller started reviewing patents in the 1950s, and his team investigated the patents one by one. Over 200,000 patents were investigated. TRIZ offers a systematic creativity process built primarily on the concept of abstraction. From a specific problem to a generic framework, out of which comes a generic solution requiring translation back to the specific.

## *2.2 TRIZ Problem Identification*

The first step is to perform Function Analysis on the identified system for describing the system containing the problem(s) in all its detail by listing all the components and all their interactions. These interactions are called functions. The Engineering System also interacts with the components which are outside its system and are called supersystem components. These supersystem components can influence the Engineering System although they were not designed as part of the system.

**Fig. 1** Line styles for each types of function

| Function | Line Style |
|---|---|
| Normal | ⟶ |
| Insufficient | - - - - - - - → |
| Excessive | ➡ |
| Harmful | ⟶ |

TRIZ Function Analysis is different from other forms in that it includes all the negative, ineffective and excessive interactions in the system, and this is key to proper understanding of the problems and the potential solution points in the system.

All interactions are expressed as subject-action-object and it is important to use simple, clear language in order to be able to structure the problem correctly. In terms of graphical standardization, the subject has a rectangular outline while the product has an oval outline for easier differentiation. The function is shown as an arrow. These functions can be either useful or harmful functions. Useful function can be further categorized under normal, insufficient or excessive. In order to differentiate the different function, different line styles are used as shown in Fig. 1.

The second step is to perform Cause and Effect Chain Analysis that will help identify the fundamental root cause(s) of the problem. If we identify the wrong root cause, most probably the solutions generated would not work. The process of generating the Cause and Effect Chain Analysis is very similar to the "5 Whys" approach. Starting with the original problem, a brainstorming is conducted to identify the high-level cause for this original problem. For each high-level cause, the next-level cause is determined by asking "why" it happened. Continue asking "why" until the foundation laws of Physics or Chemistry are found or the limits of the Engineering System emerge. They are then the fundamental root causes of the original problem.

Next, each of these fundamental root causes is reviewed and then either discarded if not plausible or considered as the potential root cause of the original problem. Each of these potential root causes is then considered in the revised problem which needs to be resolved. In most cases, resolving them is not straightforward but requires resolving a contradiction.

## *2.3 Braking Performance*

The performance of a brake system is referring mainly to the vehicle's stopping distance which is the function of vehicle velocity at the start of braking and deceleration. The stopping distance that is needed to stop a moving car can be divided into two parts called the thinking distance and the braking distance. The thinking distance is the distance that the car travels during the driver's reaction time.

Reaction time is the duration of time between the moments when the driver realizes that the car must be stopped and when he or she applies the brakes. The distance taken to stop once the brakes are applied is known as the braking distance.

# 3 Results

## 3.1 Original Problem to Resolve

This research is an actual case study of an automotive manufacturer, where the problem was determined and decided by the company. It focuses on improving the brake system of the passenger car by using the Theory of Inventive Problem Solving (TRIZ).

For this research, the selected car has disc brakes in the front wheels and drum brakes in the rear wheels. The brake is actuated by the hydraulic pressure which is called the hydraulic brakes.

## 3.2 Function Analysis

Function Analysis helps to clearly identify the interaction between the system, subsystem and supersystem. In this case, the system is the braking system. Two subsystems are involved in that system which are actuation and foundation. For the actuation subsystem, the components include the brake pedal, master cylinder and proportioning control valve, while for the foundation subsystem, the components are the disc brake and drum brake.

The braking system also interacts with the components which are outside its system and are called supersystem components. These supersystem components can influence the braking system although they were not designed as part of the system. In this case, there are supersystem components which are the ground and the driver. The Function Analysis for a braking system is shown in Fig. 2.

When the brake pedal is pressed to apply the brakes, the piston in the master cylinder forces the brake fluid that will increase the pressure of fluid. This pressure is transmitted to all the pipes and up to the brake calliper and wheel cylinder according to Pascal's law. The brake fluid acts as a link to transmit pedal force from the master cylinder to the brake calliper in the disc brake and wheel cylinder in the drum brake.

The drum brake system slows down a car through friction, by pressing the brake shoes against a drum. A drum brake mechanism consists of different components which are the drum, brake lining, return spring, piston, wheel cylinder, brake shoe, brake pads, wheel hub and stud. A vehicle can be stopped with the help of all these components by just putting a small pressure on the brake pedal.

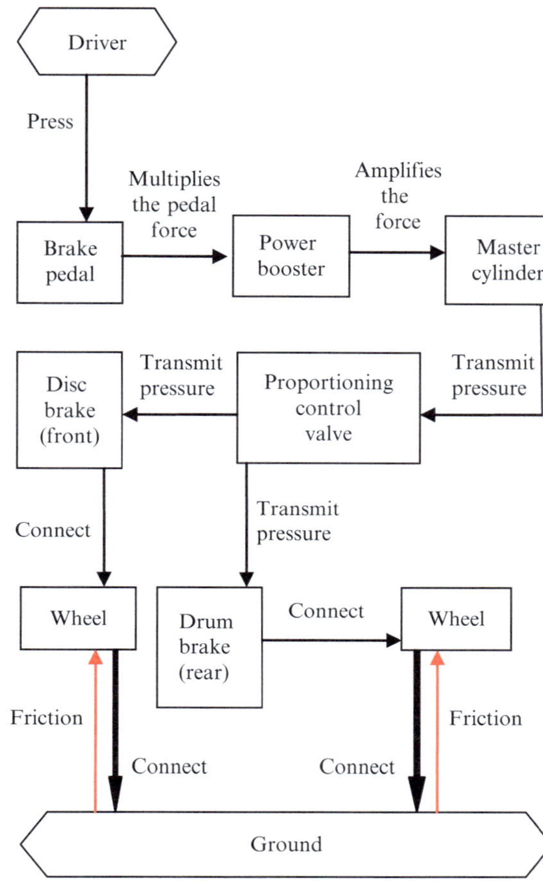


**Fig. 2** Function Analysis for the braking system

Another system is the disc brake system which is a lot simpler than the drum brake and more effective as well. It is a round, flat piece of metal, pressed against the wheel to slow or stop the car with the main components being the brake line, splash shield and disc brake.

## *3.3 Cause and Effect Chain Analysis*

A Cause and Effect Chain Analysis is used to determine the root cause of poor braking performance. The question "why" is asked in order to reach to the next level and eliminate those which are validated to be not contributing to the problem. Once the fundamental laws of Physics or Chemistry or limits of the Engineering System are reached, these would be the potential fundamental root causes. Once this root cause is identified, determine whether a contradiction needs to be resolved. The Cause and Effect Chain Analysis for inefficient braking performance is shown in Fig. 3.

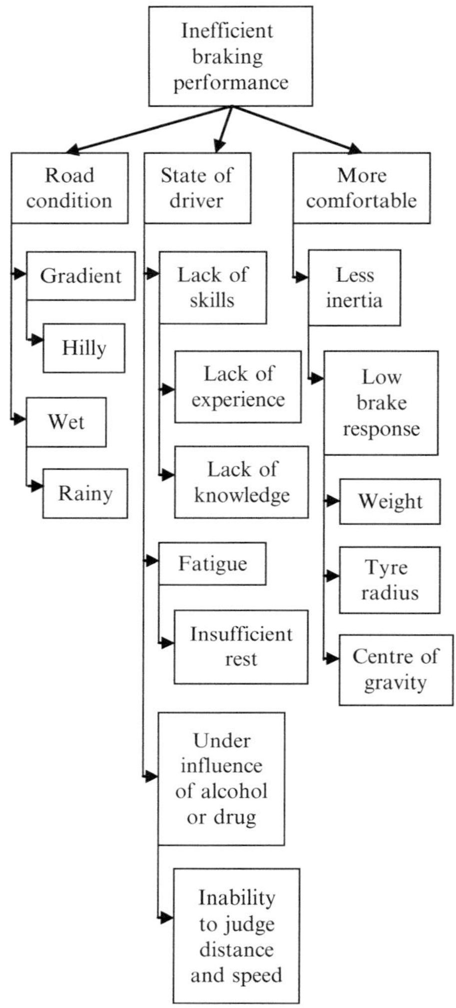


**Fig. 3** Cause and Effect Chain Analysis for inefficient braking performance

In order to find out the root cause of poor braking performance, each potential cause should be tested for its possibility. Further evaluation and hypothesis validation would be required to determine the root cause of the problem.

## 4 Discussion

The TRIZ methodology is applied to identify the problem of inefficient braking performance of the passenger car. There are two tools for identifying the problem by using the TRIZ methodology: the Function Analysis and Cause and Effect Chain Analysis. Function Analysis is applied to analyse the interactions between the system, subsystem and supersystem components specifically with their type of function. For this study, the braking system has actuation and foundation subsystems and interacts with two supersystems which are the driver and ground. The next step involved is using the Cause and Effect Chain Analysis to determine the root cause of the problem by asking "why" until the fundamental laws of Physics or Chemistry or the limits of the Engineering System are reached. Further evaluation could then be undertaken to eliminate potential causes which are validated to be not contributing to the problem, and finally the root cause of the problem will be determined.

## 5 Conclusion

This paper has presented the steps for problem identification by using the TRIZ methodology. A case study of an automotive manufacturer focusing on improving braking performance of the passenger car was conducted. Two tools of TRIZ have been used which are the Function Analysis and Cause and Effect Chain Analysis. Problem identification is the first step in problem solving and plays an important role in determining the successfulness of the study. If the identified root causes are wrong, most probably the solutions generated would not work. So, the problem identification phase helps to reduce the possibility of implementing the wrong or ineffective solution. Future research will be made to determine the best approach to test the ideas generated from these analysis and continue with the next step in TRIZ for resolving contradiction and investigate the inventive solution for improving braking system of the passenger car.

## References

1. Garrett TK, Newton K, Steed W (2001) The motor vehicle, 13th edn. Society of Automotive Engineers, Warrendale, PA
2. Dainis B (2009) Research in parameters of braking for automobiles. Engineering For Rural Development, Jelgava
3. Savranksy SD (2000) Engineering of creativity—introduction to TRIZ methodology of inventive problem solving. CRC, Boca Raton
4. Mann DL (2007) Hands-on systematic innovation. IFR Press, England
5. http://www.triz-journal.com/archives/what_is_triz/, TRIZ Journal web site.

# Part III
# Social Sciences and Humanities

# High-Performance Work System and Safety Performance

**Dayang Nailul Munna Abg Abdullah and Vanissa Karupaiah**

## 1 Introduction

There has been a growing realization that human resources are better managed by high-commitment- or high-involvement-oriented strategies [1]. Instead of relying on compliance by means of rules, regulations, and monitoring to decrease costs and increase efficiency, high-commitment management creates conditions that encourage employees to identify with the goals of the organization and to exert effort to achieve them [2].

Nowadays, the term high-performance work system (HPWS) has been used to improve workforce competencies, attitudes, and motivation. In fact, organizations that utilized this concept have been found to perform better than those without such a system in numerous ways, including having more innovation, growth, flexibility, productivity, and profits and less employee turnover, absenteeism, and labor costs [3]. HPWS is also known as a group of separate but interconnected human resource practices that together recruit, select, develop, motivate, and retain employees [4, 5]. It was suggested that this concept could be achieved by ensuring that employees possess a broad range of superior skills and abilities that are used at work, which ensure that their organizations achieve "superior intermediate indicators of firm performance" [4].

The main objective of this study was to identify the relationship between the high-performance work system determinants, namely, safety training, self-managed teams, high-quality work, and safety performance among employees at

D.N.M.A. Abdullah (✉)
Faculty of Business Management, Universiti Teknologi MARA, Puncak Alam Campus, Kuala Selangor, Selangor 42300, Malaysia
e-mail: nailul@salam.uitm.edu.my

V. Karupaiah
Department of Human Resource Management, Saito College, 18, Jalan Tengah, Section 52, PJ New Town Centre, Petaling Jaya, Selangor 46200, Malaysia

K. Ab. Hamid et al. (eds.), *The Malaysia-Japan Model on Technology Partnership: International Proceedings 2013 of Malaysia-Japan Academic Scholar Conference*, DOI 10.1007/978-4-431-54439-5_14, © Springer Japan 2015

the workplace. In addition, the dominant factor of this study would also be determined to see its effect on safety performance. This study was extended to predict that HPWS will also influence safety performance, and it was hypothesized specifically that HPWS will improve workplace safety by increasing employee trust in management and perceived safety performance approaches.

## 2 Literature Review

### *2.1 High-Performance Work System*

High-performance work system (HPWS) is defined as a specific combination of HR practices, work structures, and processes that maximizes the knowledge, skills, commitment, flexibility, and resilience of employees [6]. It is known as the high involvement or high commitment by an organization that uses a distinctive managerial approach that enables high performance through people. Although different human resource management's field authors have emphasized slightly different features and management practices in describing HPWS, the essential characteristics are the seven key dimensions introduced by Jeffrey Pfeffer in The Human Equation (1998). The dimensions are employment security; selective hiring of new personnel; self-managed teams and decentralization of decision-making as the basic principles of organizational design; comparatively high compensation contingent on organizational performance; extensive training; reduced status distinctions and barriers, including dress, language, office arrangements, and wage differences across levels; and extensive sharing of financial and performance information throughout the organization.

The main idea of HPWS is to create an organization based on employee involvement, commitment, and empowerment, not on employee control. The particular set of managerial practices will vary from company to company. The small business unit that controls its own fate and involves everyone in the business is the best image for the involvement-oriented approach [7]. In these high-involvement organizations, employees feel responsible for and involved in its success. They also know more, do more, and contribute more. They have the power, information, knowledge, and rewards to perform at the highest level [8].

There were many studies that have investigated whether firms utilizing HPWS approach achieve a higher level of performance than traditionally managed firms. Previous researches claimed that high-performance work systems are very much associated with lower turnover rates [9], enhanced productivity and financial performance [10], and increased job satisfaction [11] and help in reducing fatigue among employees [12].

### 2.2 *Safety Performance*

Safety performance can be seen as a subset of the overall job performance in situations in which worker's safety is the main concern. Safety performance is therefore perhaps best defined by the extent to which an individual performs behaviors that increase the safety of the individual and organization and avoids behaviors that decrease safety of oneself and the organization. Most research on safety behavior at work has treated safety performance as one dimension encompassing a small set of behaviors that relate directly to the performance of core job tasks, such as risk-taking, engaging in behavior to protect oneself, and following safety rules [13].

Safety compliance, which refers to rule-following in core safety activities such as wearing personal protective equipment, has been shown to be distinct from safety participation, which refers to activities that enhance the safety context of the organization such as participating in voluntary safety activities [14]. The relationship between compliance and participation was also similar to those observed between general task performances, suggesting that safety performance is similar in structure to general performance in this respect.

Work behaviors relevant to safety can be conceptualized in the same way as other work behaviors that constitute work performance. In this way, the models of performance can be applied to safety performance in the workplace. The model of performance used in the present study makes a distinction among the components of performance, the determinants of performance, and the antecedents of performance [15]. The components of performance describe the actual behaviors that individuals perform at work, and the two major components are task performance and contextual performance [16], and these two components of performance can be used to differentiate safety behaviors in the workplace.

## 3 Methodology

Quantitative research approach was utilized in this study. The population consists of all employees in Carsem (M) Sdn. Bhd., Ipoh, Perak, which was about 2,500. A total of 170 respondents were selected to participate in this study using simple random sampling technique. In this study, self-administered questionnaire was used as an instrument for data collection. Questionnaire was used because a lot of primary data and useful information can be collected for the purpose of this study. The questionnaire was adapted and modified from several past researches, namely, the Barling, Kelloway, and Iverson (2003) and Pfeffer (1998). It was redesigned in order to fit in to this study which was on the relationship between HPWS and safety performance. These questions were categorized into Sections A, B, and C. Section A was designed to obtain the information regarding the demographic characteristic of the respondents like gender, age, race, and tenure in the organization; education level; and monthly salary including wages. Meanwhile, Section B consisted of questions regarding safety performance. This part was subdivided in eight items in order to gain the employees' perception toward safety performance in their organization. Section C contained statements on the safety training, self-managed teams, and high-quality work. A 5-point Likert scale ranging from Strongly Disagree with the value of 1 to Strongly Agree with the value of 5 was used in Sections B and C.

Furthermore, the primary data was analyzed by using Statistical Package for the Social Sciences (SPSS) version 17.0 for Windows. The statistical methods used to analyze the data were descriptive statistics and inferential statistics, which included Pearson correlation and multiple regression analysis.

# 4 Findings and Discussions

## *4.1 Reliability and Validity Tests*

Table 1 presented the results of reliability and validity tests. Reliability and validity tests were conducted to ensure that the survey questions were clearly phrased and would be easily answerable by the members of the sample without ambiguous, unclear, or misleading questions. The items for each variable had factor loadings of 0.4 and above, indicating that it met the acceptable minimum standard of validity analysis [17]. The KMO score for each variable was more than 0.6. These variables were significant in Bartlett's test of sphericity. All variables had eigenvalues larger than 1. Cronbach's alpha of 0.6 or higher for a component reflects that the measuring items under that particular component provide a reliable measure of internal consistency [18], and the results showed that the value of Cronbach's alpha met the acceptable standard of reliability analysis.

## *4.2 Demographic Characteristics*

The total number of respondents was 170. Majority of the respondents were female (100, 58.8 %) compared to the male (70, 41.2 %). 33.5 % (57) of the respondents were between 26 and 35 years old, followed by 28.2 % (48) of them were from the age range of 36–45. Most of the respondents were Chinese (70, 41.2 %), followed by the Malay (62, 36.5 %). Thirty percent (51) of the total respondents have attached to the organization between 1 and 5 years, and 28.2 % (48) of the respondents have been in the profession for more than 10 years. Besides, most of the respondents were bachelor's degree holders (58, 34.1 %), and only 11.8 % (20) of them were master's degree holders.

**Table 1** Results of reliability and validity tests

| Measure | Items | Factor loading | KMO | Bartlett's test of sphericity | Eigenvalue | Variance explained | Cronbach's alpha |
|---|---|---|---|---|---|---|---|
| Safety performance | 8 | 0.459–0.617 | 0.771 | 70.300, $p = 0.000$ | 1.940 | 50.841, $p = 0.000$ | 0.796 |
| Safety training | 8 | 0.401–0.628 | 0.726 | 102.583, $p = 0.000$ | 1.900 | 38.590, $p = 0.000$ | 0.898 |
| Self-managed teams | 8 | 0.462–0.560 | 0.698 | 89.981, $p = 0.000$ | 1.991 | 53.206, $p = 0.000$ | 0.886 |
| High-quality work | 8 | 0.486–0.740 | 0.824 | 115.883, $p = 0.000$ | 1.739 | 52.232, $p = 0.000$ | 0.919 |

**Table 2** Means, standard deviations, and correlation between independent and dependent variables

| Variables | Mean | Standard deviation | Significant, *p* | Pearson correlation, *r* | Results |
|---|---|---|---|---|---|
| Safety training | 3.4338 | 0.36033 | $p = 0.000$ | 0.384** | H1 Accepted |
| Self-managed teams | 3.4993 | 0.35499 | $p = 0.000$ | 0.273** | H2 Accepted |
| High-quality work | 3.5934 | 0.33669 | $p = 0.000$ | 0.330** | H3 Accepted |

*Note*: **Correlation is significant at the 0.01 level (2-tailed)

## 4.3 Results from Pearson Correlation Analysis

As shown in Table 2, the findings from this study indicated that there was a positive relationship between safety training and safety performance ($r = 0.384$, $p = 0.000$). Training allows employees to acquire greater competencies to control their work, leading to them performing their jobs more safely [19]. Besides, training could increase the problem-solving skills of employees from a human capital perspective [20]. This means that as employees receive safety training, then their ability to handle a situation during emergency can be improved which would improve the safety performance at the workplace.

Furthermore, it was argued that there are only three determinants of individual performance: knowledge, skill, and motivation [21]. Safety performance, therefore, must be determined by knowledge and skills necessary for particular behaviors and by the motivation of individuals to perform the behaviors. This supports the findings whereby safety training can be a medium to provide employees with required skills to do their job in the right manner.

Besides, it was also revealed in Table 3 that there was a significant relationship between self-managed teams and safety performance ($r = 0.273$, $p = 0.000$). These findings were supported by a research conducted on 97 manufacturing plants, in which the findings showed that the quality of supervisor–employee relationships, as well as cohesion with the work group, was the best predictor of the propensity to comply with safety rules [22]. This was in line as well with a study that claimed empowered teams or self-managed teams share "a sense of doing meaningful work that advances organizational objectives" [23]. Both of these findings indicated the role of self-managed teams in improving safety performance at the workplace which was reliant with the findings of this study.

Moreover, a study on the members of a flight crew showed that they performed more effectively as a group in dangerous situations than as a hierarchy with the captain at the top of the chain of command, presumably because team members were motivated to contribute their ideas by the belief that they must work together in achieving their goals [24].

In addition, an empowered team would have some measure of the value of its work. Without knowing its strengths, the team will not feel a sense of accomplishment. Without knowing its weaknesses, it will not know how to increase its performance. A self-managed team will find ways to evaluate its own efforts and

**Table 3** Regression results

| Model summary | | | | | | |
|---|---|---|---|---|---|---|
| Model | | $R$ | $R$ square | Adjusted $R$ square | | Std. error of the estimate |
| 1 | | 0.438[a] | 0.192 | 0.177 | | 0.30188 |
| ANOVA[b] | | | | | | |
| *Model* | | Sum of squares | df | Mean square | $F$ | Sig. |
| 1 | Regression | 3.592 | 3 | 1.197 | 13.139 | 0.000[a] |
| | Residual | 15.127 | 166 | 0.091 | | |
| | Total | 18.720 | 169 | | | |
| Coefficients[b] | | | | | | |
| | | Unstandardized coefficients | | Standardized coefficients | | |
| Model | | $B$ | Std. error | Beta | $t$ | Sig. |
| 1 | (Constant) | 1.656 | 0.301 | | 5.511 | 0.000 |
| | Training | 0.264 | 0.071 | 0.285 | 3.699 | 0.000 |
| | Teams | 0.088 | 0.076 | 0.094 | 1.155 | 0.250 |
| | Quality | 0.170 | 0.082 | 0.172 | 2.061 | 0.041 |

[a]Predictors: (Constant), Quality, Training, Teams
[b]Dependent variable: Performance

improve upon them. People who feel empowered are more likely to develop systems for evaluating themselves, to seek feedback from others, and to use the information to make positive decisions about future goals [25]. The result of evaluation and feedback will be continuously improved within the company, as the team uses its new sense of empowerment. This includes their improvement on in terms of safety management as well. Therefore, a self-managed team would open up a chance to improve the safety performance of the organization.

It was also pointed out in the findings that there was a significant relationship between the high-quality work and safety performance ($r = 0.330$, $p = 0.000$). Some researchers have included that job quality as a component of a high-performance work system [26]. Undoubtedly, the employees who are committed in doing quality works normally perform better in the jobs assigned. High-quality work provides the employee with the motivations and the opportunity to do great work. Employees will reciprocate positive attitudes and behaviors to the organization as they feel the job they are doing is safe [27].

## 4.4 Results from Multiple Regression Analysis

Researchers used multiple regression analysis to determine the dominant factor that has greatest influence on safety performance at the workplace. Based on the Table 3, the regression model explained 19.2 % of the variation in the dependent variable, safety performance, as indicated by $R$ square value. Then, the highest beta value (0.285) with $p = 0.000$ achieved by the safety training indicated as the dominant factor of this study. Therefore, the hypothesis H4 was accepted.

## 5 Conclusion

The importance of high-performance work system and safety performance in the organization is increasingly recognized by the top management because it can give profitable advantage to the organization. The HPWS can contribute toward the sustainability and the survival in the competitive environment. However, getting employees from diverse background to work together to achieve this is not always easy and straightforward. The realization of the importance of safety performance at the workplace would be the motivating aspect in implementing HPWS within the organization.

As a conclusion, research interest in the area of high-performance work systems continues to develop as well as on workplace safety. Occupational safety aspects are identified to be a motivation factor for employees to reduce turnover and increase productivity. The study on improving the safety performance through HPWS should be further justified through future researches. A slight change in the research context could bring changes in the present findings. Therefore, future researchers are welcomed to conduct similar research to give greater contribution for the same context of study.

## References

1. Lawler EE III (1996) From the ground up: six principles for building the new logic corporation. Jossey-Bass, San Francisco
2. Whitener EM (2001) Do “high commitment” human resource practices affect employee commitment? A cross-level analysis using hierarchical linear modeling. J Manag 27:515–535
3. Werner S, Schuler RS, Jackson SE (2012) Human resource management, 11th edn. Cengage Learning, South-Western, Mason
4. Way SA (2002) High performance work systems and intermediate indicators of firm performance within the U.S. small business sector. J Manag 28:765–785
5. Wood S, Wall TD (2002) Human resource management and business performance. In: Warr P (ed) Psychology at work. Penguin, London
6. Snell S, Bohlander G (2013) Principles of human resource management, 16th edn. Cengage Learning, South-Western, Mason
7. Lawler EE (1992) The ultimate advantage: creating the high involvement organization. Jossey-Bass, San Francisco
8. Lawler EE, Mohrman SA, Ledford GE (1995) Creating high performance organizations: practices and results of employee involvement and total quality management in fortune 1000 companies. Jossey-Bass, San Francisco
9. Guthrie JP (2001) High involvement work practices, turnover and productivity: evidence from New Zealand. Acad Manage J 44:180–190

10. Huselid MA (1995) The impact of human resource practices on turnover, productivity, and corporate financial performance. Acad Manage J 38:645–672
11. Berg P (1999) The effects of high performance work practices on job satisfaction in the United States steel industry. Ind Relat 54:111–134
12. Godard J (2001) High performance and the transformation of work? The implications of alternative work practices for the experience and outcomes of work. Ind Labor Relat Rev 54:776–805
13. Komaki J, Heinzmann AT, Lawson L (1980) Effect of training and feedback: component analysis of a behavioral safety program. J Appl Psychol 65:261–270
14. Neal A, Griffin MA (2006) A study of the lagged relationships among safety climate, safety motivation, safety behavior, and accidents at the individual and group levels. J Appl Psychol 91:946–953
15. Campbell JP, Gasser MB, Oswald FL (1996) The substantive nature of job performance variability. In: Murphy KR (ed) Individual differences and behavior in organizations. Jossey-Bass, San Francisco
16. Borman WC, Motowidlo SJ (1993) Expanding the criterion domain to include elements of contextual performance. Jossey-Bass, San Francisco
17. Polit DF, Beck CT (2004) Nursing research, principles and methods, 7th edn. Lippincott Williams & Wilkins, Philadelphia
18. Awang Z (2012) Research methodology and data analysis, 2nd edn. UiTM Press, Shah Alam, Selangor
19. Barling J, Kelloway EK, Iverson RD (2003) High-quality jobs, job satisfaction and occupational safety. J Appl Psychol 88:276–283
20. Osterman P (1995) Skill, training, and work organization in American establishments. Ind Relat 34:125–146
21. Campbell JP, McCloy RA, Oppler SH, Sager CE (1993) A theory of performance. In: Schmitt J, Borman WC et al (eds) Personnel selection in organizations. Jossey-Bass, San Francisco
22. Simard M, Marchand A (1997) Workgroups' propensity to comply with safety rules: the influence of micro-macro organizational factors. Ergonomics 40:172–188
23. Kirkman BL (1999) Beyond self-management: antecedents and consequences of team empowerment. Acad Manage J 42(1):8–74
24. Tjosvold D (1990) Flight crew collaboration to manage safety risks. Group Organ Stud 15:177–191
25. London M, Holt H, Larsen L, Nelleman T (1999) Relationships between feedback and self-development. Group Organ Manag 24(1):5–27
26. Bailey T, Berg P, Sandy C (2001) The effect of high-performance work practices on employee earnings in the steel, apparel, and medical electronics and imaging industries. Ind Labor Relat Rev 54:525–543
27. Tsui AS, Pearce JL, Porter LW, Tripoli AM (1997) Alternative approaches to the employee-organization relationship: does investment in employees pay off? Acad Manage J 40:1089–1121

# Determinant Factors of Insurance Demand by SMEs in Malaysia

**Junainah Junid, Siti Zahrah Buyong, and Khuzaimah Akasah**

## 1 Introduction

Small and medium enterprises (SMEs) are often referred to as the engine and the backbone of economy in Malaysia with 90 % of the businesses categorised under small and medium enterprises. Malaysian small businesses were the second most likely (after Indonesian small businesses) to grow in 2013 (63 %). Looking at the economic boom in Japan, there are 4.69 million businesses under the small and medium business enterprises (SMEs), which represent 99.7 % of the total business. However, the disasters experienced in Japan on March 11, 2012, as a result of the earthquake and tsunami caused the discontinuation of business activities. Most SMEs insured their business as a protection against the risk of business destruction. It can be seen from the significant effect where 3 months after the disaster, SMEs in the country are gradually recovering. Insurance is a method that allows the entrepreneurs to transfer the risk to the insurance company. Not only is insurance as a form of financial risk management, it can also protect a person from unprecedented disaster. According to Brown and Churchill [1], insurance reduces vulnerability as households replace the indefinite prospect of large losses with the certainty of making small, regular premium payments. Realising the importance of insurance for small businesses in Malaysia for their own protection makes insurance a demand, but yet micro businesses are unable to afford the policy due to various aspects, such as low revenue in sales and decreased awareness. The important factor that those micro businesses demand from insurance is the flexible payment method and collection of affordable premiums provided by the insurer. Since most of the micro businesses face unstable income, providing insurance for this group requires the insurer to come up with proper

---

J. Junid (✉) • S.Z. Buyong • K. Akasah
Malaysian Academy of SME and Entrepreneurship Development, University Technology MARA, Shah Alam, Selangor 40000, Malaysia
e-mail: junainahjunid@yahoo.com.my

K. Ab. Hamid et al. (eds.), *The Malaysia-Japan Model on Technology Partnership: International Proceedings 2013 of Malaysia-Japan Academic Scholar Conference*,
DOI 10.1007/978-4-431-54439-5_15, 

structuring of premium concerning the flexibility of payment and ease of collection. Hence, the timing of premium collection is structured to coincide with the cash inflows of this group [3–5]. This is also supported by [6] who emphasised that micro insurance demand for small businesses will succeed if the element of designing the premium payment system specifically for this target group is considered by the insurer. [7–9] found that the capacity to afford an insurance is directly connected to one's level of income. However, the capacity of low-income micro business operators to purchase insurance is not only related to the level of income but also due to other reasons such as proper management of their financial resources, which have a remarkable impact on their access to micro insurance [3]. Awareness on the existence and importance of insurance policy of the micro businesses is another area to look into. Low-income businesses believe that insurance is not so important. This is probably due to less confidence on the insurer and poor understanding of the risk-pooling concept in the role of helping others after an incident. Previous experience with micro insurance also shows that many businesses view insurance in a negative way; they see an insurance as something that is reserved for the rich, which is irrelevant for them, too expensive to afford, or even unfair in terms of claim made [10–13, 3, 14].

## 2 Objectives and Scope of the Study

The objectives and scope of study of this paper are explained below.

### *2.1 Objectives of the Study*

The objectives of this study are:

- To identify the factors that contribute to the demand of insurance product by SME businesses
- To examine the degree of awareness of SME business operators in Malaysia on insurance policy
- To evaluate the relationship between the premium flexibility offered by the insurer and willingness of SME business to purchase the policy

### *2.2 Scope and Method of the Study*

This study was done in 3 months, focusing on 100 SME businesses (inclusive of micro businesses) located in Klang Valley, Malaysia, which included various sectors such as manufacturing, agriculture and services. Questionnaires were

**Table 1** Descriptive analyses: demographic

| Category | Age | Frequency | Percentage |
|---|---|---|---|
| Owner's age | Below 30 | 8 | 8 |
| | 31–45 years | 17 | 17 |
| | Above 45 | 75 | 75 |
| Gender | Male | 41 | 41 |
| | Female | 59 | 59 |
| Bumiputera status | Bumiputera | 60 | 60 |
| | Non-Bumiputera | 40 | 40 |
| Year of business establishment | Below 5 years | 3 | 3 |
| | 5–10 years | 42 | 42 |
| | More than 10 years | 55 | 55 |
| Monthly revenue (level of income) | Below RM30,000 | 63 | 63 |
| | RM30,001–RM50,000 | 22 | 22 |
| | RM50,001–RM100,000 | 12 | 12 |
| | More than RM100,001 | 3 | 3 |
| Level of business owner's education | Primary/secondary school | 22 | 22 |
| | Diploma | 30 | 30 |
| | Degree | 45 | 45 |
| | Postgraduate degree | 3 | 3 |
| No. of employees | 1–5 | 35 | 35 |
| | 6–20 | 42 | 42 |
| | More than 20 | 21 | 21 |

directly distributed to all SME businesses and were then collected. A total of 100 respondents participated in this survey with a 100 % rate of return.

# 3 Result and Analysis

Descriptive analysis was chosen to analyse the demographic section where respondents were questioned on seven categories that include the owner's age, gender, Bumiputera status, year of SME business establishment, level of income (monthly revenue), level of business owner's education and number of employees in the business.

## *3.1 Frequency Distribution*

Table 1 shows the demographic profile of respondents. From the descriptive analysis, it shows that 75 % of the owners aged above 45 years will purchase an insurance policy regardless of any type of insurance product in the market. Fifty nine percent of females have insurance policies compared to males. Sixty percent of respondents have Bumiputera status. It is found that 55 % of the respondents are

**Table 2** Model summary

| Model | *R* | *R* square | Adjusted *R* square | Std. error of the estimate |
|---|---|---|---|---|
| 1 | 0.696[a] | 0.484 | 0.479 | 0.94372 |
| 2 | 0.730[b] | 0.533 | 0.523 | 0.90276 |
| 3 | 0.744[c] | 0.554 | 0.540 | 0.88665 |

[a]Predictors: (constant), level of education
[b]Predictors: (constant), level of education, premium flexibility
[c]Predictors: (constant), level of education, premium flexibility, gender
Dependent variable: determinants of insurance demand

already running the business for more than 10 years. Mostly 63 % of the businesses have an income (monthly revenue) below RM 30,000. The survey also found out that most of the business owners are degree holders from various universities and majority of them are employing between 6 and 20 employees.

### 3.1.1 Coefficient of Correlation (*R*)

*R* is the coefficient of correlation where it is able to measure the degree of association between the dependent and independent variables. The value coefficient of correlation (*R*) is denoted by the *R* range, from −1 to +1, no matter what the unit of *X* and *Y* is. The range indicates whether the correlation is strong or not. Its sign (negative or positive) indicates the direction of relationship between the variables. The value of *R* for model 1 is 0.696; model 2, 0.730; and model 3, 0.744 (Table 2). It indicates that there is a strong positive linear association between the dependent and independent variables of abnormal returns in the model. The correlation coefficient shows that free cash flow, dividend payment and abnormal returns have a strong positive relationship at 69.6, 73 and 74.4 %, respectively.

### 3.1.2 *R* Square ($R^2$)

Based on Table 2, the *R* square for model 1 is 0.484; model 2, 0.533; and model 3, 0.554. It indicates that a 48.4 % change in the determinant factors of insurance demand is due to changes in the level of education, whereas for model 2, it indicates that a 53.3 % change in the determinant factors of insurance demand is due to changes of the combination of the level of education, premium flexibility and gender. It shows that there is a strong positive linear association between three variables in the model.

**Table 3** ANOVA

| Model | | Sum of squares | df | Mean square | *F* | Sig. |
|---|---|---|---|---|---|---|
| 1 | Regression | 81.156 | 1 | 81.156 | 91.124 | 0.000[a] |
| | Residual | 86.389 | 97 | 0.891 | – | – |
| | Total | 167.545 | 98 | – | – | – |
| 2 | Regression | 89.307 | 2 | 44.654 | 54.791 | 0.000[b] |
| | Residual | 78.238 | 96 | 0.815 | – | – |
| | Total | 167.545 | 98 | – | – | – |
| 3 | Regression | 92.861 | 3 | 30.954 | 39.734 | 0.000[c] |
| | Residual | 74.684 | 95 | 0.786 | – | – |
| | Total | 167.545 | 98 | – | – | – |

[a]Predictors: (constant), level of education
[b]Predictors: (constant), level of education, premium flexibility
[c]Predictors: (constant), level of education, premium flexibility, gender
Dependent variable: determinants of insurance demand

### 3.1.3 *F*-Stats

From the result of the *f*-test in Table 3, it shows that there is a relationship between the three independent variables (level of education, premium flexibility and the gender) with the dependent determinant factors of insurance demand at significant level $p \leq 0.05$. Therefore, the null hypothesis is rejected at 5 % significance level. By accepting the alternative hypothesis, it shows that abnormal results have a significant relationship with its level of education, premium flexibility and gender. Obviously, any changes in one of these variables could lead to the changes of determinant factors of insurance demand.

### 3.1.4 Regression Model

The multiple linear regression model equation explains the coefficient analysis between the dependent variable which is the determinant factors of insurance demand and independent variables, namely, level of education, premium flexibility and gender.

Premium Flexibility

The analysis between premium flexibility and level of education found that the *t*-significant value stands at 0.005 which represents a significant value. The finding hereby rejects the null hypothesis for the level of education, which explained that the level of education did have a significant relationship with the determinant factors of insurance demand.

**Table 4** Multiple linear regression model coefficient analysis

| Model | | Unstandardised coefficients | | Standardised coefficients | | |
|---|---|---|---|---|---|---|
| | | *B* | Std. error | Beta | *t* | Sig. |
| 1 | (Constant) | 1.345 | 0.390 | 0.696 | 3.451 | 0.001 |
| | Level of education | 0.740 | 0.077 | | 9.546 | 0.000 |
| 2 | (Constant) | 0.697 | 0.426 | 0.531 | 1.638 | 0.105 |
| | Level of education | 0.565 | 0.093 | 0.275 | 6.103 | 0.000 |
| | Premium flexibility | 0.321 | 0.101 | | 3.162 | 0.002 |
| 3 | (Constant) | 0.107 | 0.502 | 0.470 | 0.213 | 0.832 |
| | Level of education | 0.500 | 0.096 | 0.247 | 5.217 | 0.000 |
| | Premium flexibility | 0.287 | 0.101 | 0.167 | 2.852 | 0.005 |
| | Gender | 0.213 | 0.100 | | 2.126 | 0.036 |

Dependent variable: determinants of insurance demand by micro businesses

Hypothesis Statement 1

$H_1$: There is a relationship between premium flexibility and the level of demand for insurance products.

Result: Accept the hypothesis since *t*-test is $0.005 < 0.05$. The result of regression in Table 4 shows that the education level is significantly related to the determinants of insurance demand.

## Gender

The analysis of the level of education, premium flexibility and gender in Table 4 shows that the large *t* stands in 0.036, which is significant. The findings hereby accept the hypothesis for gender, which explained that gender did have a significant relationship with the determinants of insurance demand.

Hypothesis Statement 2

$H_2$: There is a significant relationship between gender and the determinants of insurance demand.

Result: Accept hypothesis 2 since *t*-test is $0.036 < 0.05$. The result from the regression indicates that gender is significant with the determinants of insurance demand.

## Level of Education

The analysis between the determinants of insurance demand and level of education in Table 5 shows that the *t*-significant value stands at 0.691, which means that it is not significant. The findings hereby reject the hypothesis for the level of education, which explained that the level of education did not have a significant relationship with the determinants of insurance demand.

**Table 5** Excluded variables

| Model | | Beta In | $T$ | Sig. | Partial correlation | Colinearity statistics |
|---|---|---|---|---|---|---|
| | | | | | | Tolerance |
| 1 | Premium flexibility | 0.275[a] | 3.162 | 0.002 | 0.307 | 0.642 |
| | Level of education | 0.183[a] | 2.420 | 0.017 | 0.240 | 0.881 |
| | Gender | 0.201[a] | 2.508 | 0.014 | 0.248 | 0.782 |
| 2 | Level of education | 0.097[b] | 1.166 | 0.246 | 0.119 | 0.701 |
| | Gender | 0.167[b] | 2.126 | 0.036 | 0.213 | 0.763 |
| 3 | Level of education | 0.035[c] | 0.398 | 0.691 | 0.041 | 0.598 |

[a]Predictors: (constant), level of education
[b]Predictors: (constant), level of education, premium flexibility
[c]Predictors: (constant), level of education, premium flexibility, gender
Dependent variable: determinants of insurance demand

Hypothesis Statement 3

$H_3$: There is a significant relationship between the level of education and the determinants of insurance demand.
Result: Reject Hypothesis 3 since $t$-test is $0.691 > 0.05$.

The result from the regression indicates that the level of education is significant with the determinants of insurance demand.

## 4 Conclusion

The objective of this study has been achieved through statistical analysis. It shows that premium flexibility, level of education and gender have a significant relationship with the determinant factors of insurance demand. It means that the demand for the insurance products is affected by these three factors. It is supported by three significant relationships with the dependent variable. This is supported by the significant value for each independent variable which is less than 0.05 (5 % level of significance). Besides that, according to the model summary, the three independent variables are positively related to the dependent variable.

### *4.1 Premium Flexibility*

The result achieves the research objective which is to evaluate the relationship between the premium flexibility and willingness of consumers to buy insurance products. The findings accept the hypothesis according to the $f$-test result. It is because the insurance industry in Malaysia has created a convenient method of payment such as through direct payment, auto debit, credit card and salary deduction. Consumers can choose annual, quarterly and also monthly payments.

These factors that make the premium flexible are relevant to influence the determinants of insurance demand because people would be more satisfied with the variety of premium payments.

## 4.2 *Level of Education*

Lastly, this study also aims to identify the main reason that contributes to insurance demand. This objective was answered in the hypothesis according to the *f*-test result. It was obviously tested during this research that consumers with lower education are unable to make a wise decision on purchasing insurance products. They sometimes do not have a clear understanding of what insurance is about.

### 4.2.1 Gender

One of the objectives mentioned in the earlier chapter is to identify the factors that contribute to the demand of insurance products. The factors that were examined in this study are the level of education, gender and premium flexibility. The finding accepts the hypothesis according to the *f*-test result. The reason from the result is that women are more responsive on protection compared to men.

# 5 Recommendations

Insurance companies should be more concerned about the SME businesses' need for protection as this group of potential customers understands the importance of insurance. Therefore, insurers should recommend a relevant insurance policy which includes a flexible premium payment to enable them to strategise their premium payment. They have to be sensitive on the micro business' needs and wants towards insurance; thus, in the future, insurance companies can underwrite various products that correspond to the various types of consumer behaviour. Besides that, business owners' level of knowledge also allows them to take a wise decision on purchasing insurance products. However, insurance companies need to revise their promotion strategy to enable the client to understand the product.

# References

1. Brown W, Churchill C (1999) Providing insurance to low-income households: part I: a premier on insurance principles and products, micro enterprise best practices project. DAI/USAID, Bethesda, pp 1–92

2. Matul M (2005) Demand for micro insurance in Georgia: quantitative study results. Presented at the 8th Annual Conference on Microfinance Institution, Bucharest
3. Guha-Khasnobis B, Ahuja R (2004) Extending formal insurance to the informal economy workers in India. EGDI and UNUWIDER, Parallel Session 3.2
4. Sinha S (2002) Strength in solidarity: insurance for women workers in the informal economy. Self Employed Women's Association (SEWA), Ahmedabad
5. Tenkorang DA (2001) Health insurance for the informal sector in Africa: design features, risk protection and resource mobilization. CMH Working Papers Series, Paper No. WG3:1
6. Sebstad J, Cohen M, McGuinness E (2006) Guidelines for market research on the demand for micro insurance. For USAID under the Accelerated Microenterprise Advancement Project (AMAP)
7. Beenstock M, Dickinson G, Khajuria S (1986) The determination of life premiums: an international cross-section analysis 1970–1981. Insur Math Econ 5:261–270
8. Browne MJ, Kim K (1993) An international analysis of life insurance demand. J Risk Insur 60:616–634
9. Outreville JF (1996) Life insurance market in developing countries. J Risk Insur 63:263–278
10. McCord M, Osinde S (2005) Reducing vulnerability: the supply of health insurance in East Africa. J Int Dev 17(3):327–381
11. Cohen M, Sebstad J (2005) Reducing vulnerability: the demand for microinsurance. J Int Dev 17(3):397–474
12. Manje L, Churchill C (2002) The demand for risk managing financial services in low-income communities evidence from Zambia. Social Employment Sector, ILO, Geneva, Working Paper No. 31
13. Herrera C, Miranda M (2004) Columna Guatemala. CGAP Working Group on Microinsurance. Good and Bad Practices. Case Study No. 5, December
14. McCord M (2008) Visions of the future of micro insurance and thought on getting there, USAID/DAI. Washington, DC, Micro Insurance Note No. 9, pp 1–35

# The Perceptions of Malaysians in a Japanese Company: A Case Study in Shimano Components (Malaysia) Sdn. Bhd.

**Nor Fyadzillah Mohd Taha**

## 1 Introduction

According to Department of Statistics Malaysia (2010), Malaysia's population of 28.3 million is comprised of three main ethnic groups which are Bumiputera (67.4 %), Chinese (24.6 %) and Indians (7.3 %). Japanese companies in Malaysia were successful implement and this will prove by JETRO research in [3] which there is total of 1,412 Japanese companies from manufacturing (730) and non-manufacturing (682)—see Appendix E.

## 2 Company Selection

As discussed with Prof Emeritus Dr. Ross Mouer, the author decided to select Shimano Components (Malaysia) Sdn. Bhd. This company is located at Lot 879 & 880, Mukim Jeram Batu, 81500 Pekan Nanas, Pontian, Johor.

### 2.1 *The Company*

Shimano Components (Malaysia) Sdn. Bhd. was incorporated in 25 January 1990. In 2012, their authorized capital was nearly RM25 million (US$7.5 million) and their paid-up capital was RM16 million (US$4.8 million). The company's core activity was manufacturing bicycle components and fishing reels. The main factory

N.F.M. Taha (✉)
Graduate School of Business Administration, Meiji University, Kanda Surugadai, Chiyoda, Tokyo 101-8301, Japan
e-mail: norfyadzillah.taha@gmail.com

K. Ab. Hamid et al. (eds.), *The Malaysia-Japan Model on Technology Partnership: International Proceedings 2013 of Malaysia-Japan Academic Scholar Conference*, DOI 10.1007/978-4-431-54439-5_16, 

(bike division) processes were cold forging, CNC machining, special purpose machining, aluminium heat treatment, pad/silk printing and laser and assembly line, while reel division processes were cold forging, CNC machining, gear and shaft machining, zinc casting, aluminium heat treatment, assembly line, plastic injection, spray painting and pad printing and laser marking. Shimano Components (Malaysia) Sdn. Bhd. was certified with ISO 9001, ISO 14001, OSAS 18001, 6 sigma, Small Group Activity (SGA) and Quality Improvement Technique (QIT) and Total Productive Maintenance (TPM).

Overall, Shimano Components (Malaysia) Sdn. Bhd. builds its competitive advantage on a combination of technical, marketing and managerial skills:

1. Technical
   - Continuous strive for engineering excellence and innovation
   - Steady process improvement
2. Marketing
   - Aiming at medium/high end of market, therefore not competing directly with the top niche players and profiting from a larger market segment
   - Introduction of components branding, achieving high awareness and recognition
   - Trend creation through sponsoring of bike enthusiasts organizations at all levels, from amateurs to professionals
3. Managerial
   - Choice to manufacture in cheap labour countries
   - Strong entrepreneurial spirit
   - Capacity to extract inputs from any possible source (clients, retailers, trade fairs), allowing to keep abreast of new trends

### 2.2 *Company Organizational Structure*

Shimano Components (Malaysia) Sdn. Bhd. has a total of 2,659 employees in which 409 employees (15 %) are in the management department and 2,250 employees (85 %) in the production line (bike components and reel centre). Hence, the breakdown structure within races shows 71 % being Malaysian (1,892 employees), 28 % foreign (755 employees) and 1 % Japanese (12 employees). The foreigners are from Bangladesh, Nepal, Indonesia and Vietnam.

## 3 The Aim of Research

The aim of this research is to get an overview perception of Malaysian employees of Shimano Components (Malaysia) Sdn. Bhd. in order to determine the results of management practices among Japanese within Malaysian employees.

# 4 Methodology

This small research was conducted by using semi-structured interview while some of the research using triangulation method which is documentary analysis, participant observation and interviews [2]. The other research such as [5] conducted a various methodology in order to get the high value results of the study. They using a survey question to various respondent, observations and in depth interview (selected) with a question from survey questionnaire. To meet the research objectives [4], using a semi-structured interview format and put this interview as for them to run the pilot testing. They conduct a face-to-face interview and a telephone conversation in order to get the full involvement of respondents.

Hence, the author using a semi-structured interview by sending them (Shimano) consent letter and email conversation will determine with staff from human resources department (HRD). After a week, Shimano Components (Malaysia) Sdn. Bhd. replied to me with good news that they were willing to get involved with this small research—see Appendix B and C. By the way, the author has not conducted an interview in the office of the company but instead had it outside and during the weekend. The reason why I couldn't go to their company and conduct an interview was because the company was closed during the weekend, and 26 January 2013 was a Malaysian holiday (Thaipusam day). Therefore, the house organization was conducted at author mother in law house. The interview was conducted in a small room with a duration between 1 and 2 h every interview.

## *4.1 Collection of Information*

The research issues were determined within 3 weeks from the date of discussion with Prof. Emeritus Dr. Ross Mouer in class. After I got the company's approval, I started to find information through the company's annual reports (2011 and 2010), website, policies and handbook and newsletter from the Internet.

### 4.1.1 Semi-structured Interviews

In advance, further information was determined from the following:

1. Semi-structured interviews—interviewees were selected from the group of management people (manager from some of division—not cover all division), group of executives people (engineers, trainers, human resources staff) and group of non-executive staff (operator). The interviews were conducted outside and during the weekend (workers' day off). Each interview was conducted for almost 1–2 h.

# 5 Summary of Findings in Interviews

## 5.1 *Interviewees' Background*

This interview had 12 participants (including current and previous workers)—see Appendix A. The average age of the interviewees was 35 years old and the average length of service with the company was 11 years and 4 months. The interviewees came from various levels: management (3), executive (5) and non-executive (4). Also, the interviewees were composed of eight Malays, three Chinese and one from foreign workers who is working in the executive level (for dealing with most of foreign workers). Based on the result of the interviews, two local interviewees spoke fluent Japanese and had a working experience in Japan.

## 5.2 *Interviewees' Perception*

All interviewees were asked about their perception of a Japanese company in Malaysia, especially with the work experience they had in Shimano Components (Malaysia) Sdn. Bhd. The detail of the questions for the interview session was attached in Appendix D.

### 5.2.1 An Attraction for Working

All the interviewees had a positive answer for this question. Interviewee Mr. A and Mrs. C strongly answer with Japan country as attraction for them to working with Shimano Components (Malaysia) Sdn. Bhd. More than that, interviewees Mr. B and Mr. G had experienced working in Japan when this company sent them for exchange service in Osaka, Japan. This working experience had given them opportunities to speak fluent in Japanese.

### 5.2.2 Reputation of Parent Company

As a result, most of the interviewees knew of the company because of its links with the parent company in Japan. Shimano Components (Malaysia) Sdn. Bhd. is financially strong, well established and secure especially in producing bicycles and fishing reel in the world. When this company started their business in Malaysia in 1990, there were a lot of job opportunities, especially for local people. Interviewees Mr. B and Mr. G joined this company when it was first established in Malaysia. This gives both of them the opportunity to get a proper training in Japan (pioneer), strong career development, and well-established career enhancement.

### 5.2.3 Job Security

Most of the interviewees strongly answered with *I want to work for money and support family members*. With this answer they know that most Japanese companies would never fire its employees and they really felt secure working with Shimano Components (Malaysia) Sdn. Bhd. Interviewee Mrs. L talked about on-time payment from this company and efficiency in the production for each division. Hence, she felt secure working with Shimano Components (Malaysia) Sdn. Bhd.

### 5.2.4 Communication with Japanese

There are 12 Japanese people working as managerial officers, one as a president and the rest as general managers (GM). Interviewee Mr. A said he only meets with Japanese colleagues during monthly meetings and has bad experience in communicating with them. He said that he does not really understand when talking with them especially when they speak in English. He also adds that it is the reason why most Japanese people never talk with local people. Besides that, interviewees Mr. B and Mr. G told that they really enjoy communicating with their president because they spoke fluent in Japanese.

### 5.2.5 Chain of Command and Transparency

Interviewees Mrs. C and Mrs. D said that each production operator had their special card for expertise purposely. They were selected for production operator with their expertise and this will give them opportunities to learn any other experienced worked. They also said that from day 1 of working with Shimano Components (Malaysia) Sdn. Bhd., they already read the employees' handbook and company policies. They had gone through the training for production operator position almost 1 week before they "down to the drain".

### 5.2.6 Salary

Interviewee Mr. A was dissatisfied with the salary with this company. He said that based on the observations of some workers, they found a difference in salary between Chinese and Malay employees. He added that a Chinese employee gets 60 % more salary compared to Malay employee. This different is answer by top of management for the status of "Bumiputera" (land people) citizen especially for Malay people. He felt that this is unfair for them as Malay employees to get a different salary even though they work at the same division and position. Interviewee Ms. I said that she was never sent to any training by the company, either local or overseas, while Mrs. L said that she felt secure with the working conditions (office space, etc.) and healthy and safe working environment.

### 5.2.7 Other Financial Benefits

When I asked for other financial benefits, interviewee Mr. A said that he, as a local person especially as a Malay employee, never had a "special treatment", such as receiving a housing allowance, company cars, and a family allowance. This allowance is for Japanese people and some of the general managers (GM) and especially Chinese employees. There is no "standardization" policy for company benefits. However, Shimano Components (Malaysia) Sdn. Bhd. provided some special benefits for all employees such as an annual dinner, a grand annual dinner (once every 5 years), birthday celebrations , rewards (loyal employee—once every 5 years of service), and insurance coverage for the whole family (different for each employee—depends on the position and current working service).

### 5.2.8 Redress of Grievances

This company never had a union for employees (current). However, interviewee Mr. A said that there is one union from Malaysia Employee Union (MEU) in 1995 but this union was never success and since that, there is no union by any other employee. With this reason, I try to asking them about how they express their something "feeling" either bad or good feeling about their work, condition, environment, salary etc. Interviewees Mr. B and Mr. G said that they have a "suggestion box" for those employees who want to "say something". Some of them use this "suggestion box" to express ideas for improvement and some give a complaint. Interviewee Mrs. L said that she never had an experience of talking directly with the Japanese people, but sometimes, when her manager wants her to complete a job, she sends an email (forward) to the top management. In this case, she can just read the instruction from the top management, but she cannot reply to any of their emails (top management—Japanese). Other than that, interviewee Ms. E said that she only had opportunities to talk with Japanese during the monthly or weekly meetings. Sometimes, she talks about salary, environment, and local people with Japanese people (some general managers).

### 5.2.9 Labour Turnover

Interviewees Mrs. C and Mrs. D said that they never had a bad experience with the Japanese but had a bad experience with local and foreign workers. Both of them said that employees especially working as non-executive workers had been faced with some of the bad line leader, assistance supervisor, supervisor and senior supervisor. This problem was course for high labour turnover especially in production operator division. Moreover, interviewee Mr. F who is a foreign worker said that most of his friends quit from the job just because of discrimination of local people and foreign workers especially about relationships with colleagues.

#### 5.2.10 Perception of Japanese Colleagues

From the interview, the author found that most of the interviewees said that Japanese people working in the top management are very helpful, kind, humble and very shy. Interviewees Mr. B and Mr. G said that Japanese people will be "noisy" if "we" can talk in Japanese (*nihongo*). Some of the discussion was very clear when communicating in Japanese language.

## 6 Discussion

The bicycle industry has seen dramatic changes in the last 20 years. This reason can cause for customer preferences. The revolution of stagnant and mature business of the late 1970s by the introduction of sophisticated high-performance bikes capable of rolling down mountains and the need for high quality components, which suited this extreme usage. Hence, this revolution led to a change in the bicycle supply chain: the vertically integrated manufacturers, which dominated the market before the consumers changed their preferences, lost their dominance due to a slow adoption rate of the new trends and the lack of the technology required to manufacture components for these high-performance bicycles.

### *6.1 Problem Arise*

Since 2002, Shimano Components (Malaysia) Sdn. Bhd started to take foreign workers to work in the production line. Some of the local people had lost their job just because of low payment for foreign workers. Local people especially those who live in the Pontian area were really shocked to find "so many" foreign workers in town. And most of these foreign workers were working in Shimano Components (Malaysia) Sdn. Bhd. One of the problems is the "loving relationship" between the local people (women) and the foreign workers. Other than that social problem, other problems that arise in Shimano Components (Malaysia) Sdn. Bhd. are:

1. Poor communication between Japanese staff and local employees. Most Japanese employees refuse to talk to local people (Malays).
2. Lack of training for local workers.
3. Barriers between management, executives and non-executives. As interviewees Mrs. C and Mrs. D said, they don't have opportunities to talk with any top

management person (especially Japanese people) because before they joined the company, current employees were already asking them not to talk to any top management person (especially Japanese) except when they are given permission. They can only say "hi", "arigato", or "konnichiwa" when some of them (top management—Japanese people) start a conversation with them.
4. Bad management practices which are most of the decision-making had been made by local people (general manager—most of them are Chinese people) not Japanese people. Some of the local employees (Malays) have never been involved with decision-making (even if some of them were general managers).
5. Some of the non-executive workers are afraid to talk with Japanese people because they felt that they cannot answer if they speak in English. This is because most of them never understand English.

### *6.2 Strong and Secure Company*

Most of the interviewees said that this company was really strong and well established in order to get a secure job in future. This is because this company was successful in Malaysia since the 1990s.

## 7 Recommendation

After the results were evaluated, some recommendations can be given in order to improve the company policy, production line, etc.

### *7.1 Japanese/Malay (Bahasa Melayu)/English Course*

The most important thing among Japanese people, local employees (Malays, Chinese and Indians) and foreign workers is communication. This communication can be built up if they went to discussion or extra class for Japanese/Malay (Bahasa Melayu) or English language. With this, company will provide a "good class" in order to build a reputation in Malaysia.

### *7.2 Top Management Decision-Making*

This company is owned by Japan so that top management should take into account about decision-making made by top management (especially Chinese employee) in order to get the best decision/solution. The Japanese (president) should involve those from "Bumiputera" and the local people in the decision-making.

### 7.3 *Job Satisfaction*

Shimano Components (Malaysia) Sdn. Bhd. should improve their employees' handbook and company policies in order to improve job satisfaction among all employees. The current company policies stated the annual dinner (big event) is every 5 years—convert to 2–3 years in order to get a good end products if production.

### 7.4 *Develop a Training Program*

Management should put training as a necessary item in the company policy. The training department also can provide the "train the trainer" activity wherein the current trainer can train the other workers after they come back from training. This can also reduce the cost of the training program.

### 7.5 *Standardization of Payment*

The government of Malaysia already announced the standardization of payment in the corporate sector, wherein the minimum salary should be RM900.00. The company should take into account this new announcement and put this into the company policy. Interviewee Mr. A said that his first salary in 1995 was RM350.00.

## 8 Conclusion

As mentioned in class by Prof. Emeritus Dr. Ross Mouer about the core principles of Japanese management practices which are the lifetime employment (LTE), the seniority wages and promotion systems, enterprise unionism and company welfare services and benefits, all these are not found in Shimano Components (Malaysia) Sdn. Bhd. Local people (especially Chinese) dominate the top management positions and this look like not healthy condition whereby the decision-making was made by them not Japanese people (president). In order to improve the skills and techniques of management style, the top management should "change" the company policy into a "new and trendy" policy. This will include the key performance indicator (KPI), new salary and standardization of local employees (among Malay and Chinese employees). The general idea of recommendation can be consider as "small improvement" if they (Japanese—president) can provide it. Furthermore, the improvements of some allowance such as company car, house allowance, etc. should be revised again in order to put a standardization effort for all employees.

## References

1. Division HR (2009) Social activity report 2008 (July)
2. Foong YP, Richardson S (2008) The perceptions of Malaysians in a Japanese company. Cross Cult Manag 15(3):221–243
3. Japan External Trade Organization (JETRO) Malaysia (2011) Malaysian economy in figures 2011. http://www.jetro.go.jp/malaysia. Accessed 19 Jan 2013
4. Khatri Y, Collins R (2007) Impact and status of HACCP in the Australian meat industry. Br Food J 109(5):343–354
5. Pang F, Toh PS (2008) Hawker food industry: food safety/public health strategies in Malaysia. Nutr Food Sci 38(1):41–51
6. Rose RC, Kumar N, Rose RC, Kumar N (2007) The transfer of Japanese-style HRM to subsidiaries abroad. Cross Cult Manag 14(3):240–253

# A Conceptual Model Towards Information Security Culture in Health Informatics

**Noor Hafizah Hassan and Zuraini Ismail**

## 1 Introduction

Information security is defined as the activity to protect information from a wide range of threats in order to ensure business continuity, minimize business damage and maximize return on investments and business opportunities. With the moving clinical information in geographical locations from different types and services, technological advances constitute new opportunities among healthcare practitioners. This innovation makes established ways of doing work in e-health become outmoded. Various threat attacks have been detected in the hospital information system showing that there are security breaches [19]. This trend along with advances in health informatics is expanding the demand to build an effective information security protection for healthcare organizations. The innovation aiming towards enhancing quality of life, diagnostic and treatment options, as well as the efficiency and cost effectiveness of the healthcare system, led to the issues on information security [16]. Health informatics is perceived as being particularly useful in the operational support of the new decentralized and collaborative healthcare models being implemented in many countries. With the tendency of data breach and computer theft increasing, healthcare organizations are urged to realize that security is a vital asset in an organization. Over the past few years, the security ramification of information system in health informatics environment started to permeate the national consciousness [7, 20]. Healthcare organizations have grown rapidly in implementing an information system to support their daily operation and business process. Information security in e-health is important due to a variety of threats endangering healthcare organizations that can disclose hundreds

N.H. Hassan • Z. Ismail (✉)
Advanced Informatics School, University Technology Malaysia,
Kuala Lumpur 54100, Malaysia
e-mail: nhafizah95@live.utm.my; zurainisma@ic.utm.my

K. Ab. Hamid et al. (eds.), *The Malaysia-Japan Model on Technology Partnership: International Proceedings 2013 of Malaysia-Japan Academic Scholar Conference*,
DOI 10.1007/978-4-431-54439-5_17, 

or thousands of records. Technical issues become the main concerns among organizations as it can harm productivity and corrupt intellectual assets [27]. Despite the growing level of interest in this field, little is known about the actual issues involved in securing networks and electronic assets in the healthcare organization.

In [14], they have highlighted that non-technical issues such as human factor can create more problems for security incidents towards the organization. In [5, 24], they suggest to build an information security culture in the organization to increase the security level. However, the role of the organization in relation towards information security culture is still an area of long-standing confusion. Information security becomes vulnerable towards the organizations' concern [22]. However, most organizations found that there is an issue to maintain information security as part of organization culture to align with its holistic and integrated management approach [15]. To cater with these issues, the organization can build information security culture through behaviour [5], knowledge [23, 29] and awareness [4] of the employee. The maximum patch to ensure the effectiveness of security culture must be carefully considered in health informatics.

Creating information security culture among employees in health informatics has become top management responsibility [13]. However, little research has been conducted in addressing top management responsibility particularly in the area of health informatics. The managements become the key points to support the healthcare practitioners in adapting security culture. There are arguments on how information security culture can be built in the organization. In [13], they stated security culture needs to be separated from organizational culture for the reason that each organization has their own unique role and business purpose. As in healthcare organization, this argument must be investigated to suit with its business purpose and goal. Threats such as confidentiality of patients' information and medical error are common security risks in health informatics. Furthermore, no specific model was designed to adjust the needs and requirements of healthcare organization to promote information security culture in HIE. Security culture should be aligned with business strategy and integrated into business goals and objectives. Lack of awareness, commitment from management and not knowing the existence of policy among healthcare providers and medical students are the factors identified influencing information security culture in healthcare [9]. Therefore there is indeed a growing body of literature concerning questions of cultural influences on organizational behaviour and performance according to healthcare organization needs. This paper intends to investigate the factors influencing information security culture especially in HIE. It aims to highlight the human factor as the main factor that influences information security culture to be built in healthcare organization.

This paper is organized as follows. The first section of this paper provides an overview of information security whilst the second section reviews organizational culture model. The third section of this paper describes health informatics, followed by the health belief model. The next section presents the related work on information security culture in health informatics. The succeeding section will demonstrate the conceptual model developed from the theoretical study. The last section will conclude the paper and discuss future works of this research.

## 2 Information Security

The adoption of information security management is built on different aspects. This focus should look into human factor and technological and organizational element that challenge the information security within organizations. Human factor can be described as those related to cognition at the individual level as well as culture and the communication among people. Organizational aspects reflect those which directly involve the structure of the organization and include also the size and managerial decision among IT security, whilst technological aspects involve technical solutions such as applications and protocol.

### *2.1 Human Factor*

Information security provides policies and procedures that represent the process which need to be followed to avoid unsafe practices that would jeopardize the confidentiality, integrity and availability of data. However, there is a third component that will need to accomplish the process, that is, human [1]. Human beings were to blame for the machine's faults. There is a field of science called "human factor engineering" that seeks to address the problem. There have been some arguments on the challenges for implementing information security management. A long discussion has been thoroughly explained from [25] that stated that the ten deadly sins that led to the failure of information security management which solely highlighted human factors on each of the sins. In human cases, having the right people in charge towards information security at the right place can compensate for deficiencies in process and technology.

### *2.2 Organizational Factor*

When talking about information security, most researchers refer and relate information security of an organization tight closely with the security policy and goals of the organization. Security policies become one necessary foundation for information security [13]. Having solid security products or technology alone cannot protect an organization without a good management and policy implementation [26]. The organization learned the needs of information security strategy and the importance of it through the growth and application of IT. Understanding the threats is not an issue anymore; the organization needs to know the patterns of their information security to suit the business goals and process as well as cost effectiveness. It is possible for organizations to consolidate information security policies with human-behavioural consideration through a process called ontology which allows information security to be associated with information security properties [18].

### *2.3 Technological Factor*

The use of technologies such as firewall, encryption, disaster recovery plan, intrusion detection system (IDS), PKI, and antivirus software has been discussed and argued as the solution for information security. However, the technological complexity becomes a challenge towards security practitioners. The complexity of the technology becomes one of the big reasons for the organization not achieving full commitment of security practice. The rapid change of the new technology affects security breach whilst giving the attacker a chance to use new technology to attack the system. This complexity makes it extremely difficult for the decision-makers to manage the big picture and design security policies that cover all the security in their organization.

## 3 Organizational Culture Model

In exploring the information security culture, thus it is important to understand the concept of organizational culture. This paper uses organizational culture dimensional model from [6] in conceptualizing information security culture. Table 1 shows the organizational cultural dimension proposed in [6] with the recommended total quality management (TQM).

## 4 Health Informatics

As defined from ISO, health informatics is "A document, established by consensus and a approved by a recognized body, that provides for common and repeated use, rules, guidelines or characteristics for activities or their results, aimed at the achievement of the optimum degree of order in a given context". Information management in health informatics constitutes a major activity for the healthcare professional. The development of information systems supports the infrastructure of medicine, and a greater focus on the needs of physicians and other healthcare managers and professionals is occurring to support education, decision-making, communication and many other aspects of professional activity. Informatics can be referred as the discipline focused on the acquisition, storage and use of information in a specific setting or domain [10].

**Table 1** Organizational culture dimension [6]

| No. | Organization culture dimension | TQM value |
|---|---|---|
| 1 | The basis of truth and rationality | Decision-making should rely on factual information and the scientific method |
| 2 | The nature of time and time horizon | Improvement requires a long-term orientation and a strategic approach to management |
| 3 | Motivation | Quality problems are caused by poor systems—not the employees. Employees are intrinsically motivated to do quality work if the system supports their efforts |
| 4 | Stability versus change/innovation/ personal growth | Quality improvement is continuous and never ending. Quality can be improved with existing resources |
| 5 | Orientation to work, task and co-workers | The main purpose of the organization is to achieve results that its stakeholders consider important. Results are achieved through internal process improvement, prevention of defects and customer focus |
| 6 | Isolation versus collaboration/ cooperation | Cooperation and collaboration (internal and external) are necessary for a successful organization |
| 7 | Control, coordination and responsibility | Shared vision and shared goals are necessary for organizational success. All employees should be involved in decision-making and in supporting the shared vision |
| 8 | Isolation versus collaboration/ cooperation | Cooperation and collaboration (internal and external) are necessary for a successful organization |
| 9 | Control, coordination and responsibility | Shared vision and shared goals are necessary for organizational success. All employees should be involved in decision-making and in supporting the shared vision |
| 10 | Orientation and focus—internal and/or external | An organization should be customer driven. Financial results will follow |

## *4.1 Health Belief Model*

Health belief model is the earliest model that has been adapted for use in healthcare study. It has been widely used to study user behaviour [3]. This model aimed to predict and explain health-related behaviour by focusing on the attitude and belief of the respective persons. HBM is widely implemented to explore from low-risk to high-risk disease. HBM consists of six elements which are self-efficacy, perceived vulnerability, perceived benefits, perceived severity, perceived susceptibility and cues to action that is use to indicate patients' behaviour. HBM also has been used in information security research to study security behaviour in health informatics environment ([17] and [11]).

**Table 2** Related work of information security culture in health informatics

| No. | Author | Information security values | Key findings |
|---|---|---|---|
| 1. | Humaidi and Balakrishnan [11] | Security behaviour | Transformational leadership style, transactional leadership style, ISPs training, perceived usefulness of security, perceived usefulness of security-countermeasure and perceived ease of use of ISPs |
| 2. | Gaunt [8] | Attitudes, ignorance, conflicting demands, inadequate system, inconsistent policies | Security infrastructure, security awareness, communication, stronger security measures |
| 3. | Williams [28] | Trust | Organizational cultural and technical contribution |
| 4. | Bansal, Zahedi and Gefen [2] | Personality traits, information sensitivity, health status, prior privacy invasions, risk beliefs and experience | Integrated approach from personality perspectives and situation of the user |
| 5. | Stahl, Doherty and Shaw [21] | Information security policy | Information security policy is the key in ISMS |
| 6. | Katsikas [12] | Information security awareness | Positive feedback on security training required |

## 5 Related Work of Information Security Culture in Health Informatics

There are few related works that have been done in promoting information security culture in health informatics. This work focused on security behaviour, trust and security awareness and security policy as tabulated in Table 2.

## 6 Research Methodology

The methodology used in designing the conceptual model is from the secondary data collected. These include using search engine from the selected databases such as ACM, IEEE, Taylor and Francis, SpringerLink, ScienceDirect and Journal on HealthCare Informatics.

## 7 Model Development

From HBM, organizational culture model by [6] and related literature on information security culture, the proposed model is developed accordingly. The conceptual model development of the study showing the stated hypothesis, namely, H1a, H1b H2, H3, H4, H5, H6, H7, H8a, H8b, H9, H10 and H11 is shown in Fig. 1. This section follows by identifying each proposed hypothesis:

H1a: Security knowledge of healthcare practitioners has a positive relationship with information security culture.
H1b: Knowledge of healthcare practitioners has a positive relationship with security awareness.
H2a: Security awareness of healthcare practitioners has a positive relationship with information security culture.
H2b: Security awareness of healthcare practitioners has a positive relationship with security behaviour.

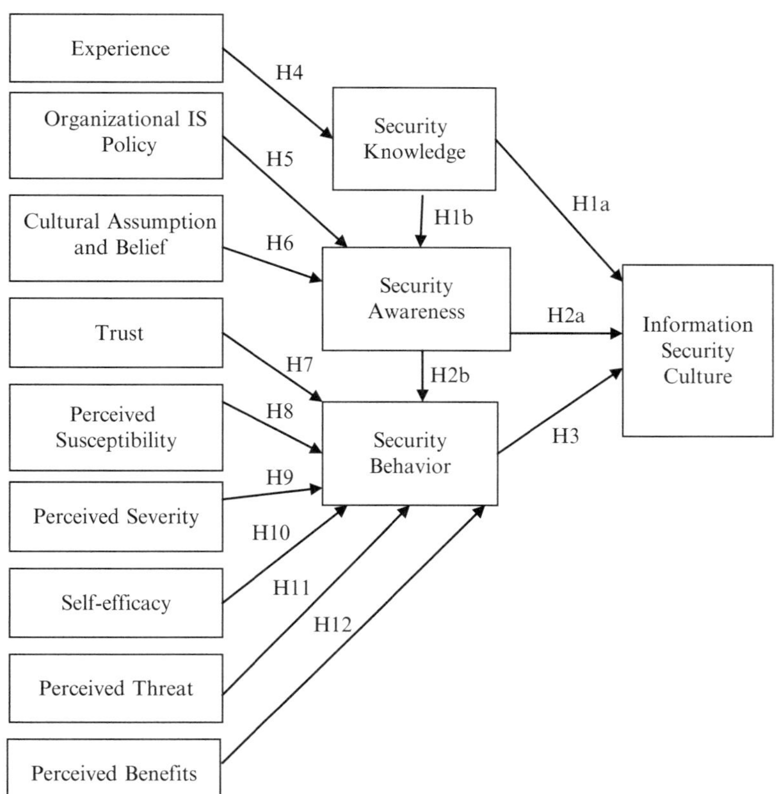


**Fig. 1** Proposed conceptual model

H3: Security behaviour of healthcare practitioners has a positive relationship with information security culture.
H4: Experience of healthcare practitioners has a positive relationship with security knowledge.
H5: Organizational IS policy has a positive relationship with security awareness.
H6: Cultural assumption and belief of healthcare practitioners has a positive relationship with security awareness.
H7: Trust of healthcare practitioners has a positive relationship with security behaviour.
H8a: The respondents' perceived susceptibility affects perceived threat.
H8b: The respondents' perceived severity affects perceived threat.
H9: Perceived threat of healthcare practitioners has a positive relationship with security behaviour.
H10: Self-efficacy of healthcare practitioners has a positive relationship with security behaviour.
H11: Perceived benefits of healthcare practitioners have a positive relationship with security behaviour.

## 8 Conclusion

This paper has demonstrated the factors that influence information security culture in health informatics. Three major factors that have been identified are knowledge, awareness and behaviour. These become the mediating factors influencing information security culture. This model will be used as guidance for the healthcare organization to manage their information security at the early stage in minimizing the risks of security incidents. It also becomes a guideline for the top management in creating information security culture among employees. This paper gives the different views on the solutions for security incidents by looking into the human factor. Hence, it becomes the responsibility of the management to ensure that they consider highlighting these factors to ensure the effectiveness of implementation of information technology in health informatics. It helps to give the patients the confidence that their data is secured. This in-progress work will look into the evaluation of the proposed conceptual model. This evaluation will be conducted with healthcare practitioners who are doctors, therapist, nurses and admin clerks. It will be held in selected healthcare organizations in Malaysia. This evaluation helps in refining and validating the conceptual model. This model also can be used to replicate the study in other organizations and other countries.

## References


1. Albrechtsen E (2007) A qualitative study of users' view on information security. Comput Secur 26(4):276–289. doi:10.1016/j.cose.2006.11.004
2. Bansal G, Zahedi F, Gefen D (2010) The impact of personal dispositions on information sensitivity, privacy concern and trust in disclosing health information online. Decis Support Syst 49(2):138–150
3. Carpenter CJ (2010) A meta-analysis of the effectiveness of health belief model variables in predicting behavior. Health Comm 25(8):661–669. doi:10.1080/10410236.2010.521906
4. Chia PA, Maynard SB, Ruighaver AB (2002) Understanding organizational security culture. In: Proceedings of Pacific Asia conference on information system 2002. Japan, pp 1–23
5. Da Veiga A, Eloff JHP (2010) A framework and assessment instrument for information security culture. Comput Secur 29(2):196–207. doi:10.1016/j.cose.2009.09.002
6. Detert JR, Schroeder RG, Mauriel JJ (2000) A framework for linking culture and improvement initiatives in organizations. Acad Manage Rev 25(4):850–863
7. Garg V, Brewer J (2011) Telemedicine security: a systematic review. J Diabetes Sci Technol 5(3):768–777
8. Gaunt N (2000) Practical approaches to creating a security culture. Int J Med Inform 60 (2):151–157
9. Gebrasilase T, Lessa L (2011) Information security culture in public hospitals: the case of Hawassa referral hospital. Afr J Inform Syst 3(3):72–86
10. Hersh W, Bhupatiraju R (2006) Adopting e-learning standards in health care: competency-based learning in the medical informatics domain. AMIA Annual Symposium Proceedings, pp 334–338
11. Humaidi N, Balakrishnan V (2012) The influence of security awareness and security technology on users' behavior towards the implementation of health information system: a conceptual framework. In: Proceeding of international conference on management and artificial intelligence, Singapore, 35:1–6
12. Katsikas S (2000) Health care management and information systems security: awareness, training or education? Int J Med Informat 2(1):129–135
13. Knapp K, Marshall T (2006) Information security: management's effect on culture and policy. Inform Manag Comput Secur 14(1):24–36. doi:10.1108/09685220610648355
14. Kraemer S, Carayon P, Clem J (2009) Human and organizational factors in computer and information security: pathways to vulnerabilities. Comput Secur 28(7):509–520. doi:10.1016/j.cose.2009.04.006
15. Lim JS, Ahmad A, Chang S, Maynard S (2010) Embedding information security culture. In: Proceedings of the PACIS 2010, pp 463–474
16. Meingast M, Roosta T, Sastry S (2006) Security and privacy issues with health care information technology. Conference proceedings. Annual international conference of the IEEE engineering in medicine and biology society. IEEE engineering in medicine and biology society. conference, vol 1, pp 5453–5458. doi:10.1109/IEMBS.2006.260060
17. Ng B-Y, Kankanhalli A, Xu Y (2009) Studying users' computer security behavior: a health belief perspective. Decis Support Syst 46(4):815–825. doi:10.1016/j.dss.2008.11.010
18. Parkin SE, van Moorsel A, Coles R (2009) An information security ontology incorporating human-behavioural implications. In: Proceedings of the 2nd international conference on security of information and networks, ACM, pp 46–55
19. Samy GN, Ahmad R, Ismail Z (2010) Security threats categories in healthcare information systems. Health Informat J 16(3):201–209. doi:10.1177/1460458210377468
20. Savastano M, Hovsto A, Pharow P, Blobel B (2008) Identity-management factors in e-health and telemedicine applications. J Telemed Telecare 14(7):386–388
21. Stahl B, Doherty N, Shaw M (2012) Information security policies in the UK healthcare sector: a critical evaluation. Inform Syst J 22(1):77–94

22. Stanton JM, Mastrangelo P, Stam KR, Jolton J (2004) Behavioral information security: two end user survey studies of motivation and security practices. In: Association for information system conference (AMCIS), pp 175–181
23. Thomson K, Von Solms R, Louw L (2006) Cultivating an organizational information security culture. Comput Fraud Secur 2006:49–50
24. Van Niekerk JF, Von Solms R (2010) Information security culture: a management perspective. Comput Secur 29(4):476–486. doi:10.1016/j.cose.2009.10.005
25. Von Solms B, Von Solms R (2004) The 10 deadly sins of information security management. Comput Secur 23(5):371–376. doi:10.1016/j.cose.2004.05.002
26. Vroom C, Von Solms R (2004) Towards information security behavioural compliance. Comput Secur 23(3):191–198. doi:10.1016/j.cose.2004.01.012
27. Whitman ME (2008) Security policy: from design to maintenance. Adv Manag Inform Syst 11 (2007):123–151
28. Williams P (2009) Capturing culture in medical information security research. Methodological Innovat Online 4(3):15–26. doi:10.4256/mio.2010.0003
29. Zakaria O, Gani A, Nor MM, Anuar NB (2007) Reengineering information security culture formulation through management perspective. Proceedings of the international conference on electrical engineering and informatics institut Teknologi Bandung, pp 638–641

# Media Conversion of Paralinguistic and Nonlinguistic Speech Information into Animated Texts for Foreign Language Learning

**Nur Syafikah Binti Samsudin and Kazunori Mano**

## 1 Introduction

In foreign language learning, the study of pronunciation and prosodic information has become important. For example, in Japanese language learning, foreign students often greatly lose naturalness of Japanese prosody [1]. According to the previous study, due to the difficulty in conveying the correct intention and nuance for the foreign language learners' pronunciation, they focused on what kinds of accent cause the greatest loss in naturalness in terms of Japanese pronunciation. Besides, a transcription method of utterance impressions is also proposed [2, 3]. Thus, for the purpose of achieving effective foreign language learning, a good understanding of linguistic information in speech such as intonations and rhythms for the learner is indispensable.

This paper proposes a speech visualization system of linguistic information in enhancing effective and interactive language learning for foreigners. We focus on how to visualize the learners' speech paralinguistic and nonlinguistic information in the animated texts effectively. Hence, in order to develop this visualization tool, the conversion process between speech attribute information and text attribute information is considered.

The paralinguistic information is defined as direct complementary information, which can be consciously controlled by the speaker. For example, this information contains the prosodic information (intonation, accent, rhythm, pause, or speed), speaker's intention, and attitude information. In contrast, the nonlinguistic information is the speaker's individual features (voice quality) or emotions, which cannot be consciously changed by the speaker. This nonlinguistic information is also described as indirect complementary information [4].

N.S.B. Samsudin (✉) • K. Mano
Systems Engineering and Science, Shibaura Institute of Technology, Graduate School of Engineering and Science, Saitama, Japan
e-mail: mf12044@shibaura-it.ac.jp; mano@sic.shibaura-it.ac.jp

K. Ab. Hamid et al. (eds.), *The Malaysia-Japan Model on Technology Partnership: International Proceedings 2013 of Malaysia-Japan Academic Scholar Conference*, DOI 10.1007/978-4-431-54439-5_18, © Springer Japan 2015

The animated texts are described as animation graphics using various font or color effects, dimensional graphics effects, and motion or animation effects. By choosing the appropriate mapping options between speech and text attributes, the speaker's intention can be expressed lively and effectively.

The rest of the paper is organized as follows. Section 2 explains the conversion process of speech attribute information into text attribute information. Section 3 explains the proposed system methodology, while Sect. 4 presents the evaluation results. Finally, we conclude the paper in Sect. 5.

## 2 Conversion of Speech Attributes into Text Attributes

The main process in the proposed speech visualization tool is the information combination process between the attributes of speech information (e.g., accent types, pause position, pitch) and the attributes of text information (e.g., font color, size, and text position). In this process, we apply the Scheffe's paired comparison method [5]. This method is used to set priorities where there are multiple possible matched pairs between the attributes of speech information and text information. The matching process example is shown in Fig. 1.

Based on Fig. 1, each attribute of speech information can be related to one or more attributes of text information, and the result of the paired comparison will be plotted as a graph in Fig. 2. The analysis example shows that the attributes of text information for visualizing the "Pause" speech information are ranked from lowest to highest comparison values. From the result, we can conclude that the "Phrase Position" is the most suitable text attribute information and the "Thinness" is the most irrelevant text information to visualize the "Pause" speech information in the animated text version.

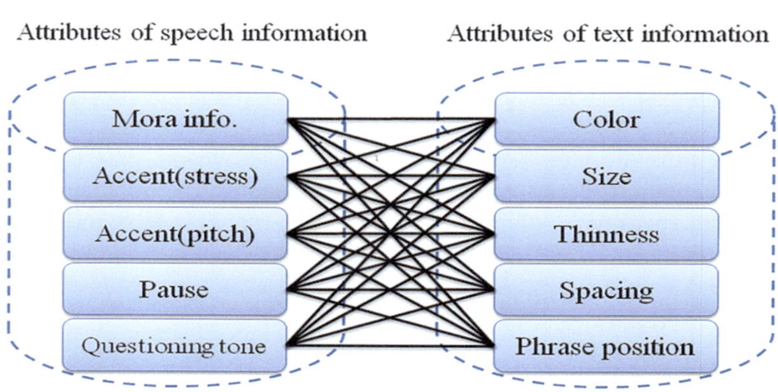


**Fig. 1** Speech-to-text mapping between attributes of speech and texts

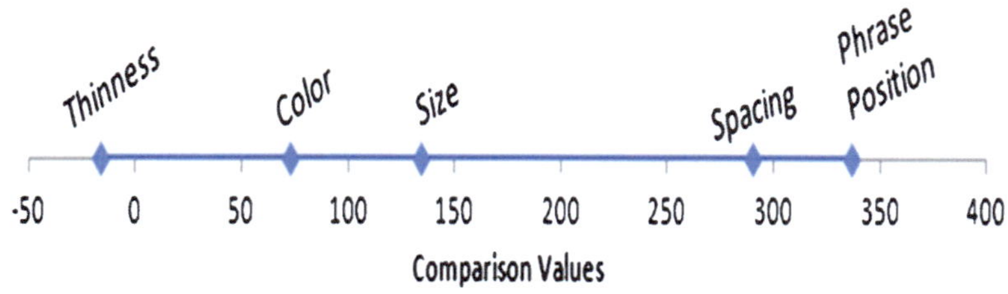


**Fig. 2** The obtained comparison values of the text attribute for "Pause" speech information

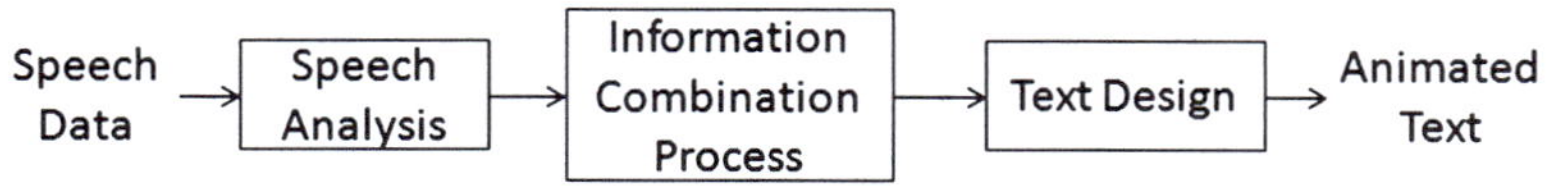


**Fig. 3** Block diagram of the proposed system

## 3 Outline of the Proposed System

The proposed text visualization system for paralinguistic and nonlinguistic information of speech is shown in Fig. 3 [6]. The speech data are used as the input of the system, and various features corresponding to speech attributes will be extracted in the speech analysis process.

In the speech analysis stage, the paralinguistic or nonlinguistic information contained in the speech will be analyzed and extracted. For example, the perception of the speaker's emotions is caused by many factors such as fundamental frequency (F0), sound intensity (SI), duration, or speech quality. Due to these factors, as the nonlinguistic information, the speaker's emotions can be analyzed into these speech features (F0, SI, etc.). Then, the analyzed speech attribute information will be matched with the suitable text attribute information in the information combination process based on the paired comparison method that has been mentioned in a previous chapter.

By using the information combination process results, the text design process will be conducted. The design will be created based on the matched text information data and will be divided into 3–5 styles for each speech attribute information. Then, as the final stage of the system, the created design will be visualized as the animated text. Furthermore, in order to evaluate the most suitable approaching text to represent the paralinguistic or nonlinguistic speech information, the user evaluation will be conducted. In the evaluation task, we are able to clarify the relation between speech information and text information.

# 4 Evaluation Experiment

## *4.1 Experimental Conditions*

An experiment was performed to evaluate the users' judgment and view of the animated text and also the effectiveness of this visualization tool. In this paper, as an example case, we focused on Japanese learning for Malaysian students because of its phonological structure which is mora-timed rhythm. The Japanese words are divided into mora, and each mora has the same length which is easy to be applied in the animated text's visualization timing. Moreover, the Japanese sentence structure occasionally has an ambiguous meaning because of dependency parsing, and due to this problem, the learners tend to misunderstand the speaker's intention. Therefore, as the experimental sample, we chose the Japanese ambiguous meaning sample speech.

Thirteen Japanese language learners (Malaysian) took part in the evaluation, nine of whom were females and four males. Their ages ranged from 20 to 26, with a mean age of 23. All the participants are foreign students in Japan with a Japanese-Language Proficiency Test N2–N4 level. We used two sample Japanese sentences that contain ambiguous meaning because of dependency parsing as described in Fig. 4. The speech paralinguistic and nonlinguistic information chosen from the sentences was mora information, accent, pause position, questioning tone, and speech intensity to determine the meaning and the correct pronunciation.

## *4.2 Animated Text Design*

Two data samples were visualized into three aspects which were accent information, pause position, and questioning tone. The evaluated design is shown in Fig. 5. For the accent element, the font color was applied in each mora on the sentences. Red color represented the high (H) pitch accent; however, blue color represented

| **Speech data1** | **Speech data2** |
|---|---|
| かのじょ　しんはつばい　かし　か<br>*彼女は新発売のお菓子とジュースを買った*<br>/ka/ /no/ jyo/ /ha/ /shi/ /n/ ha/ /tsu/ /ba/ /i/ /no/ /o/ /ka/ /shi/ /to/ /jyu/-/su/ /o/ /ka/ /tta/ | かれ　わら　はな　ゆうじん　み<br>*彼は笑いながら話す友人を見ていた。*<br>/ka/ /re/ /ha/ /wa/ /ra/ /i/ /na/ /ga/ /ra/ /ha/ /na/ /su/ /yu/ /u/ /ji/ /n/ /o/ /mi/ /te/ /i/ /ta/ |
| かのじょ　しんはつばい　かし　か<br>**a)** 彼女は新発売の / お菓子とジュースを買った。<br>(She bought the new release juice and snack.)<br>かのじょ　しんはつばい　かし　か<br>**b)** 彼女は新発売のお菓子と / ジュースを買った。<br>(She bought the juice and new snack.) | かれ　わら　はな　ゆうじん　み<br>**c)** 彼は / 笑いながら話す友人を見ていた。<br>(He saw his friend laughing while talking.)<br>かれ　わら　はな　ゆうじん　み<br>**d)** 彼は笑いながら / 話す友人を見ていた。<br>(He laughed while watching his friend talking.) |
| Different meanings : *新発売→お菓子とジュースorお菓子*<br>*new release→the juice and snack or snack* | Different meanings : *笑う→友人or彼*<br>*laughing→friend or he* |

**Fig. 4** Speech data sample

| | | |
|---|---|---|
| Accent information | Speech Data 1 | かのじょ は しんはつばい の おかし と ジュース を かった。<br>/ka//no/jyo/ /ha/ /shi/ /n/ ha/ /tsu/ /ba/ /i/ /no/ /o/ /ka/ /shi/ /to/ /jyu/-/su/ /o/ /ka/ /tta/<br>*'_' means (H) pitch accent |
| | Speech Data 2 | かれ は わらいながら はなすゆうじん を みていた.<br>/ka//re/ /ha/ /wa//ra//i//na//ga//ra/ /ha//na//su//yu//u//ji//n/ /o/ /mi//te//i//ta/ |
| Pause position | Speech Data 1 | かのじょ は しんはつばい の/ おかし と ジュース を かった。<br>かのじょ は しんはつばい の おかし と/ ジュース を かった。 |
| | Speech Data 2 | かれ は/ わらいながら はなすゆうじん を みていた?<br>かれ は わらいながら/ はなすゆうじん を みていた? |
| Questioning tone | Speech Data 1 | かのじょ は しんはつばい の おかし と ジュース を かった?<br>かのじょ は しんはつばい の おかし と ジュース を かった? |
| | Speech Data 2 | かれ は わらいながら はなすゆうじん を みていた?<br>かれ は わらいながら はなすゆうじん を みていた? |
| Questioning tone, and pause position combination | Speech Data 1 | かのじょ は しんはつばい の/ おかし と ジュース を かった?<br>かのじょ は しんはつばい の おかし と/ ジュース を かった? |
| | Speech Data 2 | かれ は/ わらいながら はなすゆうじん を みていた?<br>かれ は わらいながら/ はなすゆうじん を みていた? |

**Fig. 5** Text design for speech data 1 and speech data 2

the low (L) pitch accent. The text accents were determined according to NHK Accent Dictionary [7]. In order to identify the correct meaning of the speech data, the pause position recognition in the modification structure is important. Therefore, we located the pause position in the text caption with "/" symbol, and the last moras located at the pause position were highlighted with the same color of the pause symbol (green color). For the speech intensity, the user can identify it by viewing the pause position and also speech pitch accent elements. On the questioning tone element, we used the question symbol "?" at the end of the sentence. Lastly, we combined all three speech information elements in the same text caption.

## *4.3 Evaluation Results and Discussion*

We distributed a questionnaire among the Japanese learners (Malaysian students) to evaluate created text design based on appropriateness, effectiveness, and usefulness criteria. The learners were asked to complete sixteen questions about their understanding of the animated text design and their opinion of the various speech information elements. Six questions asked the learners about text elements such as text color, text size, symbol, and screen background based on the 5-point Likert scale where 5 was "strongly agree" and 1 was "strongly disagree." Besides, the other ten questions were asked to rate the learners' readability and understanding of

**Table 1** Means and standard deviations of technical text elements

| | | Speech data 1 | | Speech data 2 | |
|---|---|---|---|---|---|
| | | *M* | SD | *M* | SD |
| 1. | Text color | 3.85 | 0.69 | 4.00 | 0.82 |
| 2. | Text size | 3.85 | 0.69 | 3.54 | 0.78 |
| 3. | Text tones | 3.69 | 0.85 | 3.85 | 0.69 |
| 4. | Text position on screen | 4.08 | 0.76 | 3.92 | 0.76 |
| 5. | Symbol | 4.15 | 0.80 | 4.23 | 0.83 |
| 6. | Screen background | 3.85 | 0.90 | 3.77 | 0.93 |

**Table 2** Means and standard deviations of learners' readability and understanding

| | | Speech data 1 | | Speech data 2 | |
|---|---|---|---|---|---|
| | | *M* | SD | *M* | SD |
| 1. | Understanding of text element meaning (accent) | 3.92 | 0.86 | 3.85 | 0.90 |
| 2. | Understanding of text element meaning (pause position) | 4.31 | 0.75 | 4.38 | 0.77 |
| 3. | Understanding of text element meaning (questioning tone) | 4.15 | 1.07 | 4.46 | 0.66 |
| 4. | The text color effect is effective and relevant | 3.54 | 1.13 | 4.15 | 0.99 |
| 5. | Users can easily identify the modification structure (pause position) | 4.08 | 0.86 | 4.23 | 0.83 |
| 6. | Users can easily identify the stress-accented part (pause position) | 4.15 | 0.80 | 4.38 | 0.77 |
| 7. | Questioning tone symbol color | 4.08 | 0.95 | 4.00 | 0.82 |
| 8. | Question symbol size | 3.23 | 1.30 | 3.31 | 1.18 |
| 9. | The combination of each speech element is reasonable | 3.85 | 0.55 | 4.08 | 0.76 |
| 10. | The animated test can help your pronunciation learning | 4.08 | 0.95 | 4.23 | 0.83 |

the relationships between speech nonlinguistic information and text information. These questions were also based on the Likert scale with the same text element questions. From this evaluation task, we were able to clarify the relation between speech information and text information.

From the questionnaire analysis, we summarized the results in means (M) and standard deviations (SD) of the technical text elements and the learners' readability and understanding of the animated text. According to the mean and standard deviation results shown in Table 1, the questionnaire participants appeared to agree with the designed text elements on each speech data sample because each mean value score of text elements (text color, text size, etc.) is more than 3.0.

Moreover, from the learners' readability and understanding of the animated text results shown in Table 2, participants also appeared to have a good understanding and agreement in the three designed aspects. For example, as we can see in the *understanding of text element meaning* (*pause position*), the result shows the highest mean and standard deviation value ($M = 4.31$, SD $= 0.75$) in speech data 1 because participants appeared to understand the meaning of syntactic ambiguity sentence.

# 5 Conclusion

A media conversion system for foreign language learning is proposed. In the experiment, the participants examined eight animated text designs of two Japanese sentence samples those had various nuances and meanings depending on paralinguistic and nonlinguistic information. The participants' readability, impacts on the technical elements, and understanding of the animated texts were evaluated. Overall results showed that participants gave positive responses to the approach used for speech information visualization into the animated texts. Participants were able to understand the accent, pause position, and questioning tone information. This study suggests that there is more effective animated text which can be designed by considering the variation of the text information aspects and also can be evaluated by native speakers. Moreover, further research is required to evaluate the other language such as English or Malaysia language which are defined as stress-timed or syllable-timed languages in this speech visualization system.

# References


1. Kato S, Short G, Minematsu N, Hirose K (2011) Perceptual study on the effects of language transfer on the naturalness of Japanese prosody for isolated words. IEICE SP, Speech 110 (452):19–24
2. Masafumi N, Yasuo H, Shingo K, Akira I (2009) Analysis of utterance impressions in speech and text for indexing to transcriptions. IPSJ 50(2):460–468
3. Kimoto S, Taira K, Horiguchi Y, Kuroiwa S (2011) Transcription method to express utterance impressions. Acoustical society of Japan (research committee meeting), pp 439–442
4. Hirose K (2006) Prosody and spoken language information processing (in Japanese). Maruzen, p 2
5. Scheffé H (1952) An analysis of variance for paired comparison. J Am Stat Assoc 47:381–400
6. Samsudin NSB, Mano K (2013) Text visualization of paralinguistic and nonlinguistic information of voices. In: Proceedings of the 7th SEATUC Symposium, OS-2.3
7. NHK Publishing (1998) NHK new Japanese accent dictionary

# Analysis of Pancreatic Tissue Injury Caused by Ultrasonically Activated Devices

**Bik Ee Lau, Tao Gao, and Hideki Hayashi**

## 1 Introduction

In recent years, a couple of studies have shown that unexpected pancreatic fistula occurs more frequently in laparoscopic surgery compared with conventional open surgery for gastric cancer [1]. Pancreatic tissue could be injured during gastric cancer surgery because a significant part of regional lymph nodes to be resected are adjacent to the pancreas. Although minor pancreatic tissue injury rarely leads to pancreatic fistula, it could cause serious postoperative complications such as intra-abdominal bleeding and anastomotic leakages.

Causes of pancreatic fistula have been unclear in most cases of laparoscopic distal gastrectomy or total gastrectomy without splenectomy or pancreatico-splenectomy, so far. However, mechanical stress such as retraction or mobilization of the pancreas and thermal damages caused by ultrasonically activated devices (USADs) have been assumed to be associated with this complication[2]. Moreover, USAD employs ultrasonic energy to disrupt tissues by cavitation [3–5]; it could exert direct harmful influence onto the pancreatic tissues during the course of peripancreatic and suprapancreatic lymph node dissection of gastric cancer surgery. However, there have been no studies revealing the modality and severity of the pancreatic tissue damages caused by USAD in detail. In this study, we analyzed pancreatic tissue damages caused by USAD using an animal model and address the question of whether USAD could be the direct cause of pancreatic fistula after laparoscopic gastric cancer surgery.

B.E. Lau (✉)
Graduate school of Engineering, Chiba University, Chiba 2630022, Japan
e-mail: laubikee@gmail.com

T. Gao
Graduate School of Medicine, Chiba University, Chiba, Japan

H. Hayashi
Research Center for Frontier Medical Engineering, Chiba University, Chiba, Japan

K. Ab. Hamid et al. (eds.), *The Malaysia-Japan Model on Technology Partnership: International Proceedings 2013 of Malaysia-Japan Academic Scholar Conference*,
DOI 10.1007/978-4-431-54439-5_19, © Springer Japan 2015

# 2 Materials and Methods

## *2.1 Ultrasonically Activated Device*

USAD was developed by Amaral et al. and reported first in 1994. This device can be used for both cutting and coagulation of soft tissues and disseminates as a dissection modality in laparoscopic surgeries for gastrointestinal, urological, and gynecological diseases [6, 7]. USAD consists of a generator, handpiece, and blade applicator. The generator is a microprocessor-controlled high-frequency power supply device that delivers an alternating current to the handpiece. The latter houses the acoustic mount and ultrasonic transducer, which generates vibrations at its harmonic frequency of 55.5 kHz. These vibrations are transmitted along the handpiece to a blade, the tip of instrument. In this study, the scissors-type laparoscopic handpiece Harmonic Ace (ACE36J, Ethicon Endo-Surgery, LLC) was used. The output level of the generator was set to Level 5, according to the general usage of tissue coagulating dissection for clinical operations.

## *2.2 Animal Experiment and Analysis*

The experimental protocol was reviewed and approved by the Chiba University Institutional Animal Care and Use Committee. The pancreas of male, castrated 3-month-old domestic pigs weighing 30 kg was exposed under general anesthesia. In order to provide constant pressure and the contact area applied to the pancreas, a unique locking device is developed to fix the blade of the USAD securely during every creation of pancreatic damage (Fig. 1). The activation of the USAD was operated by a foot switch.

The tip or side aspect of the blade was set in contact with the pancreas, and the USAD was activated for either 1, 3, or 5 s (Fig. 2a, b). As a control, an electric scalpel was also activated with the tip of the blade in contact for the same duration as the USAD (Fig. 2c). Marking color papers were placed adjacent to the injury with a glue so as to localize the point afterward.

Total pancreatectomy was performed after attempts to make injuries on the pancreas, and the sites of injury were excised for tissue analyses. Excised specimens were fixed with 10 % formalin solution and embedded in paraffin blocks. Serial sections of 5 μm in thickness were made in the directions to reveal maximum depth of injuries (Fig. 3). Every set of two serial sections was tested for either routine checkup of the tissue structure with hematoxylin and eosin (HE) staining.

We also measured the maximum width and depth of the damaged area from the optical microscopic findings of the tissue.

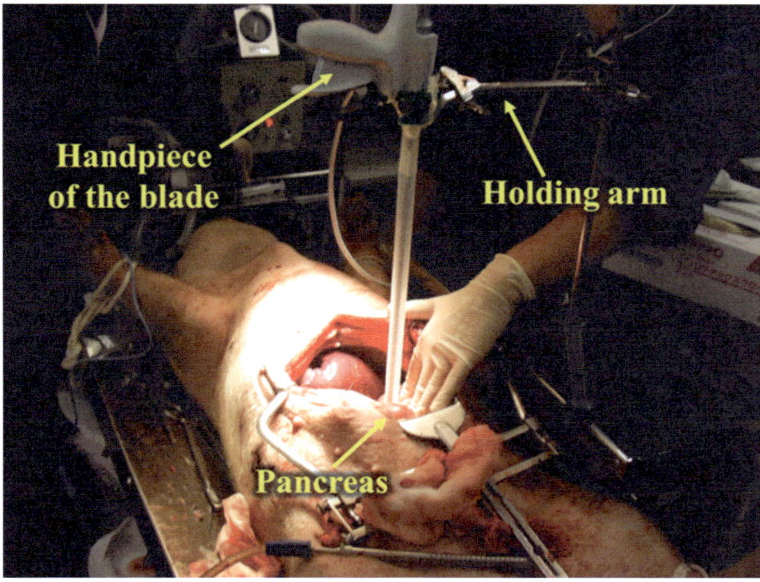


**Fig. 1** Animal experiment

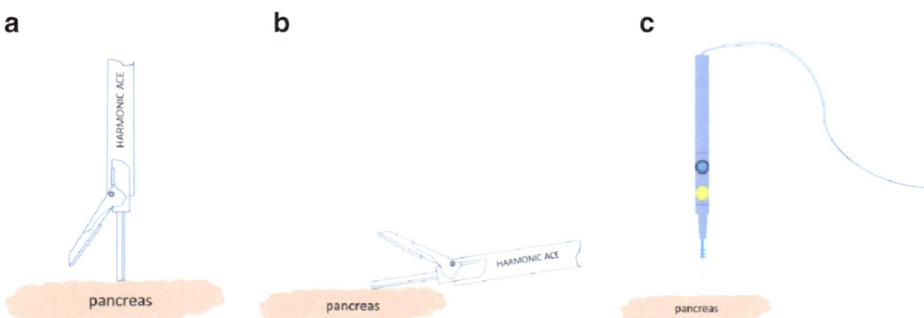


**Fig. 2** Conditions of creating tissue damage: (**a**) tip of the USAD's blade, (**b**) side aspect of the USAD's blade, (**c**) tip of electric scalpel

## 3 Results

First, macroscopic conditions of the pancreatic injuries were checked. Apparently, extent of the damages expanded as the activation time increased with both electric scalpel and USAD. However, bleeding occurred only at damaged sites made with the tip of the electrical scalpel.

Figure 4 shows the result of the histological changes made by the tip of electric scalpel with activation for 1, 3, and 5 s. HE staining showed that coagulation necrosis occurred around the site of activation in all specimens tested. Surface

Fig. 3 Setting for tissue sectioning

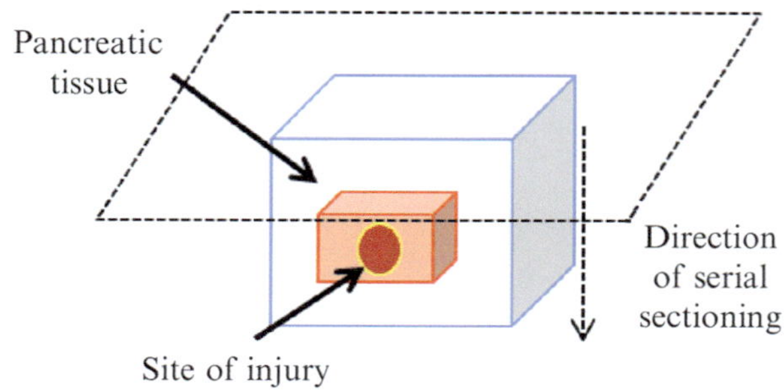


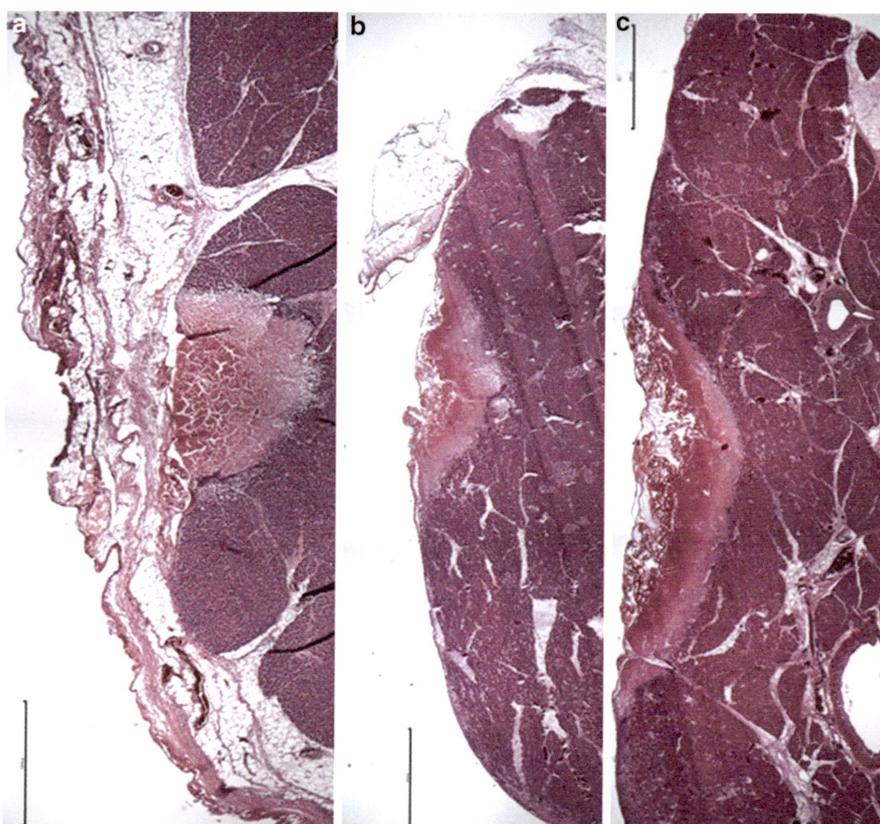


**Fig. 4** The result of the histological damage made by the tip of electric scalpel (HE staining): (**a**) 1 s, (**b**) 3 s, (**c**) 5 s

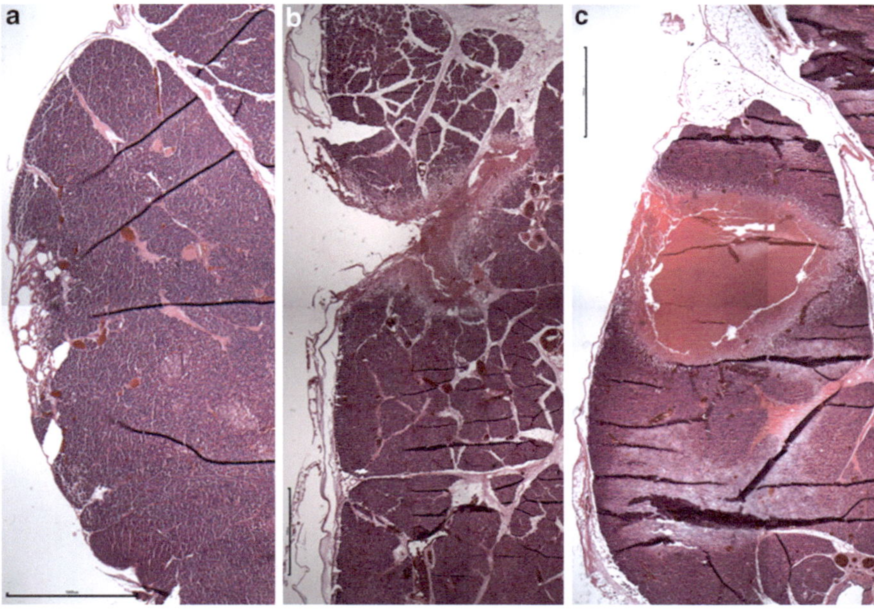


**Fig. 5** The result of the histological damage made by the tip of USAD (HE staining): (**a**) 1 s, (**b**) 3 s, (**c**) 5 s

extent of the damage expanded as the activation time increased; however, depths of the damages were similar regardless of durations of activation.

Figure 5 shows the results of the histological changes made by the tip of USAD's blade with the same conditions. Hollowing underneath the serosal layer which remained almost intact was observed in the specimen with 1 s activation. A significant part of the tissue around the activation site disappeared in the specimens with 3 or 5 s activation, and especially serous effusion was observed in the tissue defect of the specimen with 5 s activation. Coagulation necrosis was found in all specimens; however, protein clotting covering the damaged sites was hardly observed in any of these specimens. The surface width of the damages increased with the activation duration, but the depths of the damaged were similar in the specimens with either 3 or 5 s activation.

Figure 6 shows the results of the histological changes made by the side aspect of USAD's blade with the same activation durations. The serosal layer remained almost intact, and coagulation necrosis with focal effusion underneath was observed in the specimen with 1 s activation. Surface extent of the damage was much wider compared with electric scalpel or tip of the USAD, reaching 4.68 mm wide. Tissue damage observed in the specimen with 3 s activation showed similar findings, but coagulation necrosis expanded as deep as 2.53 mm from the surface with fat necrosis. The specimen with 5 s activation showed surface tissue defect along 6.88 mm and the similar degeneration finding underneath as that with 3 s activation. Protein clottings were hardly seen in all the damaged tissues, similar to the findings with specimens activated with the tip of the USAD.

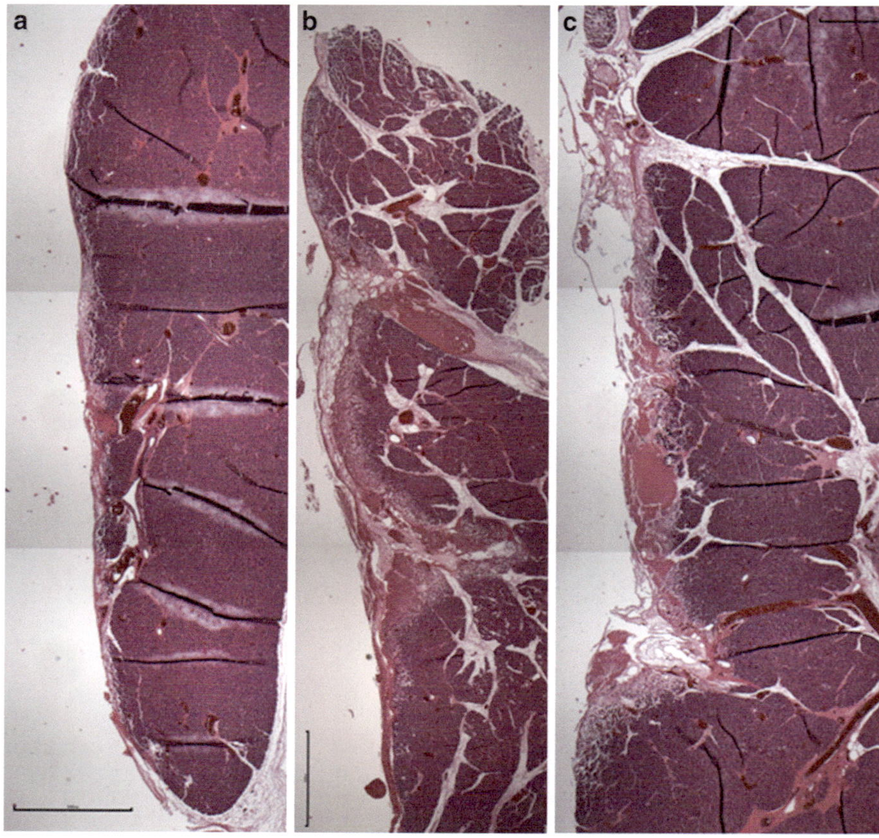


**Fig. 6** The result of the histological damage made by the side of USAD (HE staining): (**a**) 1 s, (**b**) 3 s, (**c**) 5 s

Figure 7 shows the maximum depth of the damaged areas for each activation duration setting. There were no significant differences observed in the depth of the damage by the activation durations or devices for damages. However, the surface extent of damages caused by the side aspect of USAD was the largest among all devices as shown in Fig. 8.

## 4 Discussion

Recently, the frequency of pancreatic fistula has been reported to be significantly higher after laparoscopic surgery compared with conventional open surgeries for gastric cancer. The most part of regional lymph nodes to be resected for gastric cancer are located around the pancreas; thus, energy devices, which are widely used

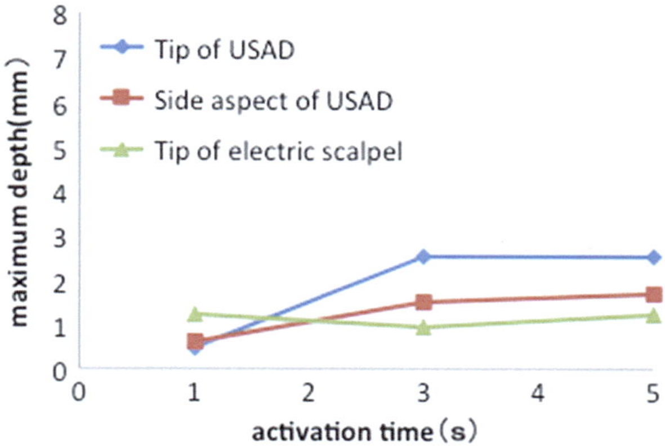


**Fig. 7** The maximum depth of damaged areas

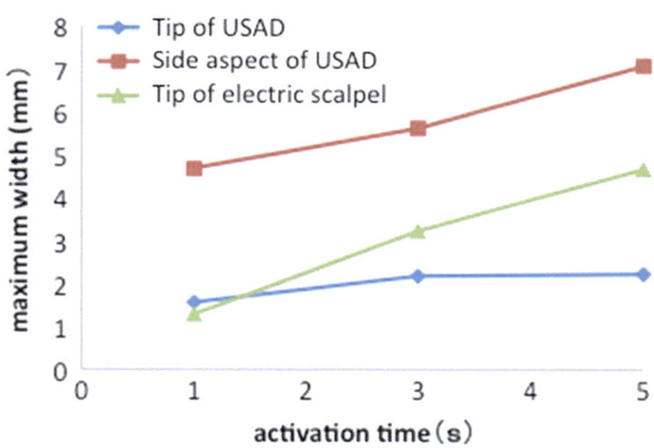


**Fig. 8** The maximum width of damaged areas

for lymphadenectomy in laparoscopic surgery, such as USADs, are allegedly one of the candidates that cause pancreatic tissue injury. However, there have been no detailed analyses to explore the direct effect of USAD activation to pancreatic tissues. In this study, we examined a direct impact of the USAD activation to the organ and compared the results with those from a conventional energy device using an animal model.

The experimental results (Figs. 4, 5, 6, 7, and 8) showed that the range of tissue damage expanded with activation duration with both devices. The operation of the electric scalpel degenerates the tissue in contact into protein clotting, and coagulation necrosis occurred in the vicinity of the tissue. On the other hand, the tissue injuries with USAD featured disruption of the tissue in contact with the blade, focal effusion in the same area, and coagulation necrosis of the surrounding tissue.

Although no disruption of pancreatic ductules was seen in the specimens, focal effusions observed in the damaged tissues with USADs presumably contain pancreatic secretion from acinar cells. In contrast, protein clotting covered the damaged tissues, and no focal effusion was seen in the specimens with activation by electric scalpel. These differences of findings could be consistent with the fact of higher incidence of pancreatic fistula after laparoscopic gastric cancer surgeries accompanied with frequent usage of USADs.

As for the USAD, surface extent of damages was much larger with the side aspect than that with the tip, because the area in contact with the tissue was larger. However, the depth of injuries was bigger with the tip than those with the side aspect of the blade. These data suggested that the power of cavitation to disrupt tissues is larger toward the tip direction than that toward lateral directions. Moreover, only a second activation results in coagulation necrosis of the pancreatic tissue, suggesting that the USAD should be used with its blade never in touch with the pancreatic tissues.

## 5 Conclusions

Biological effects on the pancreatic tissues which were caused by surgical energy devices were precisely investigated using an animal model. Consequently, differences in histological changes brought with conventional electric scalpel and USAD activation were elucidated. Further investigation is required to determine whether this characteristic of USADs leads to pancreatic fistula after gastric cancer surgeries. However, our data have shown that not only the tip but also the side aspect of the blade should be always away from the pancreas during lymphadenectomy for gastric cancer to achieve secure laparoscopic surgery.

## References

1. Obama K et al (2011) Feasibility of laparoscopic gastrectomy with radical lymph node dissection for gastric cancer: from a viewpoint of pancreas-related complications. Surgery 149(1):15–21
2. Jiang X, Hiki N, Nunobe S, Kumagai K, Nohara K, Sano T, Yamaguchi T (2012) Postoperative pancreatic fistula and the risk factors of laparoscopy-assisted distal gastrectomy for early gastric cancer. Ann Surg Oncol 19:115–121
3. Nduka CC, Poland N, Kennedy M, Dye J, Darzi A (1998) Does the ultrasonically activated scalpel release viable airborne cancer cells? Surg Endosc 12:1031–1034
4. Hashimoto S et al (2012) Analysis of tissue damage caused by ultrasonically activated device. JSES 15(2):175–181
5. Hachiya H et al (2012) Verification of generated cavitation of an ultrasonically activated scalpel. Jpn J Med Ultrason 39(2):101–111
6. Ogura G et al (2009) Development of an articulating ultrasonically activated device for laparoscopic surgery. Surg Endosc 23(9):2138–2142
7. Kinoshita T et al (1999) Experimental study on heat production by a 23.5-kHz ultrasonically activated device for endoscopic surgery. Surg Endosc 13:621–625
8. Maxwell C (1892) A treatise on electricity and magnetism, vol 2, 3rd edn. Clarendon, Oxford, pp 68–73

# The Integration of the Kano Model and SERVQUAL into the House of Quality for Developing Occupational Safety and Health Training Program

**Mohd Saiful Izwaan Saadon, Zainol Mustafa, and Khairul Za'im Kamarulzaman**

## 1 Introduction

A good employee is a worker who can perform excellently on the task given as stated in the Job Description and the Key Performance Index. In order to fulfill the Job Description and Key Performance Index, every employee should have adequate skills and knowledge. They can obtain the skills and knowledge by attending related courses which are provided by the employer or external trainer. A good course is a course that puts into consideration the needs of the customers and fulfills them. The customer in this context is the employer of the said worker. This study provides a new perspective on a new method of constructing skills and learning course based on the needs of the employer by using the house of quality (HOQ). In theory, the course model produced will grant maximum output to the employer. In order to measure the employer's satisfaction, it is suggested to use Kirkpatrick's Evaluation Model. This model measures the performance of the staff who has attended the training program, and it is also used to measure the results of the employer's investment by sending their staff to undergo the training program.

---

M.S.I. Saadon (✉)
Faculty of Science and Technology, Open University Malaysia,
40100 Shah Alam, Selangor, Malaysia
e-mail: saiful_saadon@oum.edu.my

Z. Mustafa
Faculty of Science and Technology, National University of Malaysia,
43600 Bangi, Selangor, Malaysia
e-mail: zbhm@ukm.my

K.Z. Kamarulzaman
Malaysian Institute of Industrial Technology, Kuala Lumpur University,
81750 Johor Bahru, Johor, Malaysia
e-mail: khairul.zaim@gmail.com

K. Ab. Hamid et al. (eds.), *The Malaysia-Japan Model on Technology Partnership: International Proceedings 2013 of Malaysia-Japan Academic Scholar Conference*,
DOI 10.1007/978-4-431-54439-5_20, © Springer Japan 2015

# 2 Literature Review

## 2.1 *Kano Model*

Kano et al. [20] developed a model for improvement and enhancement of a product or service. According to Kano, customer needs can be classified into three categories ([21, 22]); the categories are "must be," "one-dimensional," and "attractive." For the must be category, the need can be considered as the basic requirement for the product or services. Both the customer and the service or product provider agree on the importance of the needs in this category; they are expected but unspoken and unarticulated [7]. For this category of needs, the level of satisfaction for the customer will not raise above the normal level if the product or service provider is able to fulfill those needs, but the customer will feel very disappointed if the needs are not fulfilled. In other words while a low performance on such attributes leads to dissatisfaction in a customer, a high performance does not lead to satisfaction [6, 20, 30]. Meanwhile for the one-dimensional category, the needs that fall in this category are actually the expectation of the customer towards the product function or the services that have been provided by the service provider. These needs are very well expected, expressed, and articulated by the customer. The needs in this category can be expressed in a linear relationship, whereby if the customer needs are not fulfilled, the level of satisfaction will be low; on the other hand if the needs are fulfilled, the level of satisfaction will become high [36], and by providing for such needs, the product or service provider creates the expected quality for their product or services. It is important for the product and service provider to focus their resource to compete with their competitor in order for them to sustain themselves in the market. Last would be the attractive dimension. The needs that fall in this category are the customer needs that they themselves do not realize they need. In other words the needs are unexpected and unspoken. The relationship between the needs in this category and the customer satisfaction is a one-way linear relationship. If the product or service provider is able to provide the unexpected towards the customer, their level of satisfaction will increase; however, the satisfaction level will not be affected if the product or service provider attributes in this category are not performed well. In other words they lead to satisfaction of customers when present but do not lead to any dissatisfaction if not present [4], and by discovering such needs and expectations, and providing the customer with these, the product or service provider creates what can be called the attractive quality [20, 28]. By integrating the Kano model and SERVQUAL, we would be able to determine the service quality criteria that fall into the must be, one-dimensional, and attractive category. This information will be an important input towards creating a training course model.

## 2.2 SERVQUAL

Service quality is one of the main factors that can contribute to the success or failure of a manufacturing or service organization in today's competitive environment. Kuei and Lu [26] considered service quality as a critical determinant criterion for competitiveness. Compared to product quality whereby it can be easily determined, service quality is very intangible and qualitative. The customer has to undergo the service in order to determine the level of service provided to them. Parasuraman et al. [33, 34] suggest that in a service setting, customers judge quality by comparing their perceptions of what they receive with their expectations of what they should have received. Kim et al. [46] have determined two key elements in the attainment of high quality. The first one is the identification of customer service requirements and expectations whereby it is generally recognized that consumers evaluate the service they receive and their expectations are critically important in determining whether or not they are satisfied [5]. It can be simplified that the consumers' expectations are the key criteria to the quality of service that a firm delivers. The second key factor of service quality is customer perception. Zeithaml [48] suggests that the notion of perceived quality reflects the opinion of the customer regarding the superiority or global excellence of a product or service. Finally Parasuraman et al. [33, 34] suggest that service quality should be represented as the difference or gap between service expectation and actual service performance. They also suggest that service quality can be measure using the SERVQUAL scale that consists of a set of 22 questions built from the five SERVQUAL dimensions: reliability, assurance, tangible, empathy, and responsiveness. In this research we will use the modified SERVQUAL scale whereby we add two more dimensions to the current SERVQUAL dimensions: competence and content. These two dimensions are very crucial in determining the service quality for the training program. In the end we also add eight new questions to the 22 SERVQUAL questions making the total questions 30. We can use the set of 30 questions to determine the strength and weakness of the current training course model, and at the same time it can be integrated with the Kano model analysis to determine the training course must be, one-dimensional, and attractive criteria.

## 2.3 Quality Function Deployment (QFD)/House of Quality (HOQ)

Quality Function Deployment can be considered an outstanding matrix diagram that can be used as a powerful tool for product development. It involves the integration between different departments in an organization like the design department, quality department, manufacturing department, and even the marketing department. Griffin [13] considered QFD as an investment in people and information. It enables an organization to measure customer "wants" and map them against the engineering "how" in a way that highlights trade-offs and drives the product's design towards customer requirements [29]. QFD facilitates the growth and prosperity of a firm by developing an array of products that are attractive to existing and new customers [1, 8, 14]. Products designed with QFD may have lower production cost, shorter development time, and higher quality than products developed without QFD [12, 18, 35]. These benefits are attracting an increasing number of product development practitioners to the QFD methodology [1, 9, 11, 23]. Although manufacturing industries were the first to adopt QFD, service and government organizations are also using it in their efforts to improve performance [10, 15, 25, 41, 42]. Based on we can say that QFD is one of the most appropriate tools that can be used to develop a training course model using the customer requirements that we obtained using the integration of the Kano model and modified SERVQUAL. The new training course model will have all the necessary criteria that are needed to increase the level of satisfaction of the trainee.

## *2.4 Kirkpatrick's Evaluation Model*

Training evaluation is a very crucial step in determining the level of effectiveness for a training program. Kirkpatrick [24] had designed a model with four levels of evaluation. Those levels are reaction to the training, learning measures, behavior measures, and results. Steensma and Groeneveld [16] explain the method of evaluation for each level:

1. Reactions to the training: Trainees are asked if they enjoyed the training and if they have learned from it.
2. Learning measures: For example, if the purpose of a training program is to increase knowledge, an appropriate knowledge test should be used to determine whether the trainees have actually learned from the training. So, learning measures test retention of training materials.
3. Behavior measures: Behavior measures indicate the extent to which the training transfers to the job, to the workplace of the trainee.
4. Results: Results measures are used to show whether broad, often more long-term, organizational goals are attained through the training.

Measures used may vary from return on investment to lower sickness absenteeism or even reduction of turnover. The link between the training and such long-term results is, of course, often not clear. More often than not, long-term results are affected by multiple causes, and training may be only one of the many possible causes. Still, careful utility assessments and other large-scale evaluations are useful instruments to indicate the effectiveness of the training on this fourth level of evaluation. It is very important for us to use this evaluation model to measure the effectiveness of the newly developed training program and compare it with the results from the traditional training models.

## 2.5 *Occupational Safety and Health*

Occupational safety and health management system is an important part of ensuring that the safety and health of staffs are well maintained by the respective organizations. Nunez and Villanueva [19] find that the occupational safety and health (OSH) activities are never considered within the sources of the intellectual capital of the firm. However, it is widely recognized that safer and healthier workplaces are, along with competitive advantage, one of a firm's major objectives, as they can improve productivity, boost employees' morale, and reduce costs [43]. Normally the organization would expect that their staff would learn the aspect of OSH from their common sense and past experience. Shannon et al. [39] point out that it was the research into major catastrophes such as Union Carbide (India) or Challenger (USA), where conventional prevention worked, that highlighted an increasing need to device new workplace accident prevention models. Starbuck and Farjoun [40] clearly illustrate the need of supplementing traditional safety practices (inspections) with the management of organizational learning when describing the Columbia disaster in 2003. One of the best methods to increase the level of understanding with OSH principle is by sending organization staff to the OSH training program. There are several studies that have shown the correlation between organizational knowledge and OSH. The studies are diverse from different fields such as psychology [45, 47, 49], organizational management [3, 27, 37], or health studies [2, 44]. There are also researches and institutions that have been recently attracted by interrelations between OSH and the intellectual capital [32]. They provide some preliminary evidence on the role of organizational knowledge in building safer and healthier workplaces [19]. Schulte et al. [38] present the United States example on knowledge management in the field of OSH. They describe actions taken by some public initiatives such as the National Occupational Research Agenda (NORA) and several company-level real-life situations, for example, the Hewlett-Packard experience, where innovative knowledge management models are applied to OSH. Other significant initiatives include the Dutch knowledge infrastructure program for OSH practitioners [17] or the OSHAS 18001 standard that includes recommendations and norms for managing OSH knowledge in firms [19].

# 3 Research Outcome

A survey was conducted to 100 industrial workers in the state of Johor, Malaysia. Based on survey 1 the SERVQUAL dimensions have been modified whereby in the case of the training provider, there should be two more dimensions added to the original SERVQUAL dimensions. The new dimensions are trainer competency and course contents. Trainer competency is a very important dimension in training services. A highly competent trainer would contribute significantly towards customer satisfaction for a training program. Another dimension that needs to be focused on in training services is the course contents. It is important for the training

**Fig. 1** Modified SERVQUAL

| Dimensions | No of Questions |
|---|---|
| Tangible | 4 |
| Reliability | 5 |
| Responsiveness | 4 |
| Assurance | 4 |
| Empathy | 5 |
| Trainer Competency | 4 |
| Course Contents | 4 |

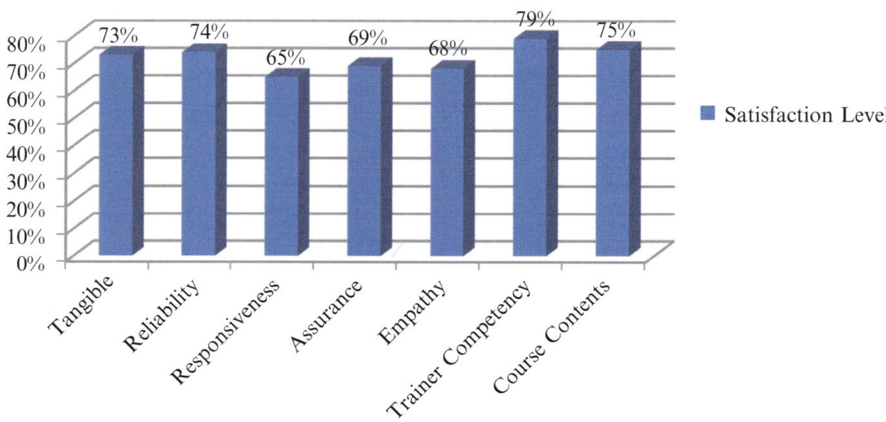


**Fig. 2** Training participant level of satisfaction for the traditional training program

provider to make sure that the training contents are aligned with the customer requirements so that the level of satisfaction for the training program would be at the maximum values. The SERVQUAL questionnaire has also been modified by adding eight more questions as a tool to measure the new dimensions. Below are the modified SERQUAL dimensions and the respective questions that are related to each one of it (Fig. 1).

The research continues by conducting another survey using the integration of the Kano model into the SERVQUAL questionnaire. The results have shown that two dimensions fall in the attractive categories (training competency and course contents), one dimension falls into the one-dimensional category (empathy), and four dimensions fall into the attractive categories (reliability, tangible, responsiveness, and assurance). The surveys also have shown the level of satisfaction for the traditional training program based on the modified SERVQUAL dimensions. The results are presented in Fig. 2: From the results we can say that the weaknesses in the traditional training program are responsiveness and empathy. These two dimensions must be improved in order to increase the level of satisfaction among the training participants. Next is the process of constructing a training program (in this case it would be an occupational safety and health training program) based on the findings above using HOQ. The program would focus on the attractive factors and

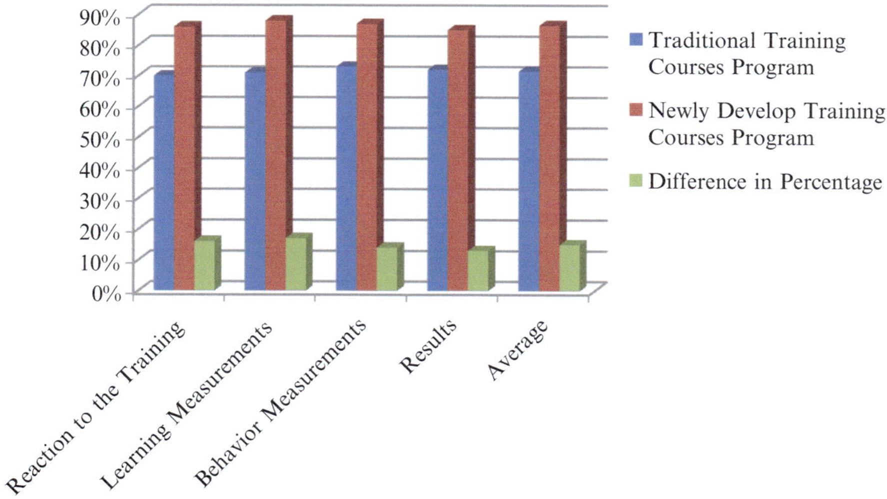


**Fig. 3** Comparison of the training model

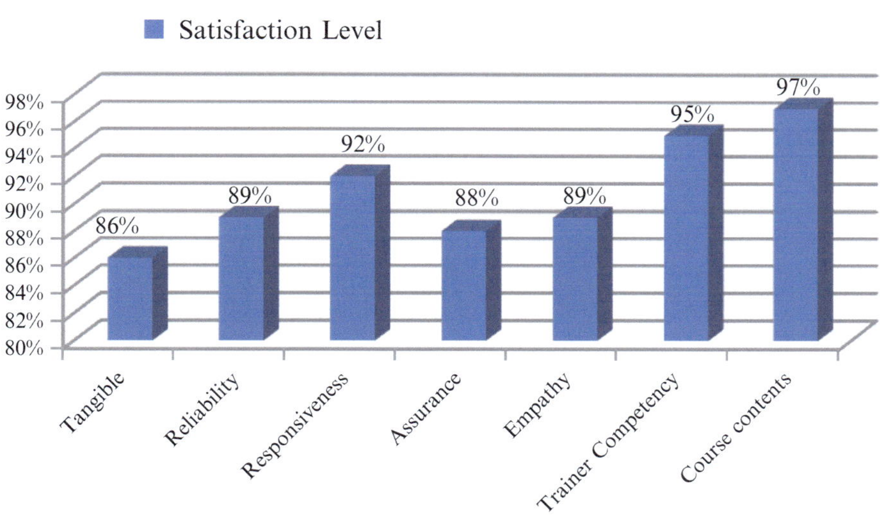


**Fig. 4** Training participant level of satisfaction for the newly developed training program

at the same time improve the weak factors. In the end a training program that can overcome the weakness of a traditional training program and provide an attractive input can be constructed and run by the training provider. An occupational safety and health training program is later conducted to 30 industrial workers. Survey 2 is then conducted after the participants have finished attending the training courses, and the results are as presented in Fig. 3. Survey 2 also needs to be able to determine the level of satisfaction for the newly developed training program based on the modified SERVQUAL dimensions. The results are presented in Fig. 4: From Fig. 3

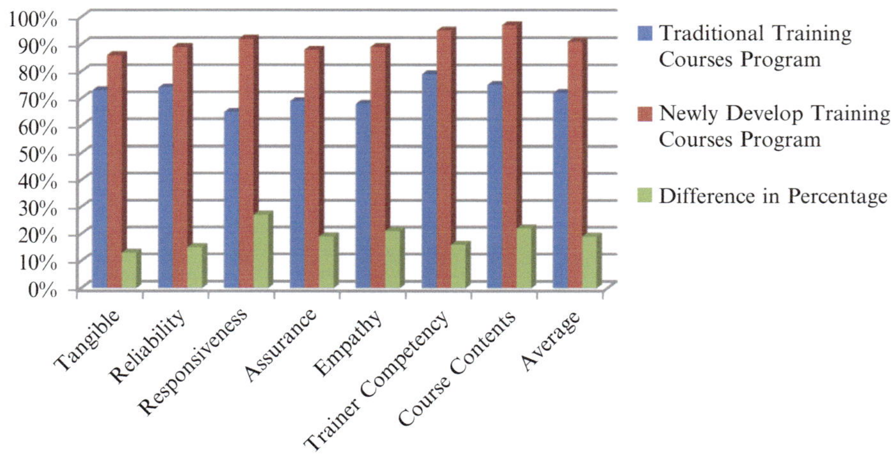


**Fig. 5** The traditional training program versus the newly developed training program

we can see clearly that the difference of performance between the ordinary training course model and the newly developed training course model is 15 %. This is a very significant value and can play a major factor for the staff performance and also for the company's return on investment in staff training. Based on Fig. 5, we may conclude that the level of satisfaction increases rapidly (19 %). This is proof that the newly developed training program is better than the traditional training program in every aspect of the modified SERVQUAL dimensions.

## 4 Conclusions

With the integration of the Kano model and SERVQUAL into HOQ, a training course model which can fulfill customer's needs and wants can be created. This new model can overcome a linear problem or the SERVQUAL model and can produce a training program which consists of unexpected aspects. This can be achieved by using the Kano model. The service provided also will obtain benefit through this model. With HOQ, the training provider can indentify needs to be prepared in order to complete the training course program. Finally we can see that by using this method, we can increase the level of reaction, learning, behavior, and results of the training participant.

# 5 Research Frameworks

Figure 6

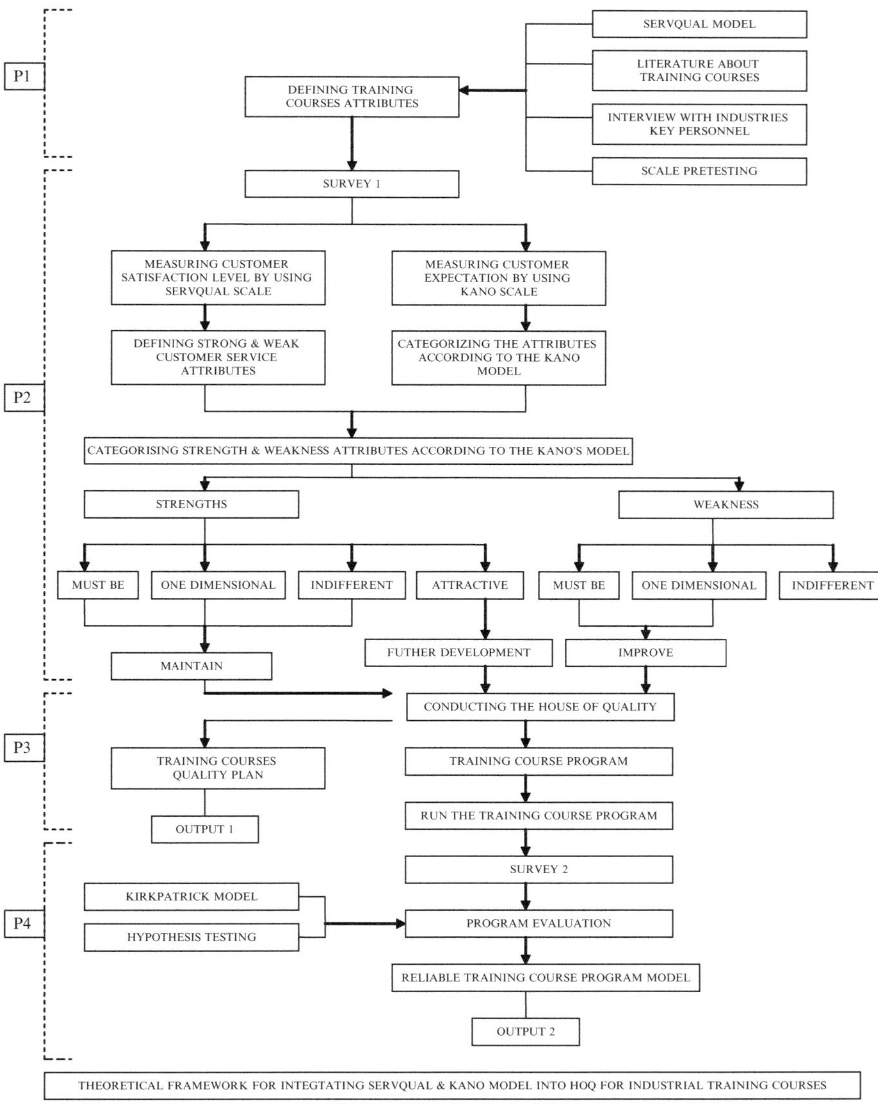


Source: Evaluation of Occupational Safety & Health Training Courses in Pasir Gudang & Senai Industrial Area, by M. S, Saadon, 2012, Johor Bahru, Johor: Universiti Teknologi Malaysia

**Fig. 6** Integration of the Kano model and SERVQUAL into HOQ to construct the training course model. *Source*: Saadon [31]

**Acknowledgment** The authors would like to acknowledge the financial support received in the form of fundamental research grant scheme FRGS/1/2013/SG04/UKM/02/4 from the Ministry of Higher Education Malaysia.

## References

1. Akao Y (ed) (1990) Quality function deployment. Productivity Press, Cambridge (English translation)
2. Amell TK, Kumar S, Rosser WJ (2001) Ergonomics, loss management and occupational injury and illness surveillance: a review. Int J Ind Ergon 28(2):69–84
3. Atherley G, Booth R, Kelly M (1975) Workers' involvement in occupational health and safety in Britain. Int Labour Rev 111(6):469–483
4. Berger C, Blauth R, Bolster C, Burchill G, Dumouchel W, Pouliot F, Richter R, Rubinoff A, Shen D, Timko M, Walden D (1993) Kano's methods for understanding customer-defined quality. Cent Qual Manag J 2(4):3–36
5. Brown SW, Swartz TA (1989) A gap analysis of professional service quality. J Market 53 (April):92–98
6. Busacca B, Padula G (2005) Understanding the relationship between attribute performance and overall satisfaction: theory, measurement and implications. Market Intell Plan 23(6):543–561
7. Cheng Lim P, Aquilano NJ, Jacobs FR (1999) An innovative framework for health-care performance measurement. Manag Serv Qual 9(6):423–433
8. Cohen L (1988) Quality function deployment: an application perspective from digital equipment corporation. Nat Product Rev 7(3):197–208
9. Ealey LA (1988) Quality by design. ASI Press, Dearborn
10. Garvin DA (1987) Competing on the eight dimensions of quality. Harv Bus Rev 6(November–December):101–109
11. Garvin DA (1988) Managing quality: the strategic and competitive edge. The Free Press Macmillan Inc, New York
12. Graessel B (1993) Using quality function deployment to improve customer service. Qual Prog 26(November):59–63
13. Griffin A (1992) Evaluating QFD's use in US firms as a process for developing products. J Prod Innov Manag 9:171–187
14. Hales R (1994) QFD: a key enabling technology in today's advanced product development environments. Ind Eng 26(12):10–11
15. Hauser JR, Clausing D (1988) The house of quality. Harv Bus Rev 66(3):63–73
16. Steensma H, Groeneveld K (2010) Evaluating training using the "Four Levels Model". J Work Learn 22(5):319–331
17. Hugenholtz NIR, Schreinemakers JF, A-Tajk MA, Van Dijk FJH (2007) Knowledge infrastructure needed for occupational health. Ind Health 45(1):13–18
18. Hunter MR (1994) Listening to the customer using QFD. Qual Prog 27(4):55–59
19. Nuñez I, Villanueva M (2011) Safety capital: the management of organizational knowledge on occupational health and safety. J Work Learn 23(1):56–71
20. Kano N, Seraku N, Takahashi F, Tsuji S (1984) Attractive quality and must-be quality. J Jpn Soc Qual Cont 14(2):39–48
21. Kano N (1995) Upsizing the organization by attractive quality creation. Proceedings of the First World Congress for Total Quality Management, 10–12 April 1995, Sheffield, pp 60–72
22. Kano N (2001) Life cycle and creation of attractive quality. Proceedings from Quality Management and Organizational Development (QMOD), Linkoping University, Linkoping
23. King B (1989) Better designs in half the time. GOAL/QPC, Lawrence, MS

24. Kirkpatrick DL (1994) Evaluating training programs: the four levels. Berrett-Koehler, San Francisco
25. Kogure M, Akao Y (1983) Quality function deployment and CWQC in Japan. Quality Progress, October, pp 25–9
26. Kuei CH, Lu MH (1997) An integrated approach to service quality improvement. Int J Qual Sci 2(1):24–36
27. Landsbergis P, Cahill J, Schnall P (1999) The impact of lean production and related new systems of work organization on worker health. J Occup Health Psychol 4(2):108–130
28. Lilja J, Wiklund H (2006) Obstacles to the creation of attractive quality. TQM Mag 18(1):55–66
29. Vonderembse MA, Raghunathan TS (1997) Quality function deployment's impact on product development. Int J Qual Sci 2(4):253–271
30. Matzler K, Hinterhuber HH (1998) How to make product development projects more successful by integrating Kano's model of customer satisfaction into quality function deployment. Technovation 18(1):25–38
31. Saadon MS (2012) Evaluation of occupational safety & health training courses in Pasir Gudang & Senai Industrial Area. Universiti Teknologi Malaysia, Johor Bahru, Johor
32. Nielsen C, Hussi T, Schunder-Tatzber S, Roslender R, Ahonen G (2007) The interrelations between health and intellectual capital. In: Johanson U, Ahonen GA, Roslender R (eds) Work health and management control. Thompson Fakta AD, Stockholm
33. Parasuraman A, Zeithaml VA, Berry LL (1985) A conceptual model of service quality and its implications for future research. J Market 49(Fall):41–50
34. Parasuraman A, Zeithaml VA, Berry LL (1988) SERVQUAL: a multiple-item scale for measuring consumer perceptions of service quality. J Retail 64(Spring):12–40
35. Raynor ME (1994) The ABCs of QFD: formalizing the quest for cost-effective customer delight. Nat Product Rev 13(3):351–357
36. Redfern R, Davey CL (2003) Supply chain market orientation in new product development in the UK: a pilot case study. J Fash Market Manag 7(1):65–77
37. Roy M (2003) Self-directed work teams and safety: a winning combination? Saf Sci 41(4):359–379
38. Schulte PA, Lentz TJ, Anderson VP, Lamborg AD (2004) Knowledge management in occupational hygiene: the United States example. Ann Occup Hyg 48(7):583–594
39. Shannon H, Robson L, Sale JM (2001) Creating safer and healthier workplaces: role of organizational factors and job characteristics. Am J Ind Med 40(3):319–334
40. Starbuck M, Farjoun M (2005) Organization at the limit: lessons from the Columbia disaster. Blackwell Publishing, Oxford
41. Sullivan LP (1986) Quality function deployment. Quality progress, June, pp 39–50
42. Sullivan LP (1988) Policy management through quality function deployment. Quality progress, June, pp 18–20
43. Thompson J (1997) Employee health programmes: a model designed for a local company. J Work Learn 9(2):83–87
44. Torp S, Moen BE (2006) The effects of occupational health and safety management on work environment and health: a prospective study. Appl Ergon 37(6):775–783
45. Van Vegchel N, De Jonge J, Landsbergis P (2005) Occupational stress in (inter)action: the interplay between job demands and job resources. J Organ Behav 26(5):535–560
46. Kim Y-P, Kim K-W, Yun D-G (2003) Exploration and development of SERVQUAL. Asn J Qual 4(1):116–130
47. Zacharatos A, Barling J, Iverson D (2005) High-performance work systems and occupational safety. J Appl Psychol 90(1):77–93
48. Zeithaml VA (1988) Consumer perceptions of price, quality and value: a means-end model and synthesis of evidence. J Market Res 52(July):2–22
49. Zohar D (1980) Safety climate in industrial organizations: theoretical and applied implications. J Appl Psychol 65(1):96–102

# An Urgent Need for Policy Change: Reform in Curriculum and Practice Through Values-Based Pedagogies and Cooperative Learning in the Curriculum in Malaysia, Japan and Thailand

**Suma Parahakaran**

## 1 Introduction

Current policies state that an integrated curriculum for education is implemented in many Asian countries, but there are no objectives and aims in the curriculum that provide students the opportunities for networking regarding social issues in their living contexts. Changes made to the curriculum are usually gradual although the input from various researches conducted by NGOs and formal bodies depicts the urgency in integrating sustainability issues for academic learning. Statistics show that 27–28 % of children in developing countries, i.e. South Asia and Saharan African, are underfed [15]. According to the report by the UNICEF, 20,000 children die out of poverty. In South Eastern Asia, emissions have doubled between 1990 and 2001 [19].

The national curriculum standards in Japan are renewed once in every 10 years, and in 2002 it was reported by Komatsu [8] that it was the seventh time the curriculum was renewed. The new basic education curriculum in Thailand was declared in June 2001 and enforced in 2002 based on the National Education Act passed in 1999. Similarly, the curriculum renewal was done in 2003 in Malaysia based on the 1996 Education Act [6].

In Thailand, educational policy emphasises on lifelong learning through teacher training, curricula, instructional media and information technology. Educational policies also include development of students with special needs and enhance learning in centres for sports and music and students and a green school. In Japan, although national curriculum standards are postulated by the National Ministry of Education, Culture, Sports, Science and Technology (MEXT), the powers have been also passed on to local authorities and schools so that they can decide what is the best curriculum for students [8]. The review of reports predicts

S. Parahakaran (✉)
Wawasan Open University, 54 Jalan Sultan Ahmad Shah, 10050 Penang, Malaysia
e-mail: sumjayan@gmail.com

K. Ab. Hamid et al. (eds.), *The Malaysia-Japan Model on Technology Partnership: International Proceedings 2013 of Malaysia-Japan Academic Scholar Conference*,
DOI 10.1007/978-4-431-54439-5_21, 

that there are studies of tangible issues related to sustainability which requires a drastic change [10] in the curriculum for helping students experience direct, experiential and problem-solving approaches.

The foundation and fundamental basis for curriculum policies in Malaysia provided the specific laws and postulates the law which is seen in the Education Act 550, Laws of Malaysia. These laws are generally applicable for all schools, and they provide directions to all curriculum documents.

## 1.1 Integrated Curriculum: Still a Remote Possibility

In the case of Malaysia, the policy stated that an integrated approach is taken in curriculum planning where content, skills and acquisition of knowledge and thinking skills are taken into account. The integration of moral values, attitudes and different ethnic languages is incorporated within the curriculum framework. The pressing issue faced by learners in all three countries is whether the curriculum gives them opportunities with real-life issues and whether it helps them in development of social processes required to meet the future needs where health, environment, economical and political issues are concerned. One of the significant facts that impacts sustainable education is the need for allocating budget for teacher training to help teachers to be equipped in skills related to teaching an integrated curriculum [12]. The report of the Curriculum Council in Japan in 1998 stated that a good relation between school and local communities is essential, thus highlighting the fact that more efforts have to be taken to link issues of local community and the school through a more integrated approach. This was one of the outputs during the proceedings of the Second UNESCO/Japan Seminar on Environmental Education in Asian-Pacific Region held in December 1998 between 9 and 11 in Tokyo.

A review of the curriculum in the Asian region revealed the fact that social issues are embedded in different ways in the curriculum design: in traditional disciplinary modes, interdisciplinary or multidisciplinary. Furthermore, analysis of the review revealed that integrated curriculum often resulted in student learning either fragmented or were not within the context. Students are then unable to have direct experiential learning with social issues in their local contexts or develop higher thinking skills for problem solving [17]. Urgent policy changes are therefore needed to be incorporated into the curriculum to *link schools and local communities* and *teach through a more integrated approach to tackle social issues*. The other pressing problem is to increase *student awareness of social issues through values-based teaching pedagogies*.

# 2 Why Values-Based Education for Sustainability?

In this article, the author stresses the need to include *values*-based sustainable education because educational policies do not include this as an agenda in their integrated curriculum. The inclusion of *values* for sustainable development in the curriculum will automatically lead to inclusion of specific pedagogies used for teaching in classrooms and ensures that students are provided learner-centred activities. Some of these pedagogies have been extensively researched in a study conducted in Thailand [5, 12], and assessment-based activities will provide evidence of sustainable practices.

## *2.1 Few Pedagogies Used Globally*

Some of the organisations have integrated teaching pedagogies into the curriculum. An example is the United Nations Educational, Scientific and Cultural Organisation Asia-Pacific Network for International Education and Values Education source book [18]. In the case of integration of values such as peace, the guidelines included integrating the cause and effects and different forms of conflict as well as religious, ethical and philosophical bases and their historical sources. The sample lesson plans included teaching pedagogies such as group activities, reflection, interactive and shared experiences. The main activity for a lesson on spirituality includes playing soft music, guided visualisation and sharing of ideas. Other activities included journal writing for helping students reflect and reinforce their ideas [18]. The teaching and learning strategies included experiential learning, storytelling, values education, enquiry learning, appropriate assessment, future problem solving, learning outside the classroom and community problem solving [18]. These different strategies have to be taken up ultimately by teachers in classrooms in order that the problems related to sustainability are addressed. Although many projects have been implemented according to the UNESCO reports, the differences in culture, religious practices and spiritual orientations have an impact on how the different pedagogies are implemented and may not be as easily extended to daily classroom teaching unless teachers themselves have personal relevance to them [12].

## *2.2 Values-Based Education for Enhancement in Personal and Community Experiences with Science, Technology and Social Issues*

In 2007, the UNESCO in cooperation with Earth Charter International discussed what values were inherent in sustainable development in the areas of learning which will encourage good behaviours. Several good practices were selected, and they included:

- Promoting learning more than teaching
- Concentrating more on the construction of meaning than on communicating the message
- Promoting the mutual transformation of teachers and learners
- Focusing on local knowledge and community-based decisions and actions
- Orienting learning more towards the process than product
- Use constructing and participating methodologies
- Use of multiple methods and integrative approaches: writing, arts, drama, debate and scientific analysis
- Context specific and action oriented
- Ongoing experiences that can also offer lessons learned

## 3 Pedagogical Implications for Sustainable Education in Multiethnic and Multicultural Asian Regions

Principles such as role modelling [5] and integrating values [5, 12] in the lessons are the starting points for bringing about change in students and an awareness of social justice as well as conservation values.

Reiss [14] stated that values education can be used broadly in two different areas. One is for personal empowerment, while the other is to help students gain values to help society and build community spirit. The aim according to Reiss [14] is to help in building autonomy and well-being and bring elements of justice in one's surroundings. In Southeast Asia, a hub of different religions and cultural practices and teachers' beliefs regarding their approaches to values-based education may have different implications for each community. Added to this, if the concepts are not clear and require time to adapt [3], teachers may lose motivation and require continuous guidance.

### *3.1 Why Values-Based Education for the Environment for Asia?*

Large organisations have implemented certain values and principles. It was noted that when specific values and principles were incorporated in the school mission and practised in an integrated mode in classrooms, there was positive transformation in a few schools in Southeast Asia [12]. For example, John Fien from the Australian School of Environmental Studies emphasised the notion of care and compassion for understanding peace and for nonhuman nature [4]. Eight values were proposed as a result of discussions by the Earth Charter, which formed the basis of 10 years of international intercultural and interfaith dialogue after Rio and is part of the 16 principles in the Earth Charter [4]. The values associated with ecological sustainability are *interdependence*, *biodiversity*, *living lightly on earth and interspecies equity* and for social sustainability are basic *human needs*, *intergenerational equity*, *human rights* and *participation*. Other organisations that include development of certain specific values have been found to be successful in the development of values such as respect, care and peace in an individual's attitudes. The first Earth Summit, the United Nations Conference on Environment and Development held in 1992, was a meeting of participative consultations of all kinds of international organisations, held in Rio de Janeiro, Brazil. The outcome of this meeting was an establishment of Agenda 21, which promulgated the principles for sustainability for the twenty-first century [20].

Of the 16 principles, Principle 1 included the following: (1) respect earth and life in all its diversity; (2) care for the community of life with compassion and love; (3) build democratic societies that are just, participatory, sustainable and peaceful and (4) secure earth's bounty and beauty for present and future generations.

The United Nations Decade of Education for Sustainable Development (2005–2014) in cooperation with the Earth Charter compiled good practices of the projects implemented by the Earth Charter, which had a successful impact in their own countries [20]. Countries such as Mexico, Spain and Brazil; Earth Scouts from the USA, Australia and Portugal and youth from Mexico, Sao Paolo, Queensland, India Florida and Seychelles were part of the working team. The activities used included pedagogies that were more reflective such as silent reflections, debates, storytelling, brainstorming and sharing ideas through participatory activities, role-play and audiovisual programmes.

The implementation of values education comes with different kinds of challenges, which are related to the training level and at the level of curricular implementation. The report of good practices as mentioned by the Earth Charter also included lessons that were learned during the project implementation [20].

In Asia, Sombath Somphone, the executive director of the PADETC, reported that some of the success factors are the inclusion of volunteer teachers from the community; integration of value concepts such as balancing head and heart, competencies, ethics and social responsibility in the curriculum and identification of real-life evidence-based learning. The sharing of teaching tools among teachers also influenced the success of the programme.

Education for Sustainable Development (ESD) reports from the Sri Lanka High Commission for UNESCO (SLNCU) in collaborative efforts with the Ministry of Education (MOE), established a project called Asia South Pacific–ESD clubs in Sri Lanka, which selected 42 Asia South Pacific schools. The project included increasing awareness in reduction of water and electricity consumption in schools and the inclusion of values such as peace and cultural diversity. However, until 2009, the report stated that the principals and teachers have not been trained and that future plans include teacher training to promote core values such as peace, respect for others, positive thinking, learning to live together and nonviolence, and the assessment for such projects is yet to be established. In Asian regions the role of assessment for affective learning for environmental education does not exist in schools because this discipline is not incorporated in the mainstream education according to the report by Bhandari and Abe [1].

## *3.2 A Need to Include Assessment for Values-Based Sustainable Education in the Curriculum*

There are mixed views regarding the effectiveness of values-based education relating to assessment procedures. Ratcliffe [13] and Hilderbrand [7] critically argued that not all teachers consider science education to integrate values in their teaching. Ratcliffe and Hilderbrand indicated that interviews of teachers in a study conducted by Levinson and Turner showed evidence that when teachers from both science and humanities backgrounds were asked their opinion regarding engaging students in discussion based on social and ethical aspects in their teaching, there were divided opinions. Half of the teachers were not ready to embrace the discussions of social conflict in their classrooms. Findings from this study showed that there was evidence that teachers' pedagogical beliefs had an impact on new reform implementations.

A report by Ratcliffe and Hilderbrand [7, 13] stated that if teachers and students were given time to reflect their own value stance and to reason out critical arguments, there may be a positive impact. Furthermore, evidence from the review of studies claims that when teachers were provided with assessment-based teaching using summative items, there was more evidence of reasoning and the provision of teachers to integrate student outcomes. According to the report, students who are provided with structured instruction using the SOLO taxonomy could perform better and reported significant differences. Reports by Ratcliffe and Hilderbrand state that there are difficulties for science teachers in modifying their pedagogies to integrate values explicitly and to include the summative and formative assessment in classroom tasks. On the other hand, Stevenson [16] argued that environmental education has to inform learners to improve the quality of life on the planet, alleviate exploitation and avoid social injustice; the notion of providing credentials to students is not effective. The reason for this is because it teaches students to excel in individual achievements with no collective responsibility when environmental agendas refer to cooperative learning and participatory approaches [16]. As a result, Stevenson argued that students remain passive spectators and are recipients of the knowledge of others rather than being rational thinkers and, instead of allowing learning and knowledge to go hand in hand, the practice is to provide theoretical knowledge to students before their actual practice. This according to Stevenson calls for changes in the structure of the organisation of schools and the professional ideologies that underlie teachers' organisation and knowledge transmission. The school curricula are predefined and serve predetermined behaviourally specific ends, which contradict the aims of environmental educational goals. Furthermore, Osterman [11] stressed that teacher accountability can impact any implementation, specifically the evaluation of affective education. From his review of studies, Osterman stated that when there are high stakes testing methods involved, teachers automatically start controlling, and this takes away the moments where teachers and students are able to respond and relate to each other for enquiry-based learning. The findings from the review of studies that involved federal, state and local policy on instructional roles revealed that teacher–student relationships were impacted and high stakes testing pressured teachers to produce high student achievement. This prompts teachers to rely on blame, praise and teacher-directed teaching that impacted student motivation.

Reflecting critically, Buissink-Smith et al. [2] stated that measuring affective learning is complex and it is still important to measure students' values, attitudes and behaviours for sustainable development. This is because it is the affective learning outcome that actually promotes values for future decision-making for students as they decide what kind of choices they need to make to contribute to society with their knowledge. Their decision-making is drawn more from their affective attributes rather than cognitive, and it has an impact on their learning outcomes and their future contribution to the society. The authors suggested that assessment could be conducted from quantitative and qualitative methods to collect data from both primary and secondary sources over a short and long time frame. This was based on earlier works of value proponents such as Krathwohl et al. [9]. These included the five hierarchical stages as proposed by Krathwohl et al. [9]. They include the ability to receive, to respond, to value, to organise and to internalise.

Some evidence is seen from reports from the Bahai community, which were funded by the EU provided below. Dahl reported on a few projects that were implemented and funded by the EU with the indigenous areas in Mexico. The project named *Development of Indicators and Assessment Tools for CSO Projects Promoting Values-Based Education for Sustainable Development* is affiliated with the Earth Charter. The report presented at the PERL International Conference at Istanbul, Turkey, stated that case studies have shown the effectiveness in contexts of indigenous children in Mexico. The Echeri Consultores is an organisation, which is affiliated with the Earth Charter and implemented the project on 9–13-year-olds in 15 schools. The activities included art workshops, guided reflection on local ecosystems and tree planting. The second sample included another multicultural group of 19 aged 12–21. The Echeri group chose 12 indicators that related to *collaboration in diversity* and 10 indicators relating to the value of *care* and *respect for community life*. There were various reasons that contributed to the effectiveness of the assessment. First, the youth publicly expressed their views by making their values visible, and the experiential learning helped them understand one another. The assessment was based on fun-filled activities, and students were given two-page questionnaires to fill out. The findings from the measurements were related to the results from a sample group of 1,500 children in schools of the area. From the review of literature on values and assessment, it can be understood that assessment for values-based education can be made effective if it is implemented in a fun-filled way for students using multiple strategies.

# 4 Conclusion

This article provides a small overview of why educational policies must incorporate curricular changes to help teachers teach for sustainable education where social issues are concerned. With rising issues in technology divide, lack of ethics and values with scientific issues, rising population and poverty in overpopulated Asia, the problems are multifaceted. The soft pedagogies will help students develop reflective practices and communal experiences which may help them in problem-solving issues in future.

# References

1. Bhandari B, Abe O (2000) Environmental education in the Asia and Pacific region: some problems and prospects. International Reviews for Environmental Strategies 1(1): 57–77. Retrieved from http://enviroscope.iges.or.jp/modules/. . ./405/. . ./057_077_bhandari_abe.pdf
2. Buissink-Smith N, Mann S, Shephard K et al (2011) How do we measure affective learning in higher education? J Educ Sustain Dev 5(1):101–114. doi:10.1177/097340821000500113
3. Clark N (2005) Water for African cities – Phase II, summary, conclusion, recommendations of the external evaluation Report of Phase 1: values based water, sanitation and hygiene strategic meeting. Libreville, Gabon, 18, 19, 20 January 2005. United Nations Human Settlements Programme (UN-HABITAT) The African Institute of Sathya Sai Education (TAISSE), pp 34–44
4. Fien J (2003) Learning to care: education and compassion. Griffith University, Brisbane. Retrieved from http://www.griffith.edu.au/__data/assets/pdf_file/. . ./breakfast-2003-fien.pdf
5. Jumsai A (2003) A development of the human values integrated instructional model based on intuitive learning concept. Doctoral Thesis, Bangkok, Chulalongkorn University
6. Hamid R (2005) Seminar on growth strategies for secondary education in Asia. Kuala Lumpur, Malaysia, 19–21, September, 2005. Co sponsored by the World Bank, the Asian Development Bank (ADB) the Southeast Asian Ministers of Education Organization (SEAMEO) hosted by the Government of Malaysia
7. Hilderbrand G (2007) Diversity, values and the science curriculum. In: Dillon J, Gunstone R, Corrigon D (eds) The re-emergence of values in science education. Sense Publishing, Rotterdam, pp 45–60
8. Komatsu S (2002) Transition in the Japanese curriculum: how is the curriculum of elementary and secondary schools in Japan determined? International Education Journal 3(5): 2002, WCCES Commission 6: Special 2001 Congress Issue, Japanese Education in Transition. http://www.flinders.edu.au/education/iej
9. Krathwohl DR, Bloom BS, Masia BB et al (1964) Taxonomy of educational objectives, the classification of educational goals – Handbook II: affective domain. McKay, New York

10. Nomura K, Osamu A (2013) Sustainability and higher education in Asia and the Pacific. Retrieved from http://www.guninetwork.org/resources/he-articles. Accessed 6 Jan, 2013
11. Osterman KF (2010) Teacher practice and students' sense of belonging. In: Lovat T, Toomey R, Clement N (eds) International research handbook on values education and student wellbeing. Springer, Dordrecht, Netherlands, pp 239–260
12. Parahakaran S (2013) Human values-based water, sanitation and hygiene education: a study of teachers' beliefs and perceptions in some Southeast Asian Countries. Doctoral Thesis. Faculty of Education and Social Work. Doctoral Thesis. University of Sydney
13. Ratcliffe M (2007) Values in the science classroom – the 'enacted' curriculum. In: Corrigan D, Dillon J, Gunstone R (eds) The emergence of values in science education. Sense Publishing, Rotterdam, pp 19–132
14. Reiss M (2007) What should be the aim(s) of school science education? In: Corrigan D, Dillon J, Gunstone R (eds) The re-emergence of values in the science curriculum. Sense Publishers, Rotterdam, pp 15–28
15. Shah A (2013) Poverty facts and stat. http://www.globalissues.org/article/26/poverty-facts-and-stats
16. Stevenson RB (2007) Schooling and environmental education: contradictions in purpose and practice. Environ Educ Res 13(2):139–153, http://dx.doi.org/10.1080/13504620701295726
17. Po S-c, Lo T-Y (2008) Developing an integrated approach in education: a case study of two junior secondary schools in Hong Kong. Complications and implications. In: Grossman DL, Lo JT-Y (eds) Social education in Asia: critical issues and multiple perspectives. Information Age Publishing, Charlotte
18. UNESCO-APNIEVE (1998) Learning to live together in peace and harmony: values education for peace, human rights, democracy and sustainable development for the Asia Pacific Region: a UNESCO-APNIEVE sourcebook for teacher education and tertiary level education. Bangkok: UNESCO PROAP. Retrieved from http://unesdoc.unesco.org/images/0011/001143/114357eo.pdf
19. United Nations (2007) The Millennium Development Goals Report. Retrieved from http://www.un.org/millenniumgoals/pdf/mdg2007.pdf
20. UNESCO-Earth Charter International (2007) United Nations decade of education for sustainable development: international implementation scheme, Paris: UNESCO. Retrieved from http://www.earthcharterinaction.org/invent/images/uploads/Good%20Practices.pdf

# Part IV
# Electronic and Computer Engineering

# The Effects of Doping Bismuth on the Thermoelectric Properties of $Zn_4Sb_3$

Afiqa Binti Mohamad and Atsumi Hisao

## 1 Introduction

Thermoelectric materials have attracted substantial interest since the last decade due to the growing demand for cooling electronic devices and reuse of waste heat for power generation. $\beta - Zn_4Sb_3$, with a complex hexagonal structure, is one of the promising thermoelectric materials. The figure of merit $ZT(=S^2T/\rho\lambda)$, where $S$, $T$, $\rho$ and $\lambda$ are the Seebeck coefficient (or thermopower), the absolute temperature, the electrical resistivity and the total thermal conductivity, respectively, of $\beta - Zn_4Sb_3$ is the highest among known thermoelectric materials in the temperature range of 323–573 K [1].

Doping is an effective way to improve the thermoelectric properties by changing the carrier concentration and/or reducing the thermal conductivity. Many empirical attempts have been made to improve the properties by doping guest atoms into the structure. The initial report of $Zn_4Sb_3$ thermoelectric properties by Caillat et al. reported an improved thermoelectric performance of a Cd-doped sample with ZT reaching 1.4 at 523 K for $Zn_{3.2}Cd_{0.8}Sb_3$ [2]. Tsutsui et al. [3] found that indium is able to substitute for part of Zn to form $(Zn_{1-x}In_x)_4Sb_3$ such that the carrier concentration of $\beta - Zn_4Sb_3$ is decreased by the In doping. Both the electrical resistivity and Seebeck coefficient of the In-doped $\beta - Zn_4Sb_3$ sample increase with increasing In content, while the thermal conductivity is reduced. As a result, $(Zn_{0.995}In_{0.005})_4Sb_3$ shows a higher ZT than the binary $\beta - Zn_4Sb_3$, indicating an improvement of the thermoelectric property with slight amount of In doping. Qin et al. [3] further investigated the thermoelectric properties of $(Zn_{0.98}M_{0.02})_4Sb_3$ (M = Al, Ga and In) at low temperatures ranging from 5 to 310 K. Their results indicated that the substitution of M (=Al, Ga and In) for Zn

A.B. Mohamad (✉) • A. Hisao
Kinki University, 3-4-1 Kowakae HigashiOsakashi, Osaka 577-8502, Japan
e-mail: fiqa0707@yahoo.com

K. Ab. Hamid et al. (eds.), *The Malaysia-Japan Model on Technology Partnership: International Proceedings 2013 of Malaysia-Japan Academic Scholar Conference*, DOI 10.1007/978-4-431-54439-5_22, © Springer Japan 2015

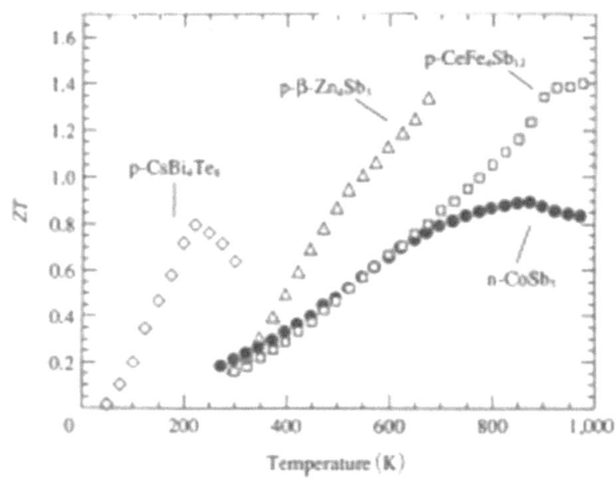


**Fig. 1** $Zn_4Sb_3$ alloy has the best figure of merit in the temperature range between 200 and 600 K

in $Zn_4Sb_3$ led to a decrease in electrical resistivity, thermopower and thermal conductivity. They also found that the $\beta \rightarrow \alpha$ phase transition of $Zn_4Sb_3$ was prohibited completely by doping with either Ga or In. The anomalies that appeared in the plots of $\rho \rightarrow T$ and $S \rightarrow T$ for $Zn_4Sb_3$ associated with the structural transition from β- to α-phase were weakened by doping with Al. Although the thermoelectric figure of merit of $\beta - (Zn_{0.98}Al_{0.02})_4Sb_3$ was reported to be slightly enhanced as compared to that of $\beta - Zn_4Sb_3$ at room temperature, further improvement in the thermoelectric performance is still of vital importance to achieve better ZT values. So looking at Fig. 1 shows that $Zn_4Sb_3$ has the best figure of merit in the temperature range between 200 and 600 K [4].

In this study, a series of bismuth compounds were fabricated. The substitution on Sb site was effective to modulate the power factor and it was expected to reduce thermal conductivity. The effect of electronic Bi at Sb site on the thermoelectric properties has been investigated in the temperature range from 323 to 573 K.

## 2 Experimental Procedures

Bismuth-doped compound $Zn_4(Sb_{3-x}Bi_x)$ ($x = 0.1, 0.2, 0.3, 0.4, 0.5$) was prepared from elements of zinc (grains, 99.9 %), stibium (grains, 99.999 %) and bismuth (shot, 99.9 %) in stoichiometric proportions. The mixtures of constituent elements were sealed into the glass ampule under vacuum for 1 hours. The glass ampule was heated in an electric furnace and held at 973 K for 1 h and cooled down to room temperature spontaneously. The X-ray diffraction was used to check the phase purity. The powders were then pressed 4 t/cm$^2$ for 5 min in an oil press machine to form bulk samples. The samples were then put in the aluminium board and inserted into the electric furnace for 1 h. The samples were measured for the electrical conductivity, Seebeck coefficient and thermal diffusivity.

## 3 Results and Discussion

The temperature dependence of Seebeck is shown in Fig. 2; the positive values indicate that all the samples are p-type conduction. The α-values of $Zn_4Sb_3Bi_{0.1}$ are very close to that of the undoped one in the temperature range 323–423 K, while the α-values of $Zn_4Sb_3Bi_{0.1}$ are the highest in the temperature range 463 K.

The temperature dependence of the electrical resistivity is shown in Fig. 3. The electrical resistivity decreases with increasing temperature. The electrical resistivity of the compound $Zn_4Sb_3Bi_{0.1}$ is close to that of the undoped $\beta - Zn_4Sb_3$ from the temperature range 323–523 K.

Based on the data of the Seebeck coefficient and electrical resistivity achieved above, the thermoelectric figure of merit ZT is shown in Fig. 4. At the temperature range 323–433 K, $Zn_4Sb_3Bi_{0.1}$, $Zn_4Sb_3Bi_{0.3}$ and undoped $\beta - Zn_4Sb_3$ are close, but at temperature 483 K, $Zn_4Sb_3Bi_{0.1}$ suddenly increases due to the improvement of the thermoelectric power factor and decrease of electrical resistivity. The $Zn_4Sb_3Bi_{0.1}$ reaches a highest ZT at the temperature 573 K.

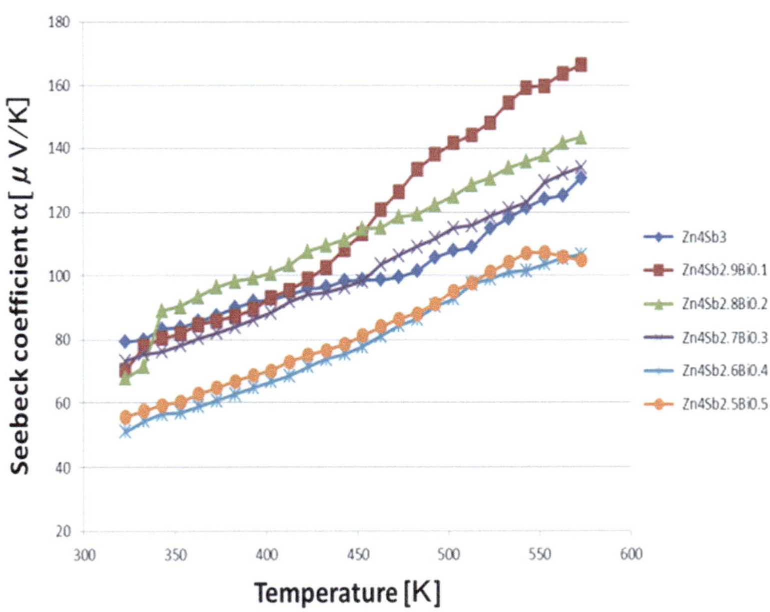


**Fig. 2** The result of the figure of merit on the $Zn_4(Sb_{3-x}Bi_x)$ ($x = 0.1, 0.2, 0.3, 0.4, 0.5$)

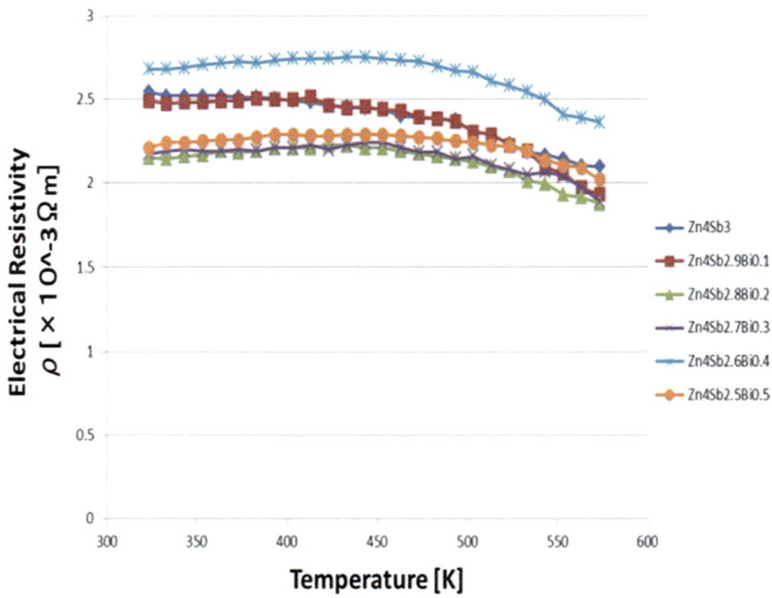


**Fig. 3** The result of electrical resistivity on the $Zn_4(Sb_{3-x}Bi_x)$ ($x = 0.1, 0.2, 0.3, 0.4, 0.5$)

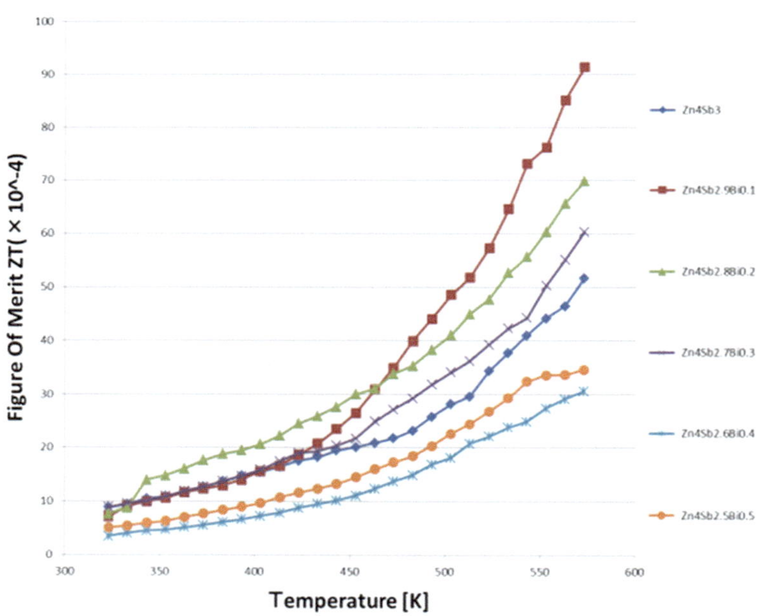


**Fig. 4** The result of electrical resistivity on the $Zn_4(Sb_{3-x}Bi_x)$ ($x = 0.1, 0.2, 0.3, 0.4, 0.5$)

## 4 Conclusions

In this study, $Zn_4(Sb_{3-x}Bi_x)$ ($x = 0.1, 0.2, 0.3, 0.4, 0.5$) are synthesized through melting method followed by pressing method. The electrical conductivity of the doped $\beta - Zn_4Sb_3$ is found to be reduced. The Seebeck coefficient of the Bi-doped samples is larger than that of undoped $\beta - Zn_4Sb_3$. The maximum ZT is obtained from the sample $Zn_4Sb_3Bi_{0.1}$ at 573 K, which shows that it optimizes the thermoelectric properties.

## References

1. Caillat T, Fleurial JP, Borshchevsky A (1997) J Phys Chem Solids 58:1119e25
2. Tsutsui M, Zhang LT, Ito K, Yamaguchi A (2004) Intermetallics 12:809e13
3. Kim HJ, Božin ES, Haile SM, Snyder GJ, Billinge SJL (2007) Phys Rev B 75:134103
4. Takenobu K, Seijiro S, Jun M (2004) The new edition of thermoelectric energy conversion systems. p 85

# Study of Transmission and Reflection Characteristics of Microstrip Line During Application of Noise Suppression Sheet by Adhesive

**Lim Yang Wei, Kyota Otsuka, and Takanobu Ohno**

## 1 Introduction

With increasing use of electronic devices in the microwave region, the density of electromagnetic waves from other electronic devices is the main reason of problems like undesirable response, malfunctioning, and degradation in the performance of electronic equipment. This brings extra challenges in the field of electromagnetic compatibility and electromagnetic interference that demand new solutions—noise suppression sheets (NSS). Noise suppression sheets are composites of magnetic and dielectric materials that exhibit significant magnetic losses in a wide frequency range. Hence, in telecommunication industry, NSS with low-pass characteristic is applied to transmission line. NSS are designed to pass low-frequency signals and attenuate signal higher than cutoff frequency which are considered as noise. NSS low-pass signals before cutoff frequency and attenuated signals after the cutoff frequency. Therefore, the highest transmission at low frequency and lowest transmission at high frequency are ideal for NSS. Besides, the lower reflection of signal represents higher transmittance and less return loss of signal. Hence, low reflection characteristic is required for NSS.

Characteristics of small thickness, sheet form, and flexibility enable NSS to be applied on diverse system. Today, NSS are widely used in cellular phones, laptop, and digital cameras. One of the most common ways to apply noise suppression sheets is to mount them onto circuit boards with double-sided adhesive tape. The most used adhesive for NSS application is acrylic-based adhesive tape. When adhesive is used, it might vary the suppression effect of noise suppression sheet. Thus, the relationship between thickness of adhesive and suppression effect is investigated [1–4].

---

L.Y. Wei (✉) • K. Otsuka • T. Ohno
Kisarazu National College of Technology, Kisarazu 292-0041, Japan
e-mail: yangweilim@gmail.com

K. Ab. Hamid et al. (eds.), *The Malaysia-Japan Model on Technology Partnership: International Proceedings 2013 of Malaysia-Japan Academic Scholar Conference*,
DOI 10.1007/978-4-431-54439-5_23, 

In this paper, we report the effects of different thickness of adhesive tape on electromagnetic noise suppression by using a microstrip line (MSL) through experiment and numerical simulation. We also compare transmission and reflection characteristic of NSS when acrylic-based adhesive is replaced with air gap and different adhesives.

# 2 Transmission and Reflection Characteristics of Microstrip Lines with Integrated Noise Suppression Sheet and Adhesive

## 2.1 Simulations

### 2.1.1 Simulation Model

Figure 1 shows the top and side view of the simulation model, including a microstrip lines (MSL) based on IEC (International Electrotechnical Commision) standard [5]. The MSL with characteristic impedance of 50 Ω. Inner conductor of rectangular coaxial line is connected with a strip conductor which acts as transmission line. The thickness of substrate and width of transmission line are 1.6 and 4.4 mm, respectively. Complex permittivity of dielectric material in MSL is fixed at $\dot{\varepsilon}_r = 2.55 - j0.05$.

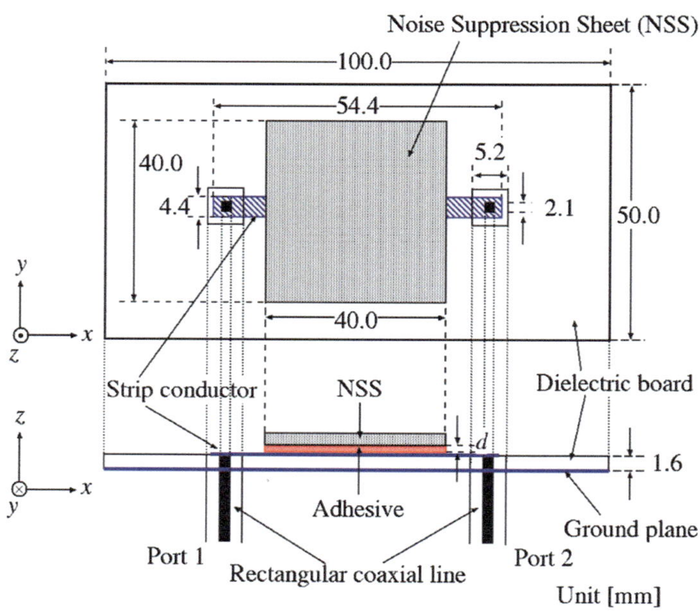


**Fig. 1** *Top* and *side* view of simulation model

**Table 1** List of material constant

| Material | $f$ (GHz) | $\dot{\varepsilon}_r$ | $\dot{\mu}_r$ |
|---|---|---|---|
| NSS | 0.5 | 172.5−$j$13.8 | 10.5−$j$5.0 |
| | 2.0 | 167.5−$j$26.2 | 5.6−$j$5.6 |
| | 3.0 | 156.3−$j$43.8 | 3.6−$j$5.5 |
| Adhesive (acrylic base) | – | 2.69−$j$0.02 | 1.0−$j$0 |
| Air | – | 1.0−$j$0 | 1.0−$j$0 |

The transmission and reflection characteristics of the MSL attached with NSS are simulated by using finite difference time domain (FDTD) method. Varying thickness of adhesive, *d* [mm], is inserted between MSL and NSS. Mur first-order absorbing boundary condition (ABC) is used in this FDTD simulation. The total number of FDTD cells in the three-dimensional FDTD grid is set to $261 \times 136 \times 171$ with cell size of $0.4 \times 0.4 \times (0.025{-}0.9)$ [mm].

Table 1 lists the material constant of noise suppression sheet at varying frequencies and adhesive tape. The complex permittivity and complex permeability of the NSS vary with increasing frequency. Adhesive tapes with varying thickness from 0.1 to 0.4 mm were placed between NSS and strip conductor. During the simulations, signal is transmitted from port 1 to port 2, ranging from 0.5 to 3.0 GHz. The frequency dependences of the reflected ($|S_{11}|$) and transmitted ($|S_{21}|$) scattering parameters were recorded. $|S_{21}|$ refers to the power of signal transferred to port 2 relative to the power input port 1, while $|S_{11}|$ represents the ratio of reflected signal power to the incident signal power at port 1. $|S_{11}|$ and $|S_{21}|$ are then converted into power by the equation below:

$$\text{Power reflection}, \frac{P_{\text{INC}}}{P_{\text{IN}}} = 10^{\frac{S_{11}}{10}} \times 100\,(\%)$$

$$\text{Power transmittance}, \frac{P_{\text{out}}}{P_{\text{IN}}} = 10^{\frac{S_{21}}{10}} \times 100\,(\%)$$

To clearly understand the variations of noise suppression effect of NSS, another similar simulation is run when adhesive tape is displaced by air gap [6, 7].

### 2.1.2 Simulation Results

Figure 2a, b compares $|S_{11}|$ and $|S_{21}|$ and characteristic of the MSL with varying thickness of adhesive. Figure 3a, b shows the frequency dependences of the s-parameters when adhesive tape is displaced by air gap. Comparison of the results for the scattering parameters with and without adhesive tape indicates difference in Figs. 2 and 3. Figure 2a shows $|S_{11}|$ decreased as the thicker adhesive tape is attached, indicating the reflectance of the signal decreased as the thickness of adhesive tape is increased. A closer examination also showed that the maximum reflectance without adhesive tape ($d = 0.0$ mm) between NSS and strip conductor that occurred at 0.5 GHz is 23.98 % (−6.2 dB). When thickness of adhesive is

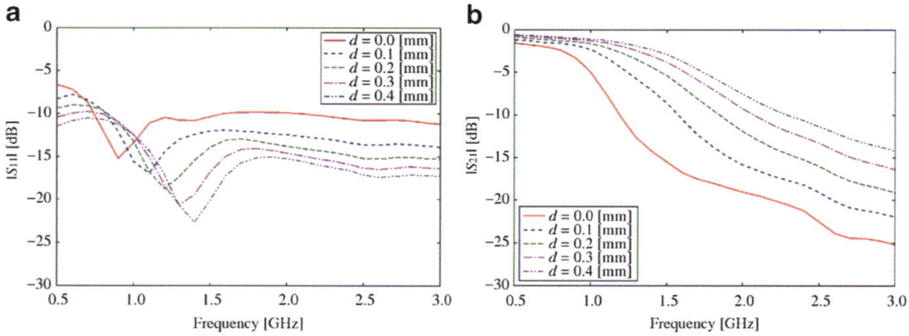


**Fig. 2** Simulated (**a**) reflection and (**b**) transmission characteristic of MSL with varying thickness of acrylic-based adhesive. *Left figure* (**a**) and *right figure* (**b**)

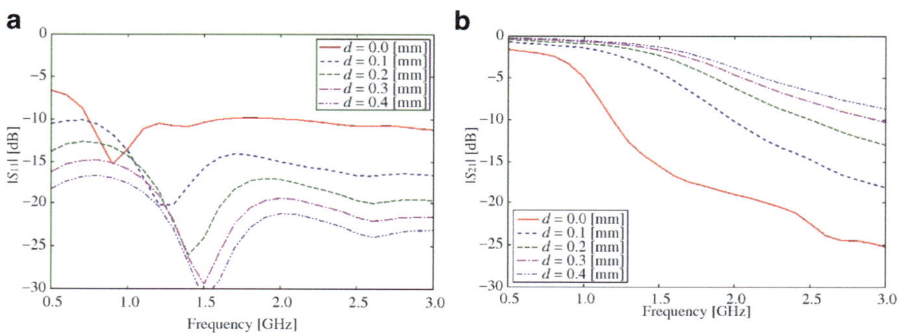


**Fig. 3** Measured (**a**) reflection and (**b**) transmission characteristic of MSL with varying distance of air gap. *Left figure* (**a**) and *right figure* (**b**)

increased to 0.1, 0.2, 0.3, and 0.4 mm, the maximum reflectance is decreased to 16.59 % (0.6 GHz), 12.30 % (0.7 GHz), 10.23 % (0.75 GHz), and 8.91 % (0.8 GHz), respectively. This means that the return loss decreased with increasing thickness of adhesive. This behavior can also be observed when adhesive is replaced with air gap as shown in Fig. 3.

Figure 2b shows that when thickness of the adhesive is increased from 0.1 to 0.4 mm, the $|S_{21}|$ is clearly increased. This behavior is more accentuate at frequency higher than 1 GHz. This indicates that transmittance increased with increasing thickness of adhesive. As shown in Fig. 2b, transmitted electric power at 1.5 GHz is 2.75 % without adhesive. At same frequency, the transmittance is increased to 13.8, 29.51, 41.69, and 89.13 % when the thickness of adhesive is 0.1, 0.2, 0.3, and 0.4 mm, respectively. Although the transmitted electric power is increased at frequency lower than 1 GHz, the low-pass characteristic of noise suppression sheet is degenerated at the same time.

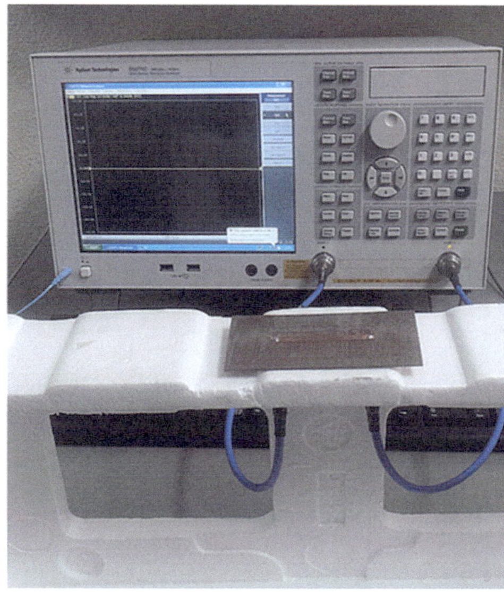

**Fig 4** Measuring system

## 2.2 *Experiment*

### 2.2.1 Measurement System

An experiment is conducted to verify the simulation results. Figure 4 shows the picture of the measurement system, including the microstrip line (MSL) with complex permittivity at $\varepsilon_r = 2.55 - j0.05$. The MSL was connected to E5071C network analyzer *Agilent Technologies* with a synthesized sweep oscillation and a scattering S-parameter test set. $|S_{21}|$ and $|S_{11}|$ are measured when signal is transferred through the MSL. The material constant of NSS and adhesive used are similar to simulated samples. Both are commercially available.

NSS is attached onto bare transmission line ($d = 0$ mm) and followed by adhesive from 0.1 to 0.4 mm inserted between NSS and strip conductor. We measured the transmission and reflection of the electric power ranging from 0.5 to 3.0 GHz. Adhesive is removed and replaced by different thickness of punched paper (air gap) to control the distance between NSS and strip conductor.

### 2.2.2 Experimental Results

Figure 5a, b shows $|S_{21}|$ and $|S_{11}$ characteristics of the MSL with varying thickness of adhesive tape, respectively. Results of another similar experiment when adhesive is displaced by air gap are shown in Fig. 6a, b. Comparison of the numerical simulation results and the experimental results shows that the results of simulation are similar to those of the experiments.

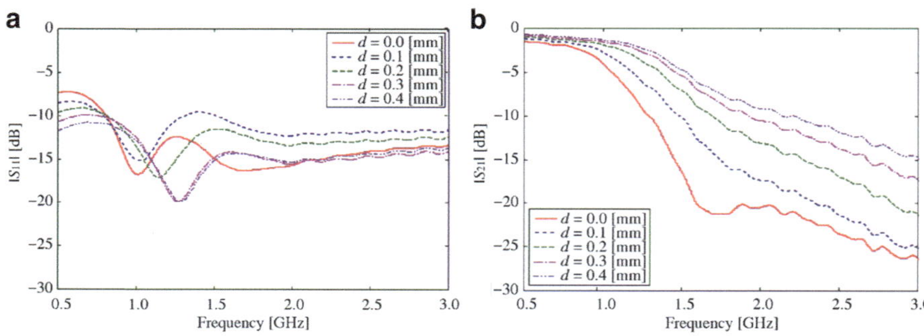


**Fig. 5** Simulated (**a**) reflection and (**b**) transmission characteristic of MSL with varying thickness of acrylic-based adhesive. *Left figure* (**a**) and *right figure* (**b**)

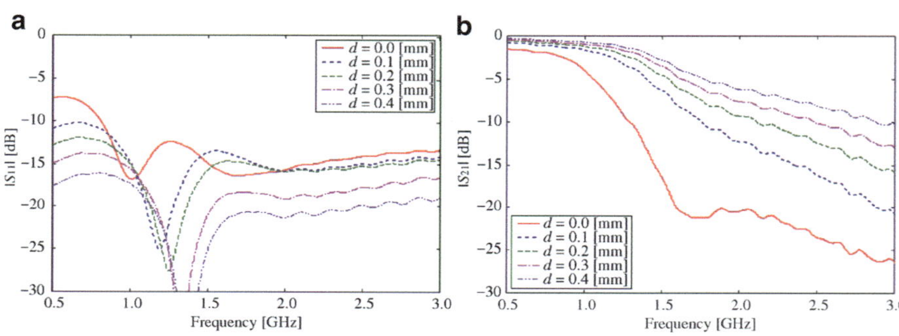


**Fig. 6** Measured (**a**) reflection and (**b**) transmission characteristic of MSL with varying distance of air gap. *Left figure* (**a**) and *right figure* (**b**)

## 2.3 Discussions

Both air gap and adhesive will affect the reflection and transmission characteristics of NSS on bare transmission line; however, adhesive clearly shows less effect compared to air gap. From Fig. 2b, the difference of transmitted electric power between 0.1 mm adhesive and without adhesive at 1.5 GHz is 4.36 % (6.4 dB). The difference is increased to 14.79 % (11.7 dB) when adhesive is replaced by air gap. The effect to transmission characteristic is clearly decreased when adhesive is used compared to the air gap. This behavior can also be observed from the reflection characteristic. The variation of transmission characteristic of adhesive and air gap is caused by their differences in complex permittivity $\dot{\varepsilon}_r$ as shown in Table 1. The effect is decreased when the dielectric constant (real part of $\dot{\varepsilon}_r$) of material between NSS and strip conductor is increased.

From the results above, we suggested another type of adhesive which has a higher dielectric constant of material—polyvinyl chloride (PVC). Complex permittivity of PVC is $\dot{\varepsilon}_r = 5.0 - j0.05$ [8]. Figure 7 showed the simulation results of

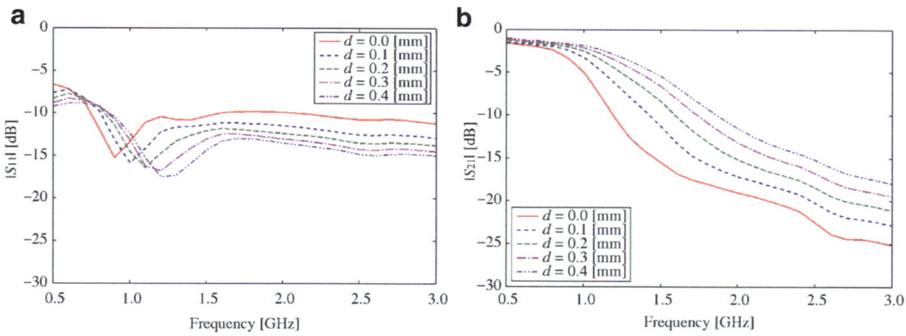


**Fig. 7** Simulated (**a**) reflection and (**b**) transmission characteristic of MSL with varying thickness of PVC-based adhesive. *Left figure* (**a**) and *right figure* (**b**)

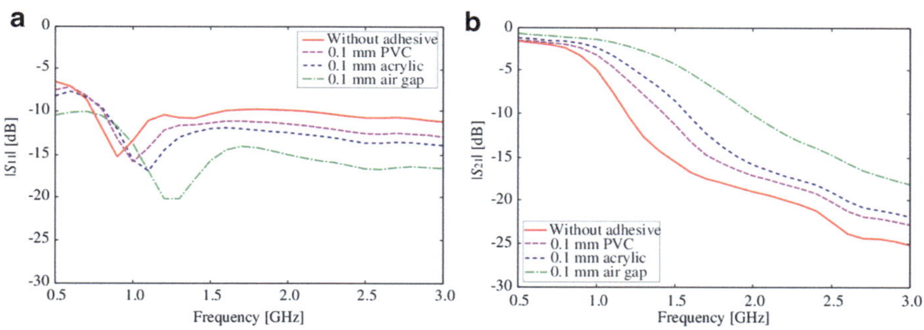


**Fig. 8** Simulated (**a**) reflection and (**b**) transmission characteristic of MSL between 0.1 mm of varying adhesive and air gap. *Left figure* (**a**) and *right figure* (**b**)

frequency dependences of transmittance and reflectance when PVC is used as the adhesive. The frequency dependences when 0.1 mm of acrylic-based adhesive, air gap, and PVC-based adhesive are compared to those without adhesive are shown in Fig. 8.

From Fig. 8a, the reflected power increased when PVC-based adhesive is used instead of acrylic. When 0.1 mm of PVC and acrylic-based adhesive is used, the maximum reflectance is 18.92 and 16.83 %, respectively. At 1.5 GHz, the difference of reflected power with and without adhesive is 2.25 and 2.96 %, when PVC- and acrylic-based adhesives are used, respectively. The return loss is increased when higher dielectric constant of adhesive is used for NSS. Hence, lower dielectric constant of adhesive decreased the reflected power.

Table 2 compares the transmitted power of 0.1 mm of acrylic-based and PVC-based adhesive at 1 and 2 GHz from Fig 8b. At 1 GHz, acrylic-based adhesive showed 10.9 % greater transmittance than PVC-based adhesive. However, when PVC-based adhesive is used, difference of transmitted power with and without adhesive at 2 GHz is decreased to 0.69 %, compared to the difference of 1.38 %

**Table 2** Comparison of transmitted power between acrylic-based and PVC-based adhesive

| | | Transmitted power @1 GHz | | Transmitted power @2 GHz | |
|---|---|---|---|---|---|
| Condition between NSS and strip conductor | *d* (mm) | (dB) | (%) | (dB) | (%) |
| Without adhesive | 0.0 | −5.0 | 31.4 | −19.0 | 1.25 |
| Acrylic-based adhesive | 0.1 | −2.3 | 58.8 | −15.8 | 2.63 |
| PVC-based adhesive | | −3.2 | 47.9 | −17.1 | 1.94 |

when acrylic-based adhesive is used. As a result, although acrylic-based adhesive showed greater power transmission than PVC-based adhesive at frequency before cutoff frequency, however it also degraded the low-pass characteristic of NSS after the cutoff frequency. This indicated that higher electric constant of adhesive is more suitable during application of NSS.

## 3 Conclusion

NSS was placed on a microstrip line (MSL) based on IEC standard. Transmission and reflection of electric power through the MSL were studied through simulation and experiment when varying thickness of acrylic-based adhesive is placed between NSS and strip conductor of MSL. When thickness of adhesive is increased, transmission characteristic is slightly improved at frequency lower than the cutoff frequency; however, transmission characteristic after the cutoff frequency is worsened simultaneously. Besides, the improvement of reflection characteristic can be observed with increasing thickness of adhesive. The same behavior was also observed when the adhesive is replaced by air gap. It showed prominent power transmission at the frequency higher than the cutoff frequency. Transmission and reflection characteristics are simulated when higher dielectric constant of PVC-based adhesive is used. When adhesive with higher dielectric constant is used, relatively small increase of power transmission can be observed at frequency higher than the cutoff frequency; however, the power transmittance decreased at the frequency before the cutoff frequency.

As a result, thickness and dielectric constant of adhesive have to be considered during application of NSS on electronic devices.

## References

1. Lee JW, Kong YK, Kim K, Joo J, Yoon YW, Kim SW, Kim YB, Kim KY (2006) RF and microwave noise suppression in a transmission line using Fe-Si-Al/Ni-Zn magnetic composite films. J Korean Phys Soc 48(6):1534–1538
2. Ott HW (1988) Noise reduction techniques in electronic systems, 2nd edn. Wiley, New York, pp 56–93

3. Burrell J (2003) Disruptive effects of electromagnetic interference on communication and electronic systems
4. International Standard IEC62333. Noise suppression sheet for digital devices and equipment Part 1, Part 2
5. Sohn J, Han SH, Yamaguchi M, Lim SH (2006) Electromagnetic noise suppression characteristic of a coplanar waveguide transmission line integrated with a magnetic thin film. J Appl Phys 100:124510
6. Yamaguchi M, Kim KH, Kuribara T, Arai KI (2002) IEEE Trans Mag 38:3183
7. Ogiso T, Ohta Y (1996) Electrical properties of the plasticized high polymer insulators. J Facul Eng Shinshu University 20:89–107
8. Kim KH, Yamaguchi M (2004) RF integrated magnetic thin film noise suppressor on coplanar transmission line. Phys Status Solidi B 241(7):1761–1764

# An Improved VEPSO Algorithm for Multi-objective Optimisation Problems

**Kian Sheng Lim, Salinda Buyamin, Anita Ahmad, Sophan Wahyudi Nawawi, Zuwairie Ibrahim, Faradila Naim, Kamarul Hawari Ghazali, and Norrima Mokhtar**

## 1 Introduction

Multi-objective optimisation (MOO) problems have multiple objective functions that are minimised (or maximised) simultaneously by evaluating the fitness of the functions with respect to any possible solutions. However, the objective functions usually conflict with each other and results in a set of multiple trade-offs, or non-dominated solutions, rather than a single solution.

The algorithms commonly optimise an objective function using the guidance from the non-dominated solutions found during the search process [1]. For example, all non-dominated solutions are grouped based on their locations in the objective space in the multi-objective PSO (MOPSO) algorithm [2, 3]. Then, one of the non-dominated solutions is randomly selected from the group that has the fewest solutions to be used as the particle guide. Other than that, the primary mechanism of non-dominated sorting genetic algorithm-II [4] is used in the non-dominated sorting PSO (NSPSO) algorithm [5] where one non-dominated solution is randomly selected based on the niche count and nearest-neighbour density estimator, to guide the particles. Furthermore, the optimised MOPSO (OMOPSO) algorithm [6] selects one of the non-dominated solutions as the guide for each particle based on the crowding distance mechanism in binary tournaments.

K.S. Lim (✉) • S. Buyamin • A. Ahmad • S.W. Nawawi
Faculty of Electrical Engineering, Universiti Teknologi Malaysia,
81310 Johor Bahru, Malaysia
e-mail: kiansheng860524@yahoo.com; sophan@fke.tm.my

Z. Ibrahim • F. Naim • K.H. Ghazali
Faculty of Electrical and Electronic Engineering, Universiti Malaysia Pahang,
26600 Pekan, Malaysia
e-mail: zuwairie@ump.edu.my

N. Mokhtar
Department Of Electrical Engineering, Faculty of Engineering, Universiti Malaya,
50603 Kuala Lumpur, Malaysia

K. Ab. Hamid et al. (eds.), *The Malaysia-Japan Model on Technology Partnership: International Proceedings 2013 of Malaysia-Japan Academic Scholar Conference*,
DOI 10.1007/978-4-431-54439-5_24, © Springer Japan 2015

Besides, Abido [7] proposed an algorithm which optimises the problem using two non-dominated solutions, a local set and a global set. Each particle is guided by the non-dominated solution that has the smallest distance between the particle and both non-dominated solution sets.

On the other hand, the vector evaluated particle swarm optimisation (VEPSO) [8] algorithm has been used to solve various MOO problems [9–13]. This algorithm solves MOO problem by assigning one swarm of particles to optimise an objective function using the guidance from the best solution with respect to a single objective function, found by another swarm. However, the non-dominated solutions are either equal or better compared with the best solution used in VEPSO. In other words, the non-dominated solution which is better with respect to all objective functions was not used to guide the particles in other swarm.

Thus, the superiority of the non-dominated solutions motivates the use of non-dominated solutions to guide each swarm for improving the VEPSO algorithm. Therefore, in this study, the guidance of a swarm is selected from the non-dominated solution which is best with respect to a single objective function, optimised by the other swarm.

## 2 Multi-objective Optimisation

For generality, consider a minimisation of a multi-objective problem:

$$\text{Minimise fitness } \vec{F}\left(\vec{x}\right) = \left\{f_m\left(\vec{x}\right) \in \mathbb{R}^M, m = 1, 2, \ldots, M\right\}$$

$$\text{subject to } \begin{cases} g_j\left(\vec{x}\right) \leq 0, & j = 1, 2, \ldots, J \\ h_k\left(\vec{x}\right) = 0, & k = 1, 2, \ldots, K \end{cases}$$

where $\vec{x} = \left\{x_n \in \mathfrak{R}^N, n = 1, 2, \ldots, N\right\}$ is the $N$-dimensional vector of decision variables that represent the possible solutions, $M$ is the number of objective functions, $f_m \in \mathbb{R}^M$ is the objective function and $g_j$, $h_k \in \mathbb{R}^N$ are the inequality and equality constraint functions, respectively.

In explaining the concept of *Pareto optimality*, consider two vectors $\vec{F}^a, \vec{F}^b \in \mathbb{R}^M$. $\vec{F}^a$ *dominates* $\vec{F}^b$ (denoted as $\vec{F}^a \prec \vec{F}^b$) if and only if $f_m^a \leq f_m^b$ for $m = 1, 2, \ldots, M$ and $f_m^a < f_m^b$ at least once. Hence, a vector of decision variables $\vec{x}^a$ is a *non-dominated solution* if and only if there is no other solution $\vec{x}^b$ such that $\vec{F}\left(\vec{x}^a\right) \prec \vec{F}\left(\vec{x}^b\right)$. The non-dominated solution is also known as the Pareto optimal solution. Thus, the set of non-dominated solutions of an MOO problem is known as the *Pareto optimal set*, $\mathcal{P}$. The set of objective vectors with respect to $\mathcal{P}$ is known as the *Pareto front*, $\mathcal{PF} = \left\{\vec{F}\left(\vec{x}\right) \in \mathbb{R}^m \middle| \vec{x} \in \mathcal{P}\right\}$.

# 3 Particle Swarm Optimisation

## 3.1 Original Particle Swarm Optimisation Algorithm

The social behaviour of birds flocking has inspired the particle swarm optimisation (PSO) which is a population-based stochastic optimisation algorithm introduced by Kennedy and Eberhart [14, 15]. Thus, the algorithm contains individuals referred to as particles that encode the possible solutions to the optimisation problem using their positions. These particles fly in a defined search space to look for solutions that better satisfy the objective function of the optimised problem. Particles collaborate with each other during the search process by comparing its current position with the best position that it and the other particles in the swarm have found [16].

For the PSO algorithm, consider a minimisation problem with single objective function: there are $I$-particles flying around in an $N$-dimensional search space, where their positions, $p_n^i(i = 1, 2, \ldots, I;\ n = 1, 2, \ldots, N)$, represent each dimension of the possible solutions. In early part of the algorithm, all particles are randomly positioned in the search space and assigned random velocities, $v_n^i(t)$. Later, the fitness, $\vec{F}^i(t)$, for each particle is evaluated by evaluating the objective functions with respect to $p^i(t)$. Next, each particle's best position, $pBest^i(t)$, is initialised to its current position. Meanwhile, the best among all $pBest^i(t)$ is set as the swarm's best position, $gBest(t)$.

After initialisation, the algorithm iterates until the stopping condition is met, i.e. either the maximum number of iterations is exceeded or the minimum error is attained. In iteration, each particle's velocity and position are updated using Eqs. (1) and (2), respectively:

$$v_n^i(t+1) = \chi \begin{bmatrix} \omega v_n^i(t) + c_1 r_1 \{pBest_n^i - p_n^i(t)\} \\ + c_2 r_2 \{gBest_n - p_n^i(t)\} \end{bmatrix} \tag{1}$$

$$p_n^i(t+1) = p_n^i(t) + v_n^i(t+1) \tag{2}$$

where $\chi$ is the constriction factor and $\omega$ is the inertia weight. $r_1$ and $r_2$ are both random values between zero and one. $c_1$ and $c_2$ are the cognitive and social coefficients, respectively. Next, the $\vec{F}^i(t)$ for each particle is evaluated again. With the newly updated $\vec{F}^i(t)$, $pBest^i(t)$ is updated with the more optimal between the new position of the $i$th particle and $pBest^i(t)$. Then, the $gBest(t)$ is updated with the most optimal $pBest^i(t)$ among all the particles. In the end of algorithm, $gBest(t)$ represents the optimum solution found for the problem.

## 3.2 Vector Evaluated Particle Swarm Optimisation Algorithm

For solving MOO problem, Parsopoulos and Vrahatis [8] introduced the VEPSO algorithm. This algorithm was inspired by the multi-swarm concept of the VEGA algorithm [17] where each objective function is optimised by a swarm of particles using the *gBest*(*t*) from another swarm. The *gBest*(*t*) for the *m*th swarm is the $pBest^i(t)$ that has most optimal fitness with respect to the *m*th objective, among all $pBest^i(t)$ from the *m*th swarm.

Generally, the PSO and VEPSO algorithms have similar process flows, except that all processes are repeated for *M* swarms when optimising problems with *M* objective functions. Since each swarm optimises using *gBest*(*t*) from another swarm, in VEPSO, the velocity is updated using Eq. (3). The velocity equation for particles in the *m*th swarm updates with $gBest^k(t)$, where *k* is given in Eq. (4):

$$v_n^{mi}(t+1) = \chi \begin{bmatrix} \omega v_n^{mi}(t) + c_1 r_1 \{pBest_n^{mi} - p_n^{mi}(t)\} \\ +c_2 r_2 \{gBest_n^k - p_n^{mi}(t)\} \end{bmatrix} \tag{3}$$

$$k = \begin{cases} M, & m = 1 \\ m-1, & \text{otherwise} \end{cases} \tag{4}$$

Besides, all non-dominated solutions found during the optimisation are stored in an archive each time after the fitness is evaluated. The archive contains non-dominated solutions only, so the $\vec{F}^i(t)$ of each particle is compared, based on the *Pareto optimality* criterion, to other particles before it is compared to the non-dominated solutions in the archive. In the end of algorithm, all non-dominated solutions in the archive represent possible solutions to the MOO problem.

## 3.3 The Improved VEPSO Algorithm

In VEPSO, each particle of a swarm collaborates with the *gBest*(*t*) from the other swarm that is optimal with respect to the objective function optimised by the other swarm, for example, an MOO problem with two objective functions; the first swarm updates its $gBest^1(t)$ when a new solution has better fitness with respect to the first objective. Hence, the $gBest^1(t)$ is not updated when the new solution has equal fitness with respect to the first objective and better fitness with respect to the second objective, as shown in Fig. 1. Thus, all particles from the second swarm will be guided by $gBest^1(t)$ from the first swarm but not the better solution, non-dominated solution.

In order to overcome this limitation, the $gBest^1(t)$ should be updated with the non-dominated solutions, which has equal fitness with respect to the optimised

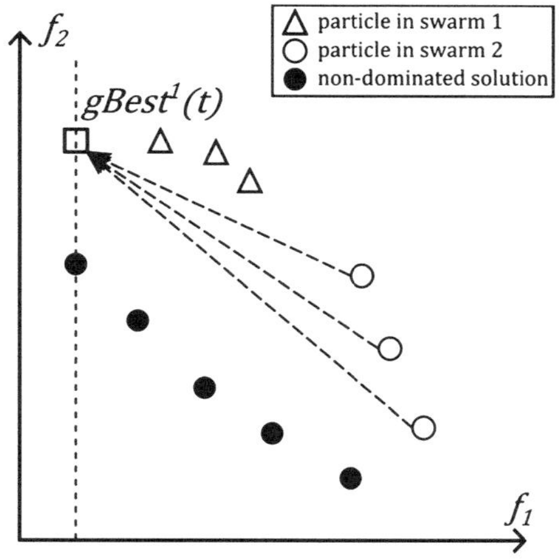


**Fig. 1** Particles guided by the best solution from the other swarm

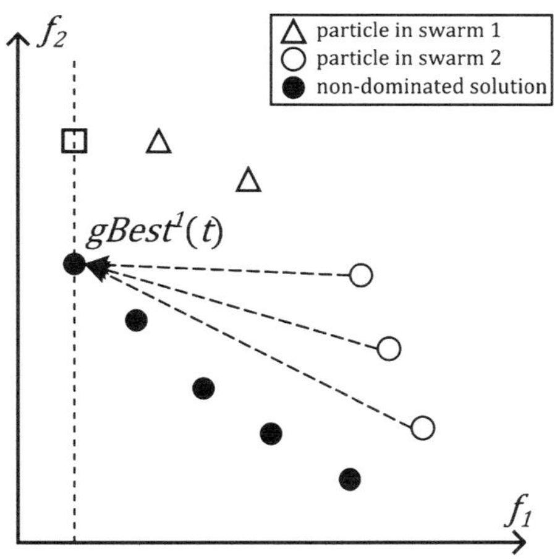


**Fig. 2** Particles guided by a non-dominated solution

objective function and better fitness with respect to the other objective. Without lost of generality in VEPSO, the *gBest*($t$) of a swarm is the best non-dominated solution with respect to the objective function optimised by the swarm. Thus, in a two-objective problem as shown in Fig. 2, the particles from the second swarm are guided by the $gBest^1(t)$ which is now a non-dominated solution that is best with respect to the first objective function. This improved algorithm is called vector evaluated particle swarm optimisation incorporate non-dominated solutions (VEPSOnds).

The particles in PSO algorithm have the potential to stick in local optimal solutions [18, 19]. Therefore, particle position is subject to polynomial mutation at certain probability to further improve the VEPSOnds algorithm. In this work, ten percent of particles are subjected to mutation.

# 4 Experiments

## 4.1 Performance Measure

Several quantitative performance measures are used to analyse the performance of the improved algorithm. First, calculate the total number of non-dominated solutions found, namely, the number of solution (*NS*). Besides, the generational distance (*GD*) [20] measures convergence or the average Euclidean distance between the obtained Pareto front and the true Pareto front. In addition, the *Spread* [4] is used to measure the diversity of the obtained Pareto front. Sometimes, GD and Spread are unfair for comparison, so the hypervolume (*HV*) [21] is included to evaluate the total space or area enclosed by the obtained Pareto front and a reference point, which is a vector constructed from the worst objective value from the true Pareto front. From these measures, larger values in *NS* and *HV* while smaller values in *GD* and *Spread* indicate better performance.

## 4.2 Test Problems

Since this work focused on continuous search space problems, five of the benchmark test problems from ZDT [22] are used to evaluate the performance of the algorithm. ZDT5 problem is a discrete problem. The parameter values are according to the paper [22]. The database from jMetal is used for evaluation.

## 4.3 Evaluation of VEPSO Algorithms

In this experiment, the VEPSO is compared to the VEPSOnds1 and the VEPSOnds2 algorithms which are the VEPSOnds without and with mutation, respectively, in order to differentiate that the performance of the improved VEPSO is not influenced by the polynomial mutation. Besides, all algorithms will have 100 numbers of particles which are equally divided among all swarms. The archive is limited to record a total of 100 non-dominated solutions and remove solutions with smallest crowding distance [4]. For fair comparison, each algorithm will be iterated for 250 cycles before stopping the search process. Besides, the inertia weight is linearly degraded from 1.0 to 0.4. $c_1$ and $c_2$ are random values between 1.5 and 2.5. For algorithm with mutation, the distribution index is set to 0.5. For statistical analysis, all algorithms will be repeated for 100 runs on each test problem.

The performance of each algorithm tested on the various test problems is presented in Table 1. In ZDT test problem, both improved algorithms, VEPSOnds1 and VEPSOnds2, show large improvement against the VEPSO algorithm in all

**Table 1** Average performance for algorithms tested on various problems

| Measure | | VEPSO | VEPSOnds1 | VEPSOnds2 |
|---|---|---|---|---|
| ZDT1 | NS | 30.220000 | 100.000000 | 99.790000 |
| | GD | 0.295865 | 0.022637 | 0.002194 |
| | Spread | 0.834481 | 0.729350 | 0.571807 |
| | HV | 0.001886 | 0.428153 | 0.631216 |
| ZDT2 | NS | 8.070000 | 38.120000 | 97.490000 |
| | GD | 0.766956 | 0.039653 | 0.002003 |
| | Spread | 0.944524 | 0.947356 | 0.687560 |
| | HV | – | 0.137784 | 0.296372 |
| ZDT3 | NS | 35.150000 | 99.600000 | 99.400000 |
| | GD | 0.173060 | 0.009607 | 0.002040 |
| | Spread | 0.871146 | 1.109448 | 1.121149 |
| | HV | 0.004722 | 0.373133 | 0.471686 |
| ZDT4 | NS | 6.610000 | 95.250000 | 64.220000 |
| | GD | 5.062543 | 0.383646 | 0.349438 |
| | Spread | 0.858655 | 1.035510 | 0.962023 |
| | HV | 0.228824 | 0.399914 | 0.437755 |
| ZDT6 | NS | 76.590000 | 78.040000 | 81.030000 |
| | GD | 0.338537 | 0.260666 | 0.266259 |
| | Spread | 1.201796 | 1.276529 | 1.286909 |
| | HV | 0.304584 | 0.303381 | 0.281256 |

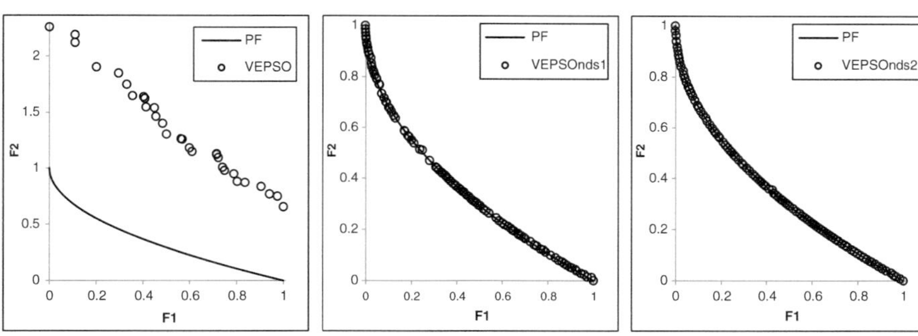


**Fig 3** Plot of non-dominated solutions returned by each algorithm for the ZDT1 test problem

performance measures. Besides, VEPSOnds2 shows much better convergence and diversity performance than VEPSOnds1. Figure 3 shows the obtained Pareto front with best *GD* value among 100 runs from each algorithm. In this figure, the large distance from true Pareto front and uneven distribution of obtained Pareto front from VEPSO explains its weak convergence and diversity.

For the ZDT2 test problem, the VEPSO still performed weakly as compared to the other algorithms. VEPSOnds1 has larger *Spread* value as compared to VEPSO. However, it is still better as indicated by its larger HV value. The obtained Pareto front for all algorithms is plotted in Fig. 4. As expected, the Pareto front obtained by VEPSO is located far from the true Pareto front. Besides, even if there is only one

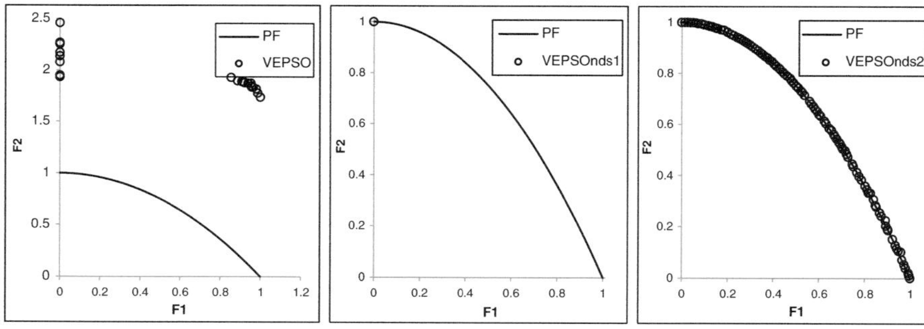


**Fig 4** Plots of non-dominated solutions returned by each algorithm for the ZDT2 test problem

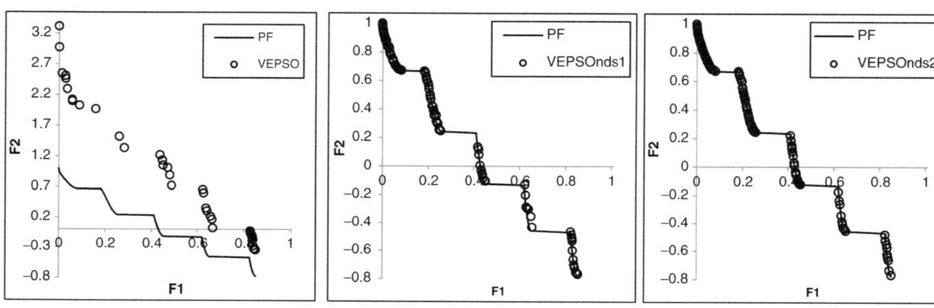


**Fig 5** Plots of non-dominated solutions returned by each algorithm for the ZDT3 test problem

non-dominated solution obtained by VEPSOnds1, it falls on the true Pareto front. Undeniably, VEPSOnds2 is very good when it exhibits well-converged and diverse Pareto front.

Again, the VEPSO exhibits weak performance in the ZDT3 test problem, as in Table 1. Even though VEPSO performs better in *Spread* measure, other algorithms actually have better solutions which are well converged and create larger *HV* value. Figure 5 also illustrates the similar analysis when only VEPSOnds1 and VEPSOnds2 are able to converge to and diverse well on the true Pareto front.

Among all test problems, ZDT4 is very challenging to PSO as there are many local optimal solutions, which could trap the particles from obtaining the best solutions. Therefore, the VEPSO exhibits poor results especially the *NS* and *GD* values. Meanwhile, the improved algorithms easily escaped from this trap. The large Spread value at both improved algorithms does not really implicate its poor performance, superiority in convergence performance. The performance for each algorithm is pretty self-explained in Fig. 6 for the ZDT4 test problem.

Finally, the ZDT6 test problem is where the density of solutions is lower when nearer to true Pareto front. However, this problem does not limit all algorithms from obtaining well-converged Pareto front, but reduce the ability to obtain more

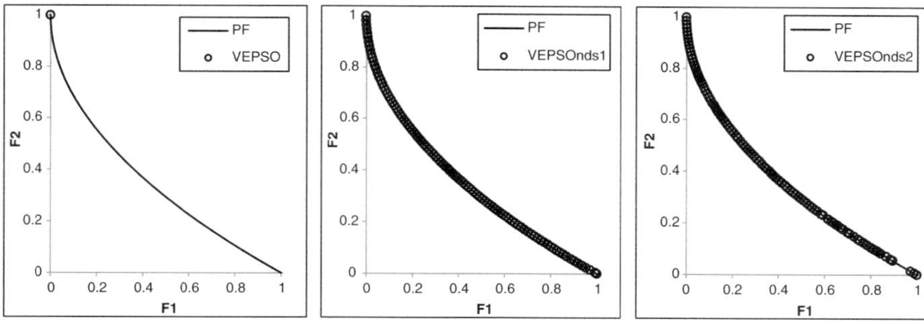


**Fig 6** Plots of non-dominated solutions returned by each algorithm for the ZDT4 test problem

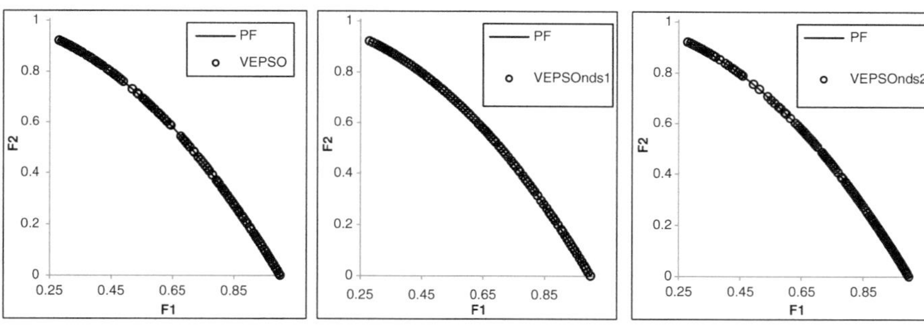


**Fig 7** Plots of non-dominated solutions returned by each algorithm for the ZDT6 test problem

non-dominated solutions. Figure 7 shows the Pareto front found by all algorithms, and they are well converged. Unfortunately, in this case, the non-dominated solutions are not well distributed over the true Pareto front for VEPSO algorithms.

In short, the improved algorithms exhibit significant improvement over the VEPSO algorithm in all test problems. This improvement is mainly contributed by guiding particles using non-dominated solutions, which is always better than the best found solution in VEPSO. In addition, the improved algorithm with mutation performs much well especially in convergence performance.

## 5 Conclusions

The VEPSO algorithm used optimises one objective function by one swarm of particles which are guided by solution from another swarm which has best fitness with respect to the objective function optimised by that other swarm. Thus, the VEPSO is improved by updating the best solution with non-dominated solution which has better fitness with respect to the optimised objective function. The rest of the algorithm is remaining the same to prevent lost of generality in VEPSO.

Based on several performance measures, the VEPSO is compared to the improved algorithm by testing on ZDT test problems. The improved algorithms show significant performance improvement especially when including position mutation to some particles.

**Acknowledgment** This work is supported by the Research University Grant (VOT 04J99) from Universiti Teknologi Malaysia, Exploratory Research Grant Scheme (RDU130605), Research Acculturation Grant Scheme (RDU121403) and MyPhD Scholarship from Ministry of Higher Education of Malaysia.

## References

1. Reyes-Sierra M, Coello CAC (2006) Multi-objective particle swarm optimizers: a survey of the state-of-the-art. Int J Comput Intell Res 2(3):287
2. Coello CA, Lechuga MS (2002) MOPSO: a proposal for multiple objective particle swarm optimization. In: Congress on evolutionary computation (CEC 2002), vol 2. IEEE, pp 1051–1056
3. Coello Coello CA, Pulido GT, Lechuga MS (2004) Handling multiple objectives with particle swarm optimization. IEEE Trans Evol Comput 8(3):256–279
4. Deb K, Pratap A, Agarwal S, Meyarivan T (2002) A fast and elitist multiobjective genetic algorithm: NSGA-II. IEEE Trans Evol Comput 6(2):182–197
5. Li X (2003) A non-dominated sorting particle swarm optimizer for multiobjective optimization. In: Cantú-Paz E et al (eds) Genetic and evolutionary computation. Springer, Berlin/Heidelberg, pp 198–198
6. Reyes-Sierra M, Coello Coello CA (2005) Improving PSO-based multi-objective optimization using crowding, mutation and ε-dominance. In: Coello Coello CA, Hernández Aguirre A, Zitzler E (eds) Evolutionary multi-criterion optimization. Springer, Berlin/Heidelberg, pp 505–519
7. Abido M (2010) Multiobjective particle swarm optimization with nondominated local and global sets. Nat Comput 9(3):747–766
8. Parsopoulos KE, Vrahatis MN (2002) Particle swarm optimization method in multiobjective problems. In: Proceedings of the ACM symposium on applied computing, ACM, Madrid, pp 603–607
9. Gies D, Rahmat-Samii Y (2004) Vector evaluated particle swarm optimization (VEPSO): optimization of a radiometer array antenna. In: IEEE antennas and propagation society international symposium, vol 3. IEEE, pp 2297–2300
10. Rao SMV, Jagadeesh G (2010) Vector evaluated particle swarm optimization of supersonic ejector for hydrogen fuel cells. J Fuel Cell Sci Tech 7(4):041014–041017
11. Omkar SN, Mudigere D, Naik GN, Gopalakrishnan S (2008) Vector evaluated particle swarm optimization (VEPSO) for multi-objective design optimization of composite structures. Comput Struct 86(1–2):1–14
12. Vlachogiannis JG, Lee KY (2009) Review: multi-objective based on parallel vector evaluated particle swarm optimization for optimal steady-state performance of power systems. Expert Syst Appl 36(8):10802–10808
13. Grobler J (2009) Particle swarm optimization and differential evolution for multi objective multiple machine scheduling in Department of industrial and systems engineering. University of Pretoria, Pretoria
14. Kennedy J, Eberhart RC, Shi Y (2001) Swarm intelligence. The Morgan Kaufmann series. In: Fogel DB (ed) Evolutionary computation. Morgan Kaufmann Publishers, San Francisco, p 512

15. Kennedy J, Eberhart R (1995) Particle swarm optimization. In: Proceedings of the IEEE international conference on neural networks, vol 4, pp 1942–1948
16. El-Sayed H, Belal M, Almojel A, Gaber J (2006) Swarm intelligence. In: Olariu S, Zomaya AY (eds) Handbook of bioinspired algorithms and applications. Taylor and Francis Group, Boca Raton, pp 55–63
17. Schaffer JD (1984) Some experiments in machine learning using vector evaluated genetic algorithms (artificial intelligence, optimization, adaptation, pattern recognition), in Faculty of Graduate School. Vanderbilt University, Nashville, p 166
18. Özcan E, Yılmaz M (2007) Particle swarms for multimodal optimization. In: Beliczynski B et al (eds) Adaptive and natural computing algorithms. Springer, Berlin/Heidelberg, pp 366–375
19. Schoeman I, Engelbrecht A (2005) A parallel vector-based particle swarm optimizer. In: Ribeiro B et al (eds) Adaptive and natural computing algorithms. Springer, Vienna, pp 268–271
20. Van Veldhuizen DA (1999) Multiobjective evolutionary algorithms: classifications, analyses, and new innovations, in Air Force Institute of Technology. Air University, Wright-Patterson AFB, Ohio, USA, p 249
21. Zitzler E, Thiele L (1999) Multiobjective evolutionary algorithms: a comparative case study and the strength Pareto approach. IEEE Trans Evol Comput 3(4):257–271
22. Zitzler E, Deb K, Thiele L (2000) Comparison of Multiobjective Evolutionary Algorithms: Empirical Results. Evol Comput 8(2):173–195

# Electrical and Photoluminescence Study of Undoped $CuGaSe_2$ Single Crystal Thin Film

**Sathiabama Thiru, Miki Fujita, Atsushi Kawaharazuka, and Yoshiji Horikoshi**

## 1 Introduction

Chalcopyrite semiconductors such as $CuInSe_2$ (CIS), $CuGaSe_2$ (CGS), and Cu $(InGa)Se_2$ (CIGS) are promising materials for efficient thin film solar cells [1]. Among these materials, CGS has superior optical properties for solar cell application. Although the bandgap energy of CGS (1.68 eV) is a little too high compared with the predicted optimal bandgap [2] of 1.4 eV, its absorption coefficient is considerably high even compared with those of CIS and CIGS. Because of the large electron mass of 0.14 $m_0$, the joint density of states of CGS becomes larger which directly results in a higher absorption coefficient. In addition, the larger electron mass and lower dielectric constant make the excitonic binding energy as high as 20 meV. Having higher excitonic binding energy makes CGS's absorption coefficient higher not only in the absorption edge but also above the bandgap energy through the Sommerfeld effect [3].

On the other end, there exist some difficulties in the electrical transport characteristics in CGS. One of the main drawbacks is the effect of grain boundaries. It has been shown that the grain boundaries of CIS and CIGS produce no fatal leakage

S. Thiru (✉)
School of Science and Engineering, Waseda University,
3-4-1, Okubo, Shinjuku, Tokyo 169-8555, Japan
e-mail: sathiabama@akane.waseda.jp

M. Fujita • Y. Horikoshi
School of Science and Engineering, Waseda University,
3-4-1, Okubo, Shinjuku, Tokyo 169-8555, Japan

CREST, JST, 4-1-8, Honcho Kawaguchi, Saitama 332-0012, Japan

A. Kawaharazuka
Waseda Institute for Advanced Study, Waseda University, 1-6-1,
Nishiwaseda, Shinjuku, Tokyo 169-8050, Japan

CREST, JST, 4-1-8, Honcho Kawaguchi, Saitama 332-0012, Japan

K. Ab. Hamid et al. (eds.), *The Malaysia-Japan Model on Technology Partnership: International Proceedings 2013 of Malaysia-Japan Academic Scholar Conference*,
DOI 10.1007/978-4-431-54439-5_25, 

paths in the solar cell device structures. This beneficial effect cannot be expected in CGS because of the energetic structure of cation-terminated grain boundaries of CGS [4]. Another difficulty of CGS is the formation of n-type material. The electrical transport of undoped CGS is dominated by high hole concentration caused by Cu-vacancy, and it is very challenging to achieve n-type material. It may be due to the fact that the formation energy of Cu-vacancy decreases with increasing in Fermi energy. Even when the donor impurity concentration is increased, the Fermi energy increases at a certain level because compensating acceptor concentration due to Cu-vacancy increases with Fermi energy. A similar mechanism is operative in CIS, but for narrow bandgap materials such as CIS, the n-type conductivity could be achieved because n-type conversion needs only a little rising up for Fermi energy to approach the conduction band edge [5].

To overcome the above difficulties, we have focused and grown high-quality CGS single crystal thin film layers on closely lattice-matched GaAs substrate by using molecular beam epitaxy (MBE). We have also, succeeded in lowering the residual hole concentration by optimizing the deposition condition. Then the impurity doping is performed to achieve n-type conductivity. No report is available for n-type doping in CGS using epitaxial growth method. However, Ge-doped n-type CGS has been reported using ion implantation [6] on growing bulk CGS. Group IV elements such as C, Si, Ge, and Sn may produce efficient donor centers in ternary CGS. Electrical and photoluminescence (PL) studies of undoped CGS single crystal thin film will be discussed here.

## 2 Thin Film Preparation

High-quality CGS single crystal layers are grown on closely lattice-matched (001) GaAs substrate by MBE at 600 °C using the migration-enhanced epitaxy (MEE) [7] deposition sequence. Since the lattice constants of GaAs and CGS are 5.653 and 5.614 Å, respectively, the lattice mismatch of this system is less than 0.7 %. Therefore, high-quality thin film CGS can be grown. Details of high-quality CGS growth process are reported elsewhere [8]. During growth by MEE, the reflection high-energy electron diffraction (RHEED) of the films exhibits streak patterns till the end of the growth run. The beam equivalent pressure (BEP) ratio of Cu and Ga during growth has been chosen at the value where no surface segregation of $Cu_2Se$ is observed. When the Cu flux is large, electrically conductive $Cu_2Se$ layers are segregated on the surface. However, by reducing the Cu flux, the segregation of $Cu_2Se$ decreases gradually and finally disappears completely. This BEP ratio is used in this experiment. Typical values of Cu-BEP and Ga-BEP are $1.1 \times 10^{-7}$ and $3.0 \times 10^{-7}$ Torr, respectively. These BEP values provide the growth rate of approximately 0.6 μm/h. Typical growth time is 2 h.

As CGS consists of alternate piling up of layers composed of Cu and Ga atoms, followed by Se atoms in the [001] direction, we have adopted MEE deposition sequence which is composed of simultaneous deposition of Cu and Ga followed by Se deposition for undoped layer growth. This ensures the stoichiometric layer of CGS.

## 3 Results and Discussion

Undoped CGS was grown at 600 °C using MEE. At this growth temperature, CGS samples showed hole concentration as high as $6.5 \times 10^{18}$ $cm^{-3}$. By lowering the substrate temperature, the residual hole concentration decreases considerably. At 530 °C the typical hole concentration reaches approximately $2 \times 10^{16}$ $cm^{-3}$. However, we have adopted the growth temperature of 600 °C because the grown layer exhibits superior PL characteristics compared with those grown at low temperatures. Figure 1 shows the PL spectrum of undoped CGS at room temperature. Two distinct peaks appear at 1.763 and 1.679 eV. Well-confined CGS/GaAs(001) has its c-axis perpendicular to the substrate plane [9]. Our PL measurement always satisfies $E_{\perp}c$ polarization configuration. Hence, these peaks are ascribed to free exciton emissions, B-exciton and A-exciton, respectively, according to the literature [9]. Peak 1.763 eV is considered to be a radiation transition associated with the $\Gamma_6$ conduction band to $\Gamma_6$ valence band (B transition). No discernible emission is observed in the lower-energy region at room temperature indicating that the purity of the undoped sample is fairly high with high crystal quality. The PL spectra measured at various temperatures are shown in Fig. 2. At 4.2 K, the undoped sample exhibits three major PL bands: (a), (b), and (c). A faint shoulder at 1.73 eV (a) is

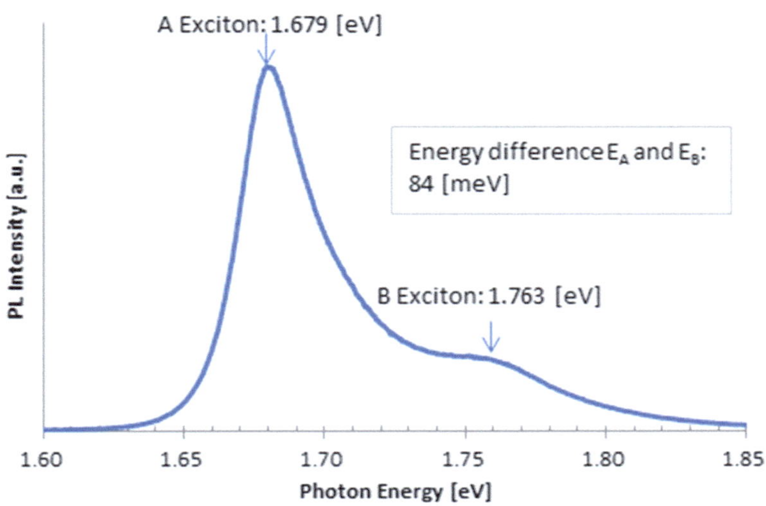


**Fig. 1** PL spectra of undoped CGS at room temperature

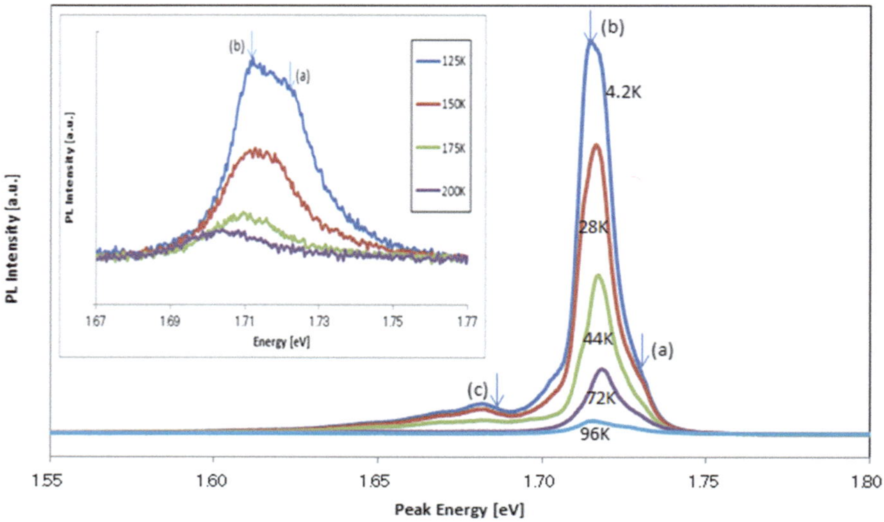


**Fig. 2** Photoluminescence spectra of undoped CGS at various temperatures

probably related to the free exciton (A-exciton), and the dominant peak at 1.715 eV (b) is related to the bound exciton, because these energies coincide well with the reported values [10]. As temperature increases, the intensity of A-exciton shoulder (a) increases and becomes dominant above approximately 150 K, while the bound exciton peak (b) becomes less intense, and the peak disappears above 125 K as demonstrated in the inset of Fig. 2. This phenomenon is explained by considering the excitation of the bound exciton to the free exciton at elevated temperatures.

Another PL peak at 1.68 eV (c) can be attributed to the band-to-acceptor radiative recombination involving a shallow acceptor level like $V_{Cu}$, according to literature [11]. The PL peak energy of emission bands (a), (b), and (c) varies with temperature as shown in Fig. 3. The Arrhenius plot for the PL intensity of band (c) indicates that its thermal activation energy is approximately 40 $\pm$ 10 meV as shown in Fig. 4. This value coincides well with the reported depth of $V_{Cu}$ acceptor, suggesting that the PL band (c) is related to band-to-acceptor transition including $V_{Cu}$ acceptor. The transition model of (a), (b), and (c) is schematically given in the inset of Fig. 4.

## 4 Conclusion

A good quality of undoped CGS single crystal thin film was successfully grown on GaAs (001) substrate using MEE. Undoped CGS showed clear free exciton emission lines (A-exciton and B-exciton) at room temperature. At 4.2 K these samples exhibit intense near band edge emissions composed of A-exciton, bound exciton,

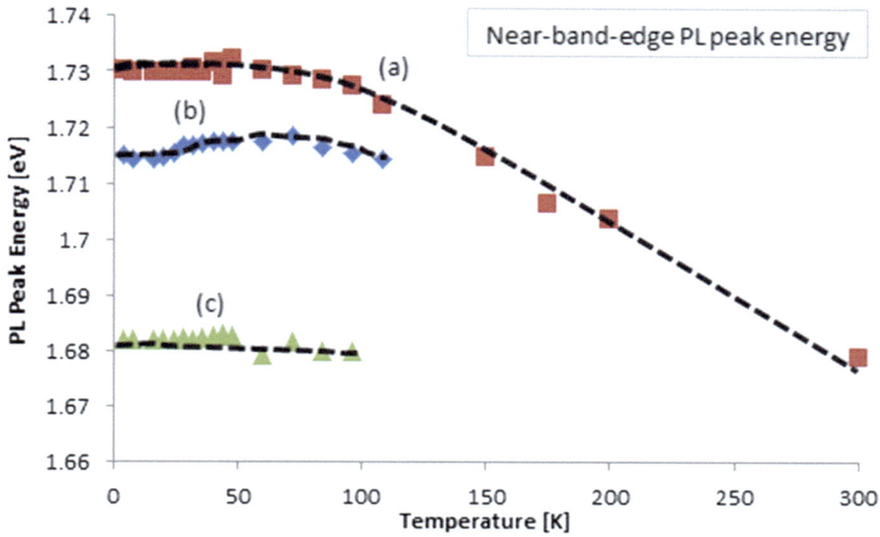


**Fig. 3** Temperature dependence of PL peak energy of near band edge emissions in undoped CGS

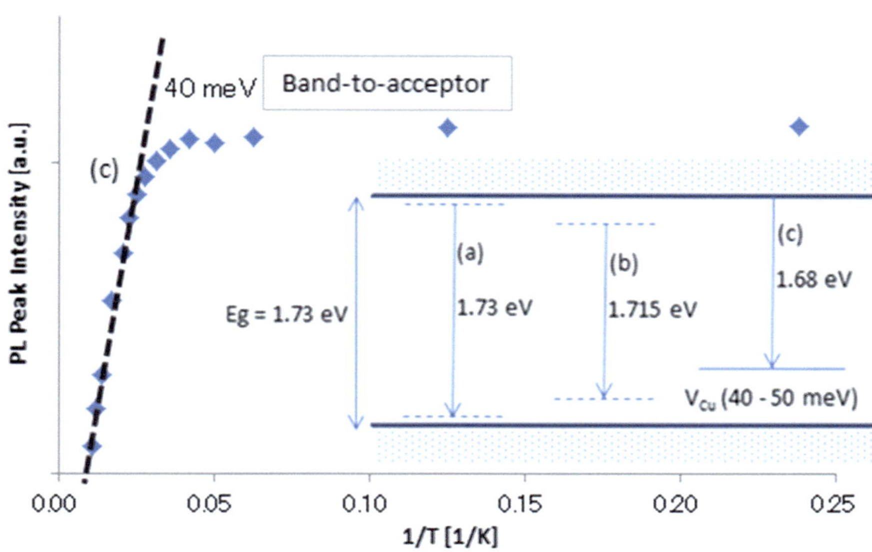


**Fig. 4** Arrhenius plot of PL peak intensity of band (c) in undoped CGS

and band-to-acceptor-related emissions. With increasing sample temperature, the intensity of the bound exciton and band-to-acceptor emission decreases, and above 125 K, they almost disappear. At higher temperature, the free exciton emission dominates until above room temperature.

**Acknowledgements** This work is partly supported by Grants-in-Aid for Scientific Research B (23360163) and Young Scientists B (20760203 and 22760233) from the Japan Society for the Promotion of Science (JSPS) and by the Global COE Program "Practical Chemical Wisdom" from the Ministry of Education, Culture, Sports, Science and Technology (MEXT).

## References

1. Ramanathan K, Contreras MA, Perkins CL, Asher S, Hasoon FS, Keane J, Young D, Romero M, Metzger W, Noufi R, Ward J, Duda A (2003) Prog Photovolt Res Appl 11:225
2. Shockley W, Queisser HJ (1961) J Appl Phys 32:510
3. Ogawa T, Takagahara T (1991) Phys Rev B 44:8138
4. Repins I, Contreras MA, Egaas B, DeHart C, Scharf J, Perkins CL, To B, Noufi R (2008) Prog Photovolt Res Appl 16:235
5. Zhang SB, Wei SH, Zunger A, Katayama-Yoshida H (1998) Phys Rev B 57:9642
6. Krustok J, Raudoja J, Schon JH (2000) Phys Status Solidi A 178:805
7. Horikoshi Y, Kawashima M, Yamaguchi H (1988) Jpn J Appl Phys 27:169
8. Fujita M, Sato T, Kitada T, Kawaharazuka A, Horikoshi Y (2012) J Vac Sci Tech B 30(2):5
9. Chichibu S, Harada Y, Uchida M, Wakiyama T, Matsumoto S, Shirakata S, Isomura S, Higuchi H (1994) J Appl Phys 76(5):3009
10. Shirakata S, Chichibu S, Miyake H, Sugiyama K (2000) J Appl Phys 87:7294
11. Vecchi MP, Ramos J, Giriat W (1978) Solid State Electron 21:1609

# Challenges in Adopting Cloud Computing for Automation Process Controller in Manufacturing

**Kean Wei Tan**

## 1 Introduction

Cloud computing centralized facilities with dynamically scalable and virtualized resource capability to provide a service over the Internet have been identified as a key technology to hit the future [1]. In today's society, industry leaders are working hard to pioneer the implementation of cloud computing to solve increasing computing, and resource problems arise in the Internet age. The main factor in providing on-demand computing services with high reliability, scalability, and availability in a distributed environment is contributing to the surge and interests in cloud computing [2].

Last year, an issue "Can cloud computing technology be adopted in process controller and automation system toward maintaining high-speed production in manufacturing?" has been raised by LinkedIn's Automation & Control Engineering Group [3]. In industry, process controller is extensively used to enable mass production of automation process. Besides, it serves as middleware for human–machine interface, alarm, or yield data collector to transfer to database server through network interface protocol [4]. Although society claims that prospects of integration of cloud computing embrace all applications in major industrial which it's no longer a mere premise, but limitation in cloud, for example, security threads, data vulnerability measurement, high performance, and real-time alarm handling and recovering to cause it's not yet mature enough to be adopted at process controller [3, 5].

The concept of cloud computing with the advantages of virtualization, cost reduction, scalability, and elasticity is one of the ideas to achieve LEAN manufacturing [8]. To hit the aim, the project is focused on every challenge especially in addressing cloud performance-scale problem and increasing data reliability with recovery contingency plan in cloud to be used for process controller.

K.W. Tan (✉)
Faculty of Computer Science, Universiti Sains Malaysia, 14000 Penang, Malaysia
e-mail: kelvin.tkw@gmail.com

K. Ab. Hamid et al. (eds.), *The Malaysia-Japan Model on Technology Partnership: International Proceedings 2013 of Malaysia-Japan Academic Scholar Conference*,
DOI 10.1007/978-4-431-54439-5_26, © Springer Japan 2015

Today, smart grid is used to improve the efficiency, reliability, and sustainability of the production and distribution of electricity [9]. Computing isn't really like electricity, but we will nevertheless see parallel evolution driven by similar forces. We can implement "cloud grid" into each client that connects to the cloud to have similar capability like power grid to achieve significant cloud computing resource management, efficient recovery mechanism, cloud sustainability, and reliability.

## 2 Definition of Cloud Computing

National Institute of Standards and Technology adds yet another definition to the already saturated list of cloud computing [6]:

> A model for enabling ubiquitous, convenient, on-demand network access to a shared pool of configurable computing resources (e.g., networks, servers, storage, applications, and services) that can be rapidly provisioned and released with minimal management effort or service provider interaction.

In this definition, first, cloud computing is a specialized distributed computing prototype with massively scalable resource and can deliver different layers of services to end user outside the cloud by encapsulating it as an abstract entity. Every service can be dynamically configured (via virtualization or other approaches) and be delivered on demand.

There are various definitions of cloud computing architecture; cloud computing provides three types of services: SaaS (software as a service), PaaS (platform as a service), and IaaS (infrastructure as a service) also defined as three-layer system structure in cloud computing (Fig. 1).

Hardware level contains the physical resources, for example, computer processor, hard disk storage, and network communication. It provides necessary raw resources to the cloud. Infrastructure as a Service (IaaS) is a layer to allow customer to provision processing and storage and to network their software, deploying and operating system execution. Platform as a service (PaaS) provides a service for user to compile, validate, and deploy their application onto cloud infrastructure. Software as a service (SaaS) delivers complete software running on platform cloud and accessible through the Internet.

Although every layer (IaaS, PaaS, and SaaS) in the cloud provides services through the standard interface, interoperability issue between clouds still exists, and it has been identified as one of the challenges to adopt cloud computing to process controller. In today's industry, a lot of applications and services emerge that required the use of multiple clouds; there will be growing incentives to adopt standard interfaces that facilitate interoperability in order to capture emerging and growing markets in a saturated cloud market [7].

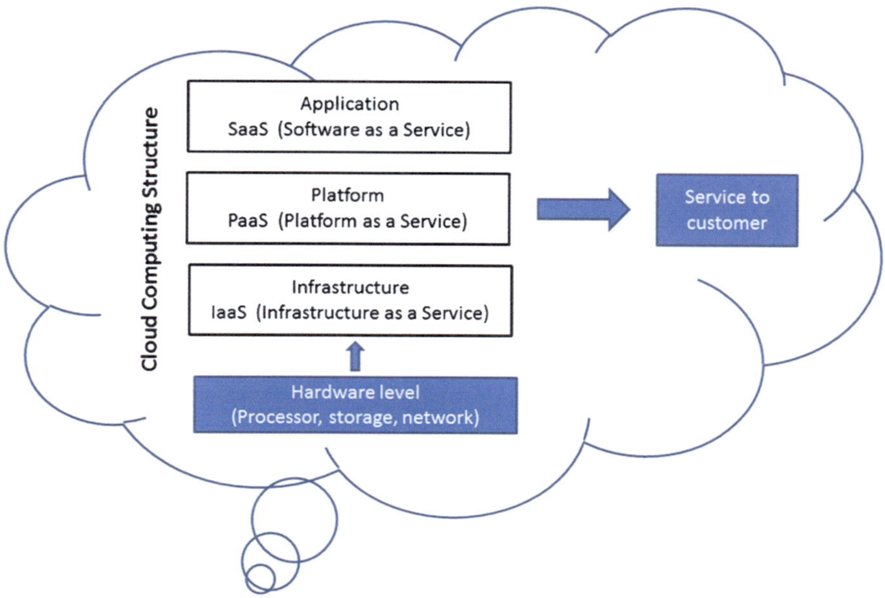


**Fig. 1** Cloud computing structure

## 3 Process Controller Application

Process control system is a general term that encompasses several types of control system, for example, supervisory control, data acquisition systems, and distributed control system used in industrial production. It normally works as middleware to receive (1) command from remote machine and (2) automated or operator-driven supervisory command to remote machine; (3) it serves to collect yield data and alarm conditions from the machines through network interface protocol (TCP/IP) [10].

The traditional process controller network setup layout (Fig. 2) is causing installation process toward high complexity and is time-consuming; a lot of efforts will be wasted on system maintenance and high capital cost to hire more manpower to operate every process controller and to purchase computer hardware and infrastructure. In conclusion, employing cloud computing could improve centralize process controller in terms of service with dynamically scalable and virtualized resource for every machine in industry which helps to reduce production cost and achieve LEAN manufacturing.

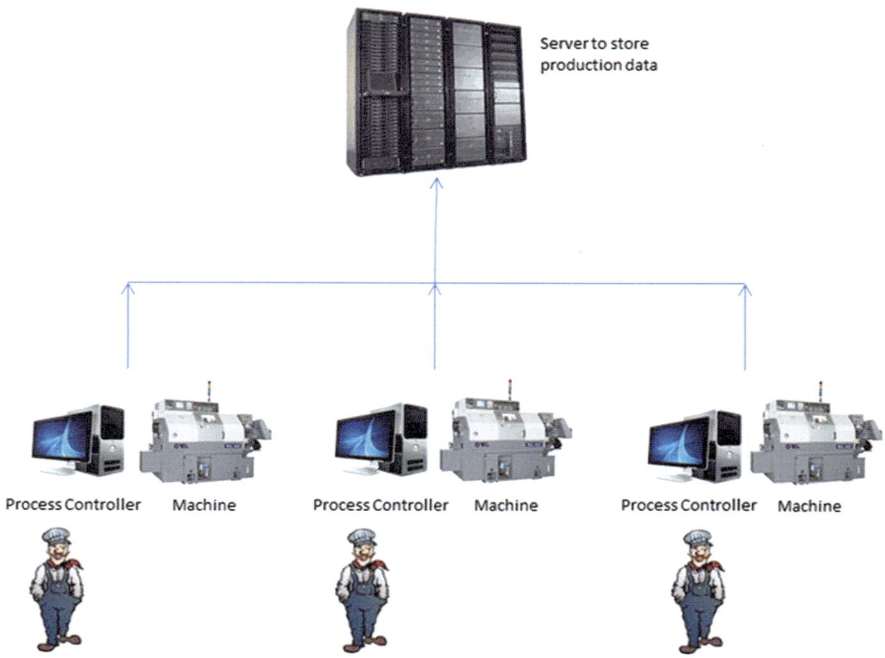


**Fig. 2** Process controller network layout in today's manufacturing

# 4 Challenges to Adopt Process Controller into Cloud

## *4.1 Performance Issue*

We understand that cloud computing talents with on-demand service can be provided easily in the Internet era that embraces simplified management features, has less maintenance, and hastens complicated application installation process in the client computer. Nevertheless, the most important is to reduce production cost such as manpower, infrastructure, and management efforts to maximize company revenue. However, performance issue that holds back some cloud computing efforts has been identified as one of the top concerns in today's IT society [11]. This happens because cloud computing works to centralize all services and resources into a server. So, the cloud server will be identified as a fundamental engine to support thousands of machines in mega industry. In addition, process controller that contains few system controllers to run synchronously in every machine is a high-density resource application, and it will absolutely create performance disaster with no way to recover. Secondly, more heavy information and data will be transmitted among core components in cloud-based application, server, and machines via network and potentially may cause network congestion. Basically, one way to resolve the problem is to raise the cloud hardware capacity such as increase computer processor, storage, and network bandwidth to support each process controller application. But that means the main objective to reduce production cost is not achieved by increasing the physical cloud computing infrastructure capacity in the cloud implementation [3, 7, 12].

Besides capacity planning, scalability is another solution to increase and preserve application performance in the cloud. The advancement of multi-core technology today brings the possibilities of having a client desktop supercomputer with thousands of core processors and bigger storage. It has driven visualization to run locally.

Finding solution to harvest and manage all unused resource from every client computing to the cloud in order to increase cloud service performance is doable. Thus, a new methodology tool—"cloud grid"—will be a new research subject with capability to monitor and allocate resource from each client computing to the cloud. As a result, cloud computing will become more powerful when more client computers are using it. If the target project successes, it will become an ultimate solution tool with significant capability to increase the performance in the cloud.

## *4.2 Security*

Apart from the performance issue, security is categorized as one of the top barriers to the cloud computing. According to the survey about the risk in cloud that had been completed by IDC, security is the number one obstacle in the cloud (Fig. 3) [13]. The unique layer virtualization in constructing cloud infrastructure has bought certain security concern to the end user. The concerns include the potential to compromise the virtualization software to cause loss of data or data theft [12]. Subsequently, everyone questions the cloud viability, authentication access support, data security, exact data location, data privacy control, and disaster recovery in the cloud and feels uncomfortable to use it until they have a solid solution to overcome it. Otherwise, security issue is setting cloud computing up for a fall although there are plenty of advantages from the cloud.

In the past few years, many companies and industries decide to forego public cloud implementation so as not to reveal their data to dirty network in public and choose instead to use private cloud which is under their firewall protection. This may be is the quick solution to erase most of the security concerns, but it doesn't mean that it is100 % safe. When we are putting everything into a single box, for example, in the cloud, just one bad apple spoils the whole box—a single department, user, or application that is not behaving as it should [3, 13].

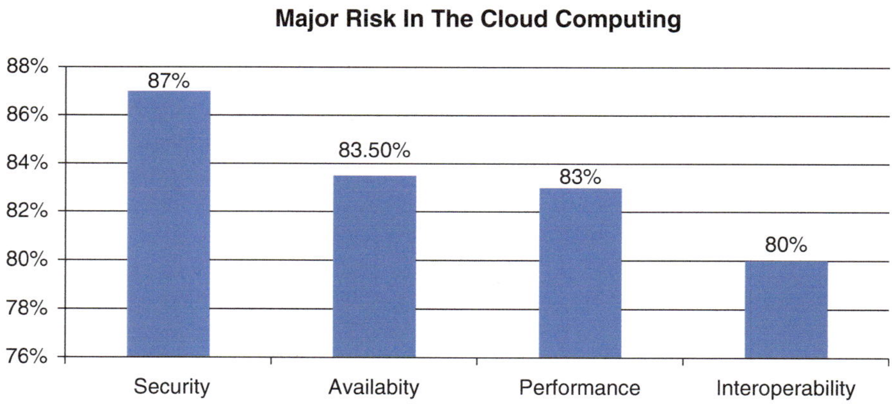


**Fig. 3** Survey of major risk in the could computing

Process controller serves as a prime system to enable mass production, and it has been claimed by LinkedIn's Automation & Control Engineering Group that it is not yet mature enough to be adopted by the cloud [3]. We cannot afford to lose all process controllers in production at the same time and have downtime for even a single second because it can cause huge losses amounting to billions when the cloud computing's security is compromised. So, it's important to research a new metrology or system to overcome the security issue in the cloud before using it for process controller. Cloud grid as my research project will serve as custom checkpoint for every client computer that connects to the cloud. It will provide identity access management to authenticate every access to the cloud and to ensure the cloud's healthiness.

### *4.3 Availability*

The availability of resource and service stored at cloud becomes another major concern for many organizations. IT society is getting worried about cloud availability gap when cloud outage was happening at Amazon and Google App Engine recently [14]. Data backup in data center that suffers downtime or going off-line, and data lock-in are the common outages in cloud. Without having a high availability from the cloud, the advantages associated with using cloud environment are diminished [15]. In addition, it may have highly dangerous consequence to the industry when the yield and production get impacted if process controller is plugged into cloud and caused by cloud availability issue.

Today, it's impossible to expect a cloud to have 100 % availability level, and IT outage will always occur in unexpected condition [16]. So, We will not aim to eliminate 100 % outages, but will concentrate on reducing number of outages to the lowest possible level without having an impact on the cloud service quality. Besides, customers should expect a certain level of downtime in cloud and have sufficient recovery mechanism in place when this happens.

There are a lot of ways that can improve the availability in cloud, such as frequent cleanup of all large, useless, and redundant data that has accumulated in cloud over time. These are the simple manual methods to increase the cloud availability.

Besides, we should make use of the multi-core processor and resources in every client computer to increase the cloud availability. We are now working on method to harvest and manage all unused resources from each client computer and create sub-cloud to run in every client computer. With this approach, service provided by the cloud, for example, process controller, is able to run in sub-cloud for each machine and will not be impacted by any outage accident from the main cloud.

### *4.4 Interoperability and Standardization*

Today's industry doesn't have the freedom to move their data and application across other cloud platforms that may impact cloud computing growth. Many international companies like Apple, Microsoft, Intel, etc., have few factories that are set up at different regions around the world. Thus, cloud computing interoperability issue will impact their business as data or resources are locked into particular cloud platform at every factory and not able to communicate and share among different clouds. As a summary, industry is losing their flexibility to use any application between their in-house infrastructure and services to communicate at different clouds [17].

Many cloud platforms don't interoperate because they use distinct hypervisor and virtual machine technologies and they store and configure operating systems and application differently. The lack of standardization in cloud interface, security feature, and management standard is the main reason contributed to the cloud computing interoperability issue [18]. In addition, cloud computing to support different standards of process controllers in production for different kinds of machines will be another challenge to adopt process controller into a cloud.

Recently, a lot of cloud computing interoperability forums have been held, for example, Cloud Computing Interoperability Forum (CCIF) and Open Cloud Consortium (OCC), to address the problem of cloud interoperability and standardization. The purpose of the forum is to discuss and come up with a generic cloud computing interface. However, while waiting for the cloud computing common interface, we can start to research on certain methodology tool to overcome the cloud's interoperability issue. Cloud grid serves as an interface between cloud and client computer that is capable to translate data and information among different clouds. This is resemble to language translator tool that we commonly search in the Internet to translate different kinds of language to the language we understand.

## 4.5 Power Consumption

The advantages of cloud computing to deliver virtual computing service that does not require physical storage and processing are impacting the PC world will be conquered by the cloud and mobile device soon. Today, mobile device is another next big wave to the computer world. But, at the same time, it also brings another potential challenge which is energy efficiency of mobile clients in cloud computing. Energy efficiency has always been identified as one of the critical items to the mobile device as use cases are moving toward always online connectivity with high-speed network communication for computation downloading from the mainstream cloud computing. Unfortunately, the development of battery technology has not been able to synchronize with the power requirements of the increasing resource demand. The amount of energy that can be stored in a battery is limited and is growing 5 % annually [19].

Besides depending on the physical battery, computation offloading between cloud and mobile client will be one of the identified improvement areas to increase overall system energy efficiency. In the recent research by the Telecommunication Nokia Group, monitoring the trade-off point for the computing-to-communication ratio, which is the critical factor for the decision between mobile client local processing and computation offloading to the cloud to depend on energy consumption of network communication and local processing, has brought a significant power reduction to the mobile client usage and mainframe cloud. Additionally, not only the amount of transferred data but also the traffic pattern is important; sending a sequence of small packets consumes more energy than sending the same data in a single burst [20]. So, we will see cloud grid to serve as a new middle tool for managing the trade-off point that strongly depends on the energy efficiency and monitoring the traffic pattern between mobile clients and cloud.

## 5 Conclusion

Although cloud computing is endowed with dynamically scalable and virtualized resource capability to provide a service over the Internet toward achieving significant manufacturing cost reduction and simplified management process which has attracted industry's attention, but cloud computing implementation is not easy and straightforward as a lot of challenges have been identified. This paper successfully illustrates sufficient information for all the challenges in the cloud computing to the industry. All these challenges should not consider as roadblocks in the pursuit of cloud computing. It's rather important to give serious consideration to these issues and the possible way out before applying process controller into cloud computing.

The comparison among all the challenges, performance and availability issue in the cloud are the first priority obstacles to be resolved in order to embrace process controller into a cloud even security has been categorized as top issue in the cloud by the society. In today's industry, they don't have the intention to expose their network to public; they prefer to hide under their own firewall protection. However, any slowness or outage issues that occur in cloud will seriously cause the company's revenue sink.

In fact, economics of computing looks similar to energy today. Either the energy or the grid computing of tomorrow will look like yesterday's electric power grid. Both will move toward a mix of micro-production and large utilities, with increasing numbers of small-scale producers (wind, solar, biomass, etc., for energy; for computing, local clusters and processors). With the new metrology tool, "cloud gird" like "smart grid" in energy field has been identified as target research project toward improving the performance, reliability, and sustainability of the cloud service and distribution of computer's resource. Table 1 shows the comparison between cloud computing and grid computing as well as cloud grid to resolve all the discussed challenges in cloud.

## 6 Future Research Direction

The target research is to find out a sufficient way of identification and allocation of unused resources in every client computer and supply for the mother cloud usage, so that it can sustain and increase the whole cloud's performance. Furthermore, we are also discovering a new methodology tool—"cloud grid"—and it has similar functionality with our smart grid today for energy field. But it is used to obtain, sustain, distribute, and exchange resources with authentication control over client computer and cloud.

Figure 4 elucidates the new cloud computing layout we have foreseen, and with the new layout we target to resolve the performance, availability, security, and interoperability issue existing in today's cloud. Nevertheless, the project final goal is to adopt process controller application into cloud computing; thus, industry can enjoy all the significant benefits from the cloud.

**Table 1** Comparison grid computing and cloud computing with cloud grid

| Category | Grid computing | Cloud computing | Cloud grid |
|---|---|---|---|
| Business goal | Enable federated resource sharing in dynamic, distributed environments for the users | A large pool of centralized computing and storage resources that delivers nontrivial application services to every client | A tool to monitor network traffic between the mainframe cloud and client computer, harvest unused resources in order to increase the entire cloud performance |
| Performance | Overall grid performance depends on the existing integration resources and control by resource access and management protocol and doesn't support interactive applications. Performance issue in grid is low | Entire performance depends on mainframe cloud's resource capacity and supports latency-sensitive application which is operating natively on cloud. Facing performance issue when cloud grows in scale and number of users | Used to harvest and manage all unused resources from every client computing that connects to the cloud in order to increase cloud performance |
| Security | Security issue in grid computing is low as grid certificate service in connectivity layer underlies every grid transaction | Security issue in cloud computing is high because of virtualization in constructing cloud infrastructure that has brought a potential data loss or data thief | Provides identity access management to authenticate every connection to the cloud in order to overcome the security issue in cloud |
| Interoperability | Always facing interoperability issue for different grid computing and has established Open Grid Forum standard to define a standard protocol to resolve the interoperability issue | Facing an interoperability issue and setup Cloud Computing Interoperability Forum (CCIF) and Open Cloud Consortium (OCC) to address the problem of cloud interoperability and standardization | Works as information translator tool between different clouds to be able to resolve the interoperability issue |
| Availability and failure management | Availability in grid computing is low, and failure management is limited. Any failure in grid computing is not recovered | Doesn't have a strong backup mechanism to apply in cloud and causes data loss or lockup when outage in cloud happens | To create a sub-cloud mechanism to run locally in every client computer so that it will not be impacted by any outage accident from main cloud |
| Power consumption | Power consumption for entire grid computing system is high as a lot of resource distribution computing and network activities are ongoing | Power consumption for cloud computing is getting higher when cloud grows in scale and a number of users are connected | To monitor trade-off point for computing-to-communication ratio between cloud and client computer in order to increase the energy efficiency for entire cloud system |

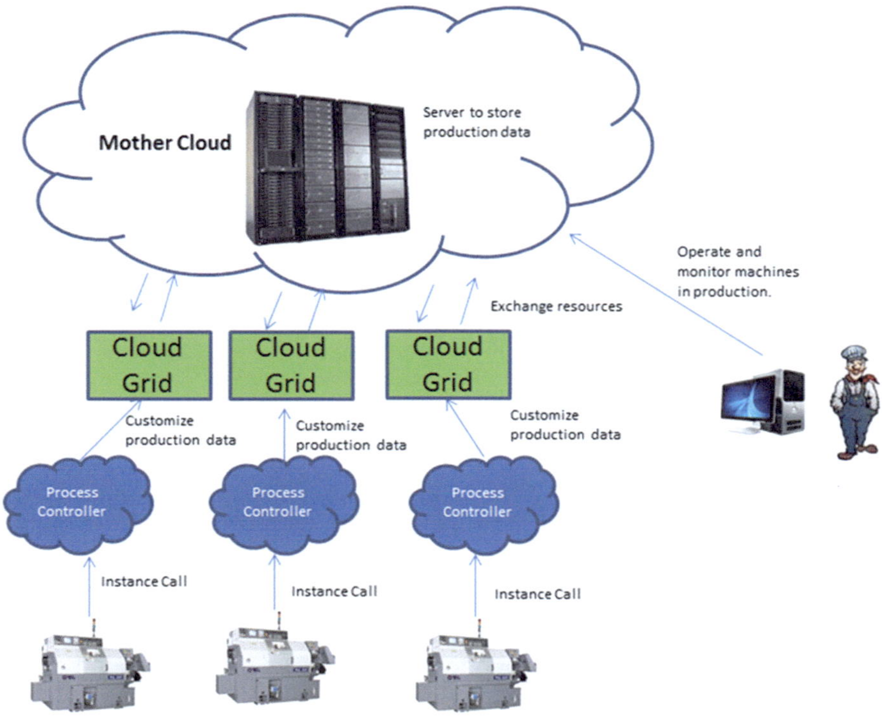


**Fig. 4** A new cloud computing layout to address performance, availability, and security issue in the cloud

## References


1. Bughin J, Chui M, Clouds Manyika J (2010) Big data, and smart assets: ten tech-enabled business trends to watch. McKinsey Quarterly, McKinsey Global Institute, USA
2. Xu X (2011) From cloud computing to cloud manufacturing. Department of Mechanical Engineering, University of Auckland, New Zealand
3. ControlEngineering.com, Katzel J (2012) Cloud computing, alarm management. http://www.controleng.com/channels/tutorials/tutorials/single-article/cloud-computing-alarm-management/cb60097162.html
4. Wikpedia.org (2013) Programmable control system. http://en.wikipedia.org/wiki/Control_system
5. Phneah E (2012) Cloud to drive industry, security concerns remain, Journalist, ZDNET
6. Mell P, Grance T (2009) Perspectives on cloud computing and standards. National Institute of Standards and Technology (NIST), Information Technology Laboratory, USA
7. Foster I, Zhao Y, Raicu I, Lu S (2008) Cloud computing and grid computing 360-degree compared. Department of Computer Science, University of Chicago, USA
8. Wikpedia.org (2012) LEAN manufacturing. http://en.wikipedia.org/wiki/Lean_manufacturing
9. Wikpedia.org (2013) Smart grid. http://en.wikipedia.org/wiki/Smart_grid
10. Wikpedia.org (2013) Industrial control system. http://en.wikipedia.org/wiki/Process_control_system

11. Cloudways.com (2013) The top 4 cloud computing concerns today. http://www.cloudways.com/blog/top-4-cloud-computing-concerns-today/
12. Wikpedia.org (2012) Cloud computing security. http://en.wikipedia.org/wiki/Cloud_computing_security
13. Kinsella J (2012) 5(more) key cloud security issues. Director, Protected Industies. http://www.csoonline.com/article/717307/5-more-key-cloud-security-issues
14. Cloudsware.com (2012) Service availability – The number one cloud computing issue. http://www.cloudswave.com/blog/service-availability-the-number-one-cloud-computing-issue/
15. Abhigna NG (2011) Why enterprises need multi-layered defense architecture. Content Editor, CyberMedia
16. Weier MH (2011) Google, workday outages show SaaS isn't perfect. Senior Managing Editor, Workday
17. Lawton G (2009) Addressing the challenge of cloud computing interoperability. IEEE Computer Society, USA
18. Claybrook B (2011) Cloud interoperability: problems and best practices, President. New River Marketing Research, USA
19. Robinson S (2009) Cellphone energy gap: desperately seeking solutions. Tech. rep., Strategy Analytics
20. Miettinen AP, Nurminen JK (2013) Energy efficiency of mobile clients in cloud computing. Telecommunication Nokia Group, USA

# Face Recognition Human–Machine Comparison Under Heavy Lighting

**Evgeny V. Mozgunov**

## 1 Introduction

There are well-known and relatively simple methods for face recognition in computer vision: Fisherfaces and Eigenfaces. Eigenface performs principal component analysis on the images represented as vectors of pixel intensities. Fisherface relies on the assumption that the subset of all pictures of the face of one person will lie close to a certain hyperplane in the pixel intensities space, unique for each human. Our goals are to see whether (1) this claim based on the theory of lighting of the convex 3D objects actually allows better recognition of faces than evolution of the human eye and brain provided and (2) the error rates (syndrome) can allow one to discriminate between recognition done by a human and recognition done by the machine.

This work is based on the following paper on Eigenfaces and Fisherfaces [1] which compares PCA and LDA for face detection as well as the original Eigenface paper [2]. More advanced face recognition algorithms like elastic graph matching [3] and using a 3D reconstruction [4] have a lot of adjustable parameters that can be tweaked to minimize the error rates for a particular set of lighting conditions. This study is restricted to the methods with effectively no adjustable parameters for a more comprehensive interpretation of the results.

E.V. Mozgunov (✉)
California Institute of Technology, Pasadena, CA 91125, USA
e-mail: emozguno@caltech.edu

K. Ab. Hamid et al. (eds.), *The Malaysia-Japan Model on Technology Partnership: International Proceedings 2013 of Malaysia-Japan Academic Scholar Conference*,
DOI 10.1007/978-4-431-54439-5_27, 

# 2 Dataset

The face dataset is taken from Yale Face database B [5]. It uses 10 test subjects, which are shown in Fig. 1. The images are arranged then into 5 different subsets. Subset 1 will be used for training, subsets 2–5 for testing. Faces are centered, no hair in view. No glasses. Some have small beards. We restrict to small changes in facial expression.

Subset 1 shows well-lit faces, subset 2 has small shadows, and subsets 3–5 are in really hard lighting conditions. Subset 5 faces are poorly recognizable. Subsets 5 and 4 are shown in Fig. 2. A task to match the images to the individuals in Fig. 1 was performed manually, and the results were that subset 4 images can all be matched after careful examination, whereas subset 5 has no more than 20 % success probability. We see how human perception shows discontinuous jump of error rate between sure recognition and impossible task.

## *2.1 Representation*

Each 50 $\times$ 50 pixel training image (from subset 1) is vectorized into a 2500-dimensional vector. The training set contains 70 images, the test set—the rest of all 270 images. Later, subsets 1 and 5 will be used for training for comparison.

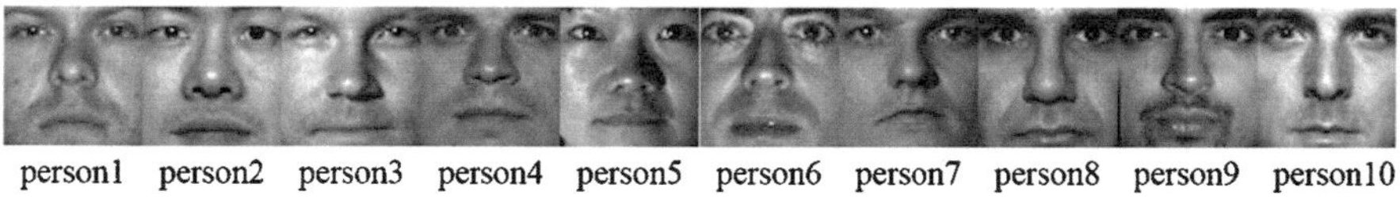


**Fig. 1** The test subjects for Yale Face database B [5]

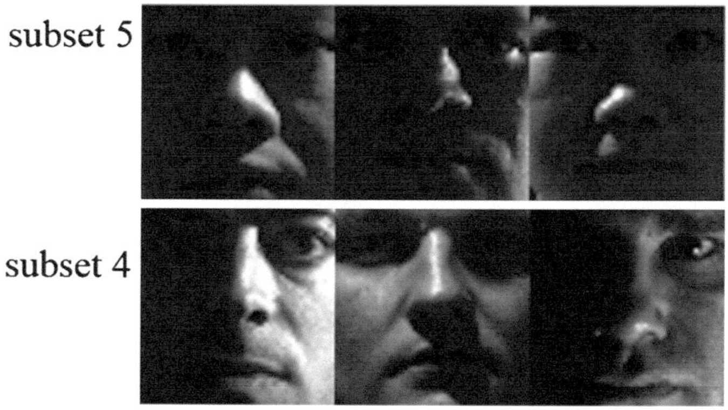


**Fig. 2** The lighting conditions of subset 5 render the person virtually unrecognizable, whereas subset 4 contains a challenging case for our eye

## 3 Eigenfaces (PCA/SVD)

The idea [2] is to perform singular value decomposition (SVD) on the matrix formed by the training set. Then principal component analysis (PCA) is performed on the entire set of training image vectors, retaining the first $d$ principal components.

MATLAB function princomp(M) is used. It runs for about a second on $2{,}500 \times 70$ matrix. The first $d = 40$ singular values in logarithmic scale are shown in Fig. 3.

The gap between the 11th and 12th singular value should not be taken seriously—the amount of eigenvalues retained directly influences the performance. As we will see later, $d = 30$ algorithm works noticeably better than $d = 9$. Thus, $d$ is the first adjustable parameter we have to keep track of when measuring the error rate syndrome.

The $d$ eigenvectors (when reshaped into $50 \times 50$ matrix again and displayed as images) are the Eigenfaces. The first 20 Eigenfaces as images are shown in Fig. 4. We normalize the peak value to 255 white and minimum value to 0 black.

We see that some of the facial features got directly into principal faces—like mustache or contrast in the eyes. That incredibly detailed contrast in the eyes took all principal components starting from 12th to describe (possibly that explains the drop in eigenvalues).

We also note that lighting is more or less uniform—that is just by the nature of subset 1.

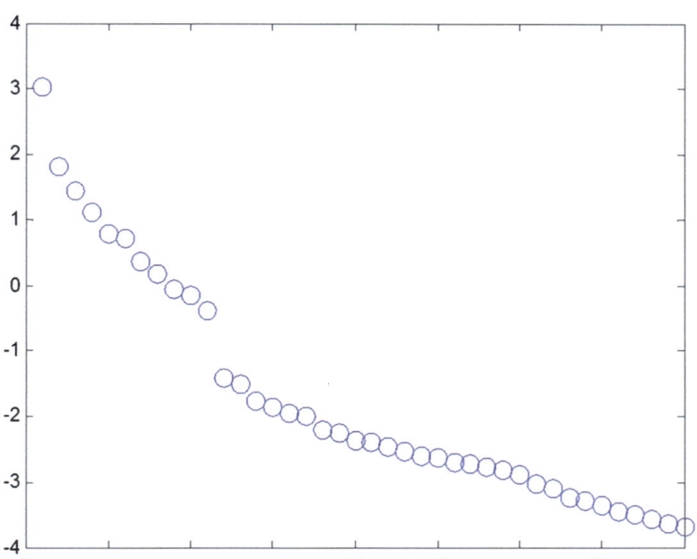


**Fig. 3** The first $d = 40$ principal values of the training set

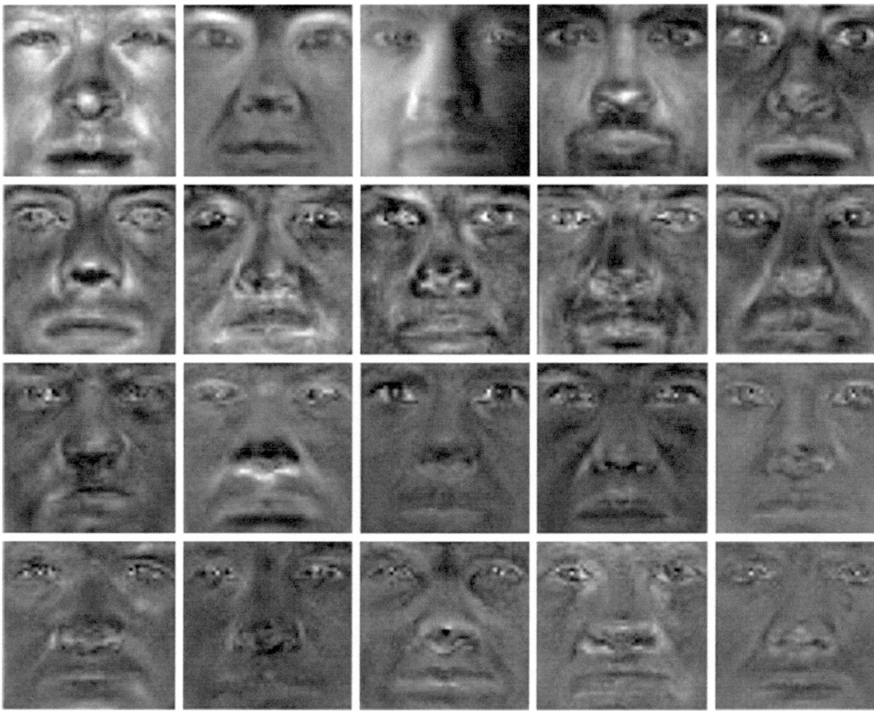

**Fig. 4** The first 20 (principal) Eigenfaces

### 3.1 Computation Speed Enhancement

When performing PCA, the number of dimensions, N, of our data (in this case 2500) is usually much larger than the number of examples, M (70). We can find the eigenvectors of the covariance matrix ($XX'$—where X is the $N \times M$ zero mean matrix of our training data and $X'$ is X transposed) without actually computing $XX'$:

Eigenvalue equation is $XX'\ v = \lambda\ v$. Consider instead an $M \times M$ matrix $X'X$. Eigenvalue equation for it is more feasible to solve (less dimension). So consider that we found all $w$'s and $\lambda$'s that satisfy $X'Xw = \lambda\ w$. Multiplying from the left by X, we obtain $XX'Xw = \lambda\ Xw$. Then we see that $v = Xw$ is a desired eigenvector. As we had M such $w$'s, we thus easily find M corresponding $v$'s. And other N-M $v$'s are not meaningful components—their $\lambda = 0$.

### 3.2 Application of the Algorithm to the Test Set

We project each of the training images to the d-dimensional eigenspace. For each test image in subsets 2–5, we project it onto the d-dimensional subspace also—then

**Table 1** Fraction of the number of correct recognitions to the size of test subset for each subset and for each $d$

| Eigenfaces | Train on 1, $d = 30$ | Train on 1, $d = 9$ | Train on 1 and 2, $d = 30$ | Train on 1 and 2, $d = 9$ |
|---|---|---|---|---|
| Total accuracy (%) | 40 | 38 | 71 | 45 |
| Accuracy subs. 2 (%) | 94 | 90 | 94 | 77 |
| Accuracy subs. 3 (%) | 51 | 43 | 51 | 23 |
| Accuracy subs. 4 (%) | 20 | 20 | 67 | 36 |

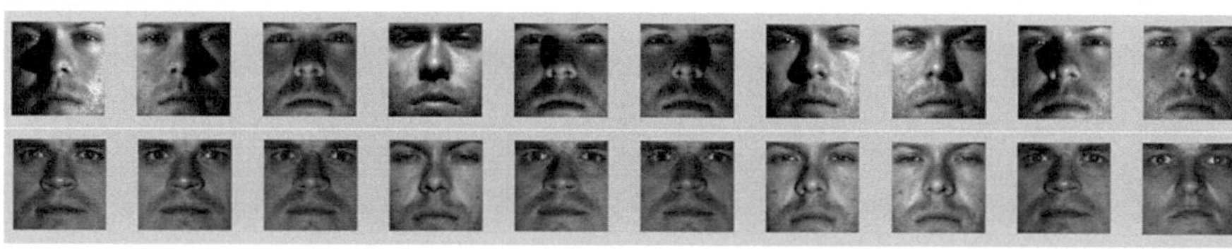


**Fig. 5** Example of false recognition

classify the test image as person 1–10 by finding its nearest neighbor (using the Euclidean distance) in the training set. In Table 1 an accuracy (number of correct recognitions over the size of test subset) for each subset and for each value of $d$ is shown. We comment on how different lighting conditions affect classification accuracy.

The last two columns show the case when subsets 1 and 5 (the hardest) are used for training. Subset 5 turns out to be very useful in improving the performance on other subsets. Using subset 5 to train raises the performance on subset 4—which drastically changes the total accuracy.

## *3.3 Examples of False Recognition*

Some examples of erroneous face recognition are shown in Figs. 5 and 6. In Fig. 6 images from subset 4 are recognized as face 4—it had mustache where shadows usually are. Such a mistake is a signature of a machine, for human would typically not recognize all the faces as one.

Then we use subset 1 and 5 for training and test on subsets 2, 3, and 4. The principal components are shown in Fig. 7. With shadows in training set, the principal faces capture face features under different light directions, which allows for more certain recognition. The overall performance becomes more humanlike.

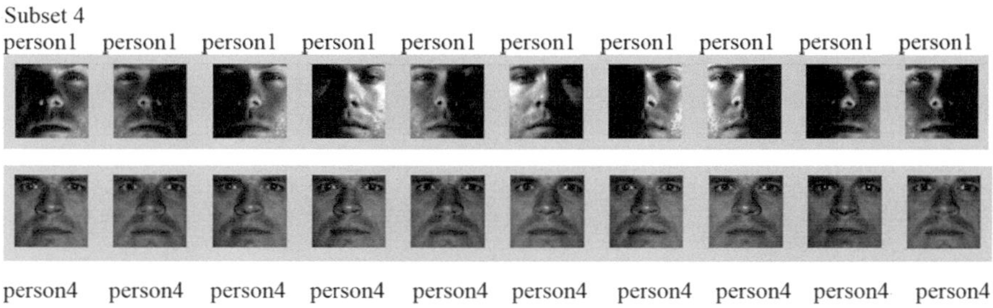


**Fig. 6** Example of false recognition

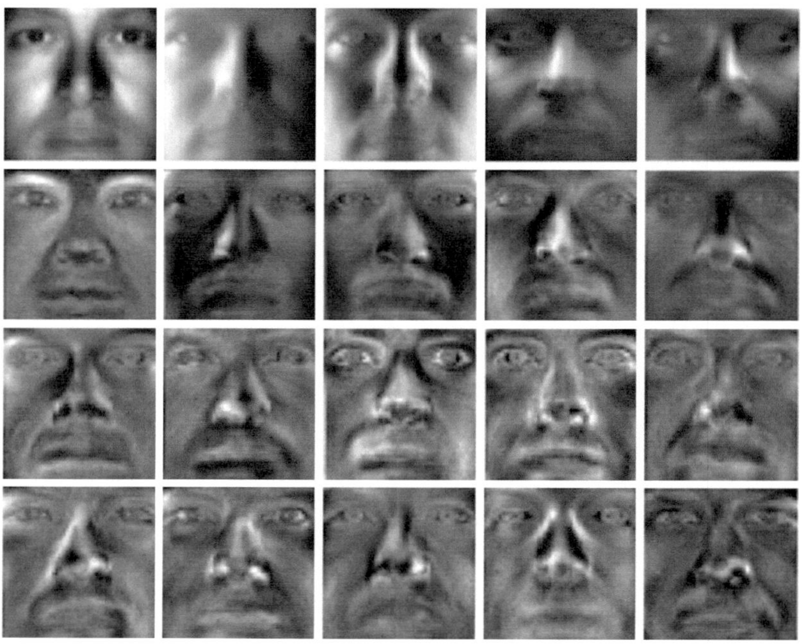

**Fig. 7** The algorithm works much better when training set contains heavy shadowing: the first 20 (principal) Eigenfaces

# 4 Fisherfaces

In the Fisherfaces [1] algorithm, we project to a lower dimensional space before performing linear discriminant analysis (LDA).

The dimensionality of this subspace is in the ideal case (at most):

$$\#\text{of training images} - \#\text{of classes} = \text{rank}\,(S_w) \tag{1}$$

where $S_W$ is the within-class scatter matrix $S_W = \sum \sum (x_i - m_i)(x_i - m_i)'$

**Table 2** Fraction of the number of correct recognitions to the size of test subset for each subset and for each $d$

| Fisherfaces | Train on 1, $d = 30$ | Train on 1, $d = 9$ | Train on 1 and 5, $d = 30$ | Train on 1 and 5, $d = 9$ |
|---|---|---|---|---|
| Total accuracy (%) | 58 | 55 | 98 | 97 |
| Accuracy subset 2 (%) | 100 | 100 | 100 | 100 |
| Accuracy subset 3 (%) | 100 | 98 | 99 | 100 |
| Accuracy subset 4 (%) | 52 | 44 | 95 | 93 |
| Accuracy subset 5 (%) | 10 | 9 | | |

To perform further analysis, we will need to invert $S_W$ on the subspace we are working in. So this subspace must be at least the dimension = rank $S_W$. It follows because we maximize $\lambda = W'SW/W'S_WW$. Taking derivative gives $SW = \lambda S_WW$. To make it an ordinary eigenvalue equation, we need $S_W^{-1}$.

The within-class scatter matrix indeed has a rank N-c (checked numerically, N-c = 60 in our case). But after leaving 60 principal components, the rank decreases (to 57). Repeated iteration converges to rank 55.

In Table 2 we see that even the advanced algorithm cannot do the impossible: faces from subset 5 are virtually unrecognizable (e.g., you see only the nose).

However, it lifted the probability on subset 4 from ~50 to ~100 %. That is why it is important to include different lighting conditions into your training set.

## 5 Speculation

In the context of face recognition with heavy lighting, the learning of a computer method struggles to outperform the human eye. In layman terms, "recognition by only the nose" seems to be impossible unless we decide to train the computer for that particular task. Such inflexibility in performance may slow down the commercial applications, but for researchers it provides invaluable opportunity to search for methods that have similar limitations as humans, as well as, possibly, methods that have different limitations than humans. Discovery of both will allow one to conjecture about algorithmic structure of our neural system. However, one has to be particularly careful in choosing the adjustable parameters that specify the method. Requirements on these parameters that would simulate the human performance with the modern face recognition methods may be a subject of further study.

## References

1. Belhumeur N, Hespanha P, Kriegman J (1997) Eigenfaces vs. fisherfaces: recognition using class specific linear projection, pattern analysis and machine intelligence. IEEE Trans 19(7):711
2. Turk M, Pentland A (1991) Eigenfaces for recognition. J Cognit Neurosci Arch 3(1):71–86
3. Shin H-C, Park JH, Kim S-D (2007) Combination of warping robust elastic graph matching and kernel-based projection discriminant analysis for face recognition. IEEE Trans Multimed 9(6):1125–1136

4. Gokberk B, Dutagaci H, Ulas A, Akarun L, Sankur B (2008) Representation plurality and fusion for 3-D face recognition, systems, man, and cybernetics, Part B: cybernetics. IEEE Trans 38(1):155–173
5. Georghiades AS, Belhumeur PN, Kriegman DJ (2001) From few to many: illumination cone models for face recognition under variable lighting and pose. IEEE Trans Pattern Anal Mach Intell 23(6):643

# Design of Passive Mark for Mobile Robot Navigation Purpose

**Khairul Salleh Mohamed Sahari and Kavinthran Devarajan**

## 1 Introduction

Mobile robots are a kind of robots which have the capability to navigate around their environment and are not fixed to one physical location in contrast with industrial robots. Navigation and motion planning in an unknown environment are important tasks for future-generation mobile robots. There are several different fields in which it is possible to apply mobile robotic systems, such as applications in industrial environments, military and security applications, underwater and mining researches, transport of dangerous substances, surveillance tasks, and many others. In all these fields, it is necessary to equip the mobile robots with a sensor system which allows the vehicle to obtain information from the environment. The processing of these data, provided by the sensor system, must be useful to facilitate actions and control planning of the mobile robot [1–6].

Navigation of mobile robots is an extensive topic, and it covers a large spectrum of different technologies and applications. Its roots include many engineering and science disciplines, from mechanical, electrical, and electronic engineering to computer science. Mobile robot navigation in an unknown environment is a very difficult problem to solve. The mobile robot must be able to safely navigate within its environment and reliably move from the starting point to the destination point. Additionally, a mobile robot which can be placed in an unknown environment must unaidedly discover and collect sufficient information about its surroundings to be able to accomplish its task. For such a robot, the most important requirement is adequate sensing [5,6].

In order to achieve accurate navigation of the robot, it is very important to equip the robot with a sensory system, permitting the acquisition of information about the

K.S.M. Sahari (✉) • K. Devarajan
Centre for Advanced Mechatronics and Robotics, College of Engineering, Universiti Tenaga Nasional, Jalan IKRAM-UNITEN, 43000 Kajang, Selangor, Malaysia
e-mail: khairuls@uniten.edu.my

K. Ab. Hamid et al. (eds.), *The Malaysia-Japan Model on Technology Partnership: International Proceedings 2013 of Malaysia-Japan Academic Scholar Conference*,
DOI 10.1007/978-4-431-54439-5_28, © Springer Japan 2015

robot's surroundings. When considering an easy task of obstacle avoidance, the use of range systems such as ultrasonic and infrared sensors may be enough. However, with the aim of acquiring more accurate information and if we need more than only range information such as size and complete position of the obstacle, the range systems will not be enough for sensing the robot's environment.

In indoor navigation, the usage of landmarks, either active or passive, is common. Passive landmarks are easier to install since they require no power supply. Many passive landmark detections are based on interpretations using vision sensors. Thus, it is very important to design a good marker which is easy to detect using vision sensors.

This paper presents a simple passive mark design for mobile robot navigation purposes. The marker can be easily detected using a webcam, with the image processing done using OpenCV. A GUI is also developed to assist the user in the robot control and also to display real-time data. Tests have been conducted to validate the performance of the marker. Results have shown that the marker has potential to be used for mobile robot navigation.

## 2 Hardware Design

This section explains about the hardware configuration used in this work. A Universal Serial Bus (USB) webcam is used to capture images of the designed markers. The images taken will be processed within a host PC using developed algorithm based on OpenCV library. A mobile robot Pioneer P3AT is used as the robot platform. The host PC is connected to the P3AT via RS232 serial cable. Pioneer P3AT is a 4WD mobile robot with a variety of unique functions. For programming purposes, P3AT is equipped with its own control library called ARIA.

The designed passive mark is generated from few conceptual designs. Design was made based on the marker's easiness to be detected by the webcam using a simple but unique algorithm. The passive mark selection was done by evaluating the errors that occurred when the dilation process is executed. The chosen design is shown in Fig. 1a. An arrow-shaped marker is chosen because it would be easier to show the direction and to set the direction when the programming is done. Square-shaped boxes were the best designs as the formula is easier to generate compared to other shapes. Based on our developed algorithm, the webcam will be able to detect the marker as shown by the red arrow in Fig. 1b.

## 3 Algorithm Design

OpenCV is used for the image processing part, CMake (Cross Platform Make) is used to create Graphical User Interface (GUI), Aria is used to control the P3AT, and Microsoft Visual Basic 2008 is used as the programming platform for this work.

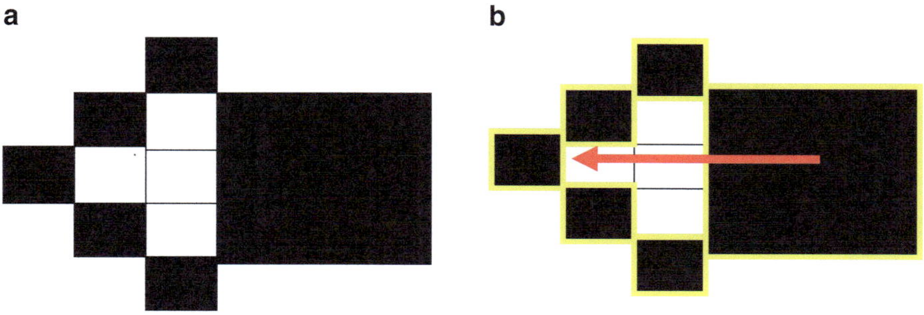


**Fig. 1** Design of passive marker. (**a**) Passive mark design. (**b**) Marker detected by camera

CMake is generally used to generate the GUI for the mobile robot under C++ algorithm. This GUI will show the result of the captured image by the webcam. Moreover, the GUI also shows all the direction and the trajectory of the mobile robot. The flow of the passive mark detection and interpretation is shown in Fig. 2.

1. Binarization: According to this flow chart, the image captured in this case the passive mark's image is used to create an algorithm so that the webcam will be able to detect the passive marks. As stated in the flow chart, the image later then undergoes the first step which is known to be the binarization process. After this process, all the images will be gray scaled so that the colors can be eliminated from the image. This process is to make the image processing much more easier because other colors can interrupt the result. As for binarization, adaptive threshold and dilation process have to be conducted. Adaptive threshold and dilation process are ways to get rid of colors in the image and make it in binary colors which in this case are black and white.
2. Adding rectangle to the edge: After the binarization process is done, adding rectangle to the edge is the next step. This adding rectangle to the edge is to eliminate the dilated image from having more shapes that can be similar to the passive mark. Adding shapes such as rectangle is a trial and error method.
3. Finding the marker squares: Finding the passive mark has to be done by generating an algorithm based on the current image of the passive mark. To perform this step, a very precise algorithm must be generated. Here, after a hard day of trying on the algorithm generation, a very precise algorithm was generated. This algorithm is based on the basic square theories. According to the passive mark design, the small squares have $2 \times 2$ cm dimension. Elsewhere, the big square has the dimension of $6 \times 6$ cm. The small squares and the big square are considered neighbors to each other. Each of the squares has an area of 4 $cm^2$ for the small squares and 36 $cm^2$ for the big square. Here, an algorithm was developed to satisfy all conditions (Fig. 3). The marker squares that fulfill these criteria will be placed in the array as the output. Figure 4 shows an example of the result.

   Explanation for the conditions in Fig. 3:

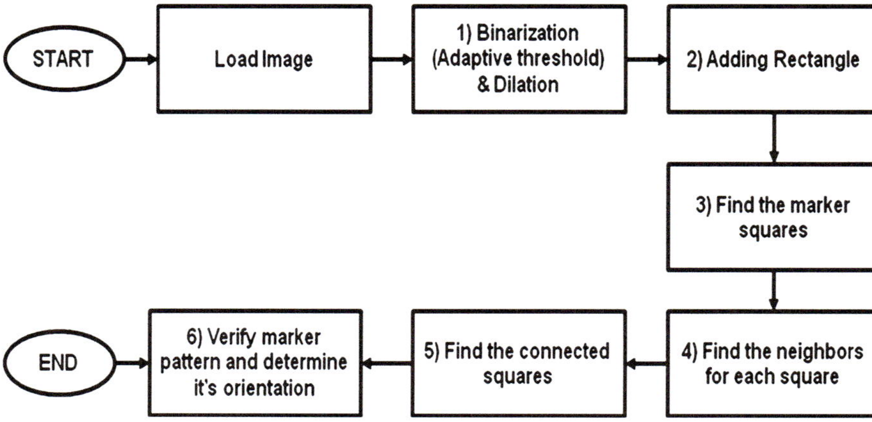


**Fig. 2** Passive mark detection flow chart

**Fig. 3** Algorithm to identify the marker squares (*note*: *p* is for perimeter)

Find the contour which fulfill these conditions:

1) has four corners
2) Is convex
3) area > min_size
4) d3 * 4 > d4
5) d4 * 4 > d3
6) d3 * d4 < area * 1.5
7) d1 >= 0.15 * p
8) d2 >= 0.15 * p

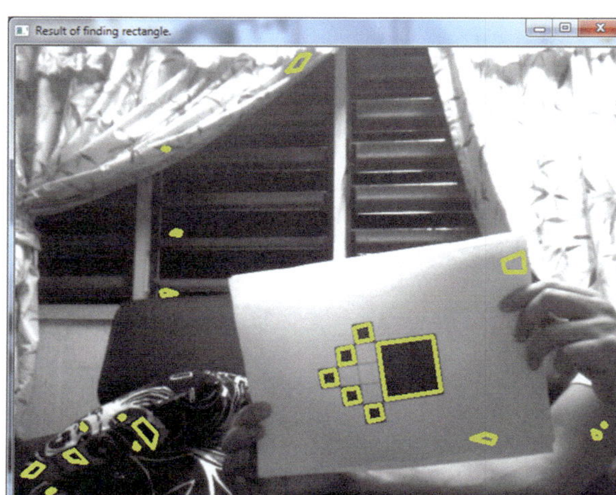


**Fig. 4** Result of finding the marker squares

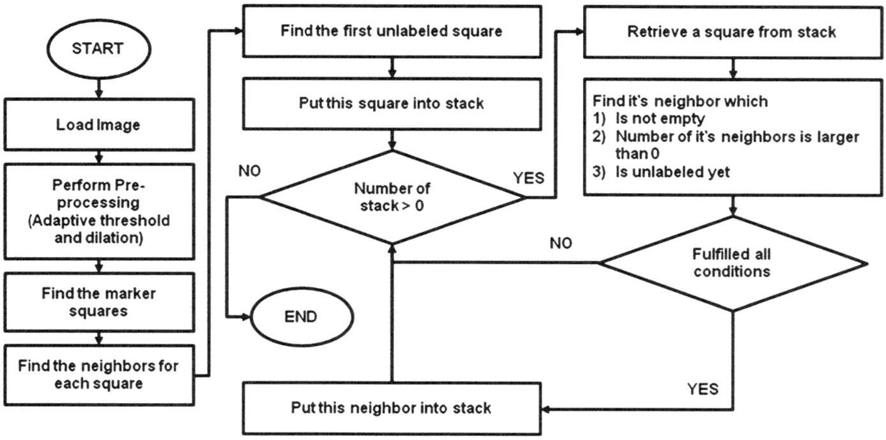


**Fig. 5** Flow chart for the neighbor finding

1. All squares have four (4) corners, so the first condition has to be fixed to four corners.
2. This condition is set to be convex because the marker will not shrink compared to other shapes that appear on the image.
3. The areas of the squares are set to be more than 4 $\text{cm}^2$ because that is the smallest area according to the marker.
4. These d3 and d4 are set based on the possible length that could be for the marker if the marker is tilted in an angle (e.g., d3 $= 2$ cm $\times$ $4 = 8$ cm $>$ d4, so d4 cannot be more than 8 cm or cannot be more than the perimeter of the square using d3).
5. The fifth condition has the same explanation as the fourth condition.
6. This condition is to make sure that d3 $\times$ d4 is not more than the area $\times 1.5$.
7. This condition is to check the hypotenuse of the square.
8. This condition is the same as the seventh condition.

4. Find the neighbors for each square: Finding neighbors for the squares is finding squares located side by side with the current square. The interconnection between these squares must be determined first before it could go for the next step, that is, to find the connected squares. Figure 5 shows the flow chart on how the neighbors are selected and stored in array. Figure 6 shows the result of this operation.
5. Verify marker pattern and determine its orientation: Finally, verification of the marker's pattern and determination of its orientation are done based on a few conditions. Those conditions are as follows:

1. All the edges must be computed to ratio method where the edge of one square is compared with other squares.
2. The compared ratio must be as follows:

- $N_0 > 2$ and $N_0 < 4$. This condition is to eliminate other squares to have the same ratio as the big square to the small square (e.g., 6 cm:2 cm, ratio $= 3$).

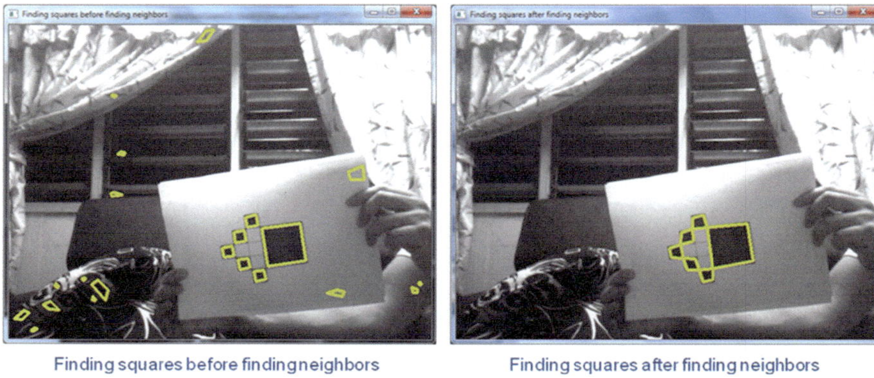


**Fig. 6** Results of before and after finding the neighbors

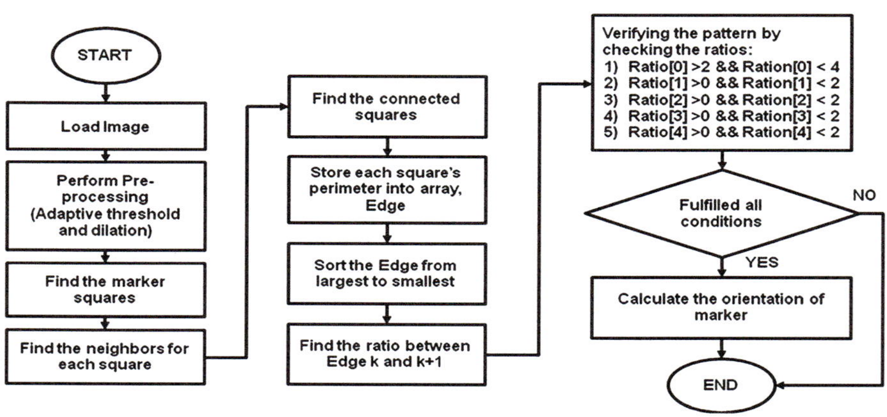


**Fig. 7** Flow chart for marker pattern verification and determination of its orientation

- $N_{1,\ 2,\ 3,\ 4} > 0$ and $N_{1,\ 2,\ 3,\ 4} < 2$. This is the ratio between the 1st small square and the 2nd small square ($N_{0,\ 1,\ 2,\ 3,\ 4}$ = ratio value for the particular squares).

The flow chart for this step is shown in Figs. 7 and 8 and shows the result of this step.

## 4 Graphical User Interface for Mobile Robot

A Graphical User Interface (GUI) is developed using Microsoft Visual Basic 2008 to incorporate all the algorithms developed on the marker detection with the codes to control the Pioneer P3AT mobile robot which is used as the base. This GUI is shown in Fig. 9.

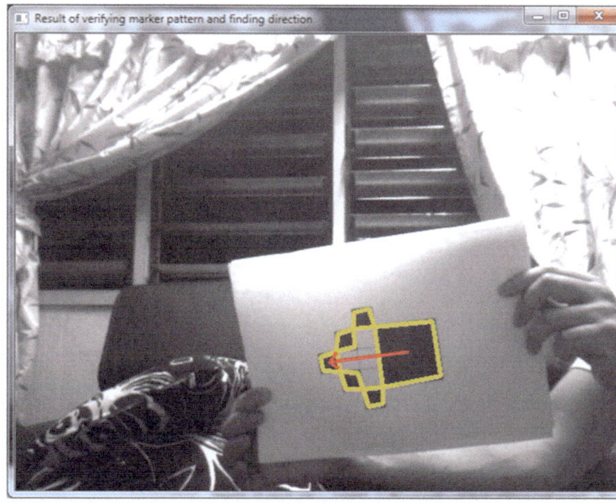


**Fig. 8** Result for marker pattern verification and determination of its orientation

# 5 Results and Discussions

The codes and GUI are tested to check the functionality of the proposed algorithm and also the integration between the image and the mobile robot. The testing was done by showing the marker to the robot, whether it is pointing up, down, left, or right. If the marker is seen pointing up, the robot will move forward; if the marker is pointing down, the robot will move backward; if the marker is seen pointing to the left, the robot will turn 360 ° counterclockwise; and if the marker is seen pointing to the right, the robot will turn 360 ° clockwise. Some sample results are shown in Fig. 9.

In general, after going through the results, it is proven that Pioneer P3AT with a webcam is capable of navigating in a given environment using passive marks. Although the software and hardware work well, there are still errors and problems encountered as follows:

1. The pictures taken have a lot of noise and error: This problem can be solved by putting more filters and using a better camera. The uniqueness of our system and method is that the webcam is basically cheap, and the setup in general can also be very cheap.
2. The marker could not be detected at an angle between 60 and 70°: This is an error due to the algorithm generated. The algorithm generated is given a range of values to work with, so if the marker is tilted at an angle between 60 and 70°, it will violate the algorithm. To overcome this problem, the algorithm can be further improved.

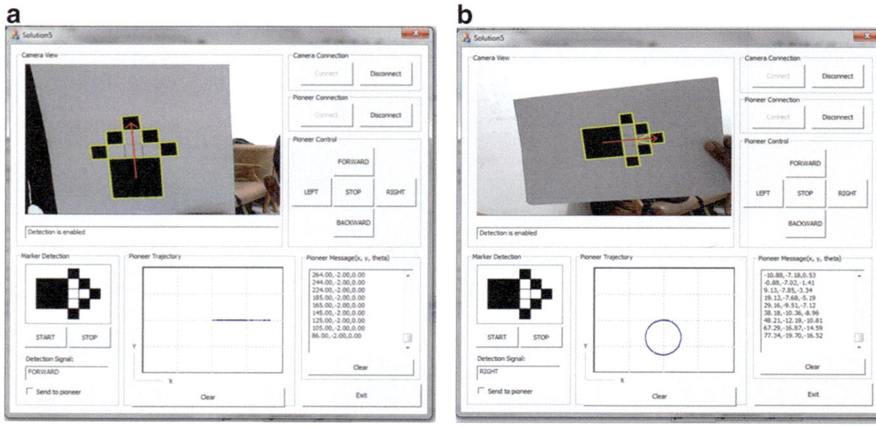


**Fig. 9** Some experimental results to test the validity of the algorithm shown in the developed GUI. (**a**) When marker is pointing up. (**b**) When marker is pointing to the right

## 6 Conclusions

As a conclusion, passive mark system for mobile robot navigation purposes was successfully developed and tested. This passive mark detection method is very useful in especially indoor environments such as homes, offices, factories, hospitals, and so on. Although the system is successful, there are still ways to make it better by eliminating the errors to bring down the percentage count.

The passive mark system for this project is done to reduce the usage of power and to sustain the resources. If the passive mark were to be active mark, the usage of power will be there as a factor of resistance in navigation. This means navigation will not take place if there is no power supply or battery source.

## References

1. Dissanayake MWMG, Newman P, Clark S, Durrant-Whyte HF, Csorba M (2001) A solution to the simultaneous localization and map building (SLAM) problem. IEEE Trans Robot Autom 17:229–241
2. Giralt G, Sobek R, Chatila R (1979) A multi-level planning and navigation system for a mobile robot: A first approach to Hilare. Proc Sixth Int Joint Conf Robot Autom
3. Greiner R, Isukapalli R (1993) Learning useful landmarks. Tech. Rep., Siemens Corporate Res, Princeton. The Twelfth National Conference on Artificial Intelligence/Sixth Conference on Innovative Applications of Artificial Intelligence, Seattle, Washington, 1994
4. Tsumura T (1986) Survey of automated guided vehicle in Japanese factory. Proc IEEE Int Conf Robot Autom 3:1329–1334

5. Elinas P, Little J (2005) sigmaMCL: Monte-Carlo localization for mobile robots with stereo vision. robotics: science and systems. Boston. Proceedings of Robotics: Science and System
6. Lorch O, Seara J, Strobl K, Hanebeck U, Schmidt G (2002) Perception errors in vision guided walking: analysis, modeling and filtering. In: Proceedings of the IEEE international conference on robotics and automation, Washington DC, USA

# Changes of Coastline: A Study Case of Carey Island-Morib Coast, Selangor, Malaysia

**Nor Lita Fadilah Rameli and Mokhtar Jaafar**

## 1 Introduction

Generally, the coast is the area that separates the coastal land and sea areas. History has proven the coast into focus as diverse as the development of settlements, fisheries, agriculture, port activities, and recreation area. Nowadays, development activities have increased geomorphological processes along coastal land and destroyed a huge area of natural coastal ecosystem. These situations have resulted in degraded coastal resources and expose the coastal landscapes to vulnerability. According Basir et al. [9], the increase in the intensity of coastal development utilizing the available resources in the coastal areas as well as the lack of awareness and involvement on conservation measures raises a lot of negative impacts on the coastal environment. This, coupled with the nature of the coastal ecosystems of the most dynamic landscape, affects land and sea and wind actions. The combination of these three agents, and accompanied by human activities, can promote changes in the coastal zone. According to Ahmad [4], changes in coastal or more accurately the changes in the coastline are also strongly influenced by the effects of sea level rise from global climate change. Global climate change is said to be able to change the process of sediment transport and the transfer of land and sea next to a beach and has implications for the increase or reduction of precipitated material in coastal zones. The direct impact of this natural process is the change in shoreline position.

Malaysia which has a length of 4,809 km coastline was not spared from natural processes in coastal zones. This could be associated with two main coastal geomorphological processes of erosion and deposition that transferred the coastal materials along the coastal zone. These physical processes have been able to change the coastline of the country [8]. Data from the Department of Irrigation and

N.L.F. Rameli (✉) • M. Jaafar
School of Social, Development and Environmental Studies, Faculty of Social Sciences and Humanities, Universiti Kebangsaan Malaysia, 43600 UKM Bangi, Selangor, Malaysia
e-mail: etafadilah@gmail.com

K. Ab. Hamid et al. (eds.), *The Malaysia-Japan Model on Technology Partnership: International Proceedings 2013 of Malaysia-Japan Academic Scholar Conference*, DOI 10.1007/978-4-431-54439-5_29, © Springer Japan 2015

Drainage (DID) show nearly 1,400 km of coastline in Malaysia experiencing critical erosion process. Many beaches in the east coast of Peninsular Malaysia are among the 65 locations experienced critical erosion. The Department of Irrigation and Drainage defined critical coastal erosion as any part along the coastal zone that suffered a serious erosion process and therefore difficult to be conserved. As a result, it's leaving a serious impact to the coastal zone and affected the stability of the natural beach profile. In the socioeconomic context, it referred to a situation that is capable of threatening the lives of coastal residents, especially the aspects of safety and economy.

Although the beaches in Malaysia have long suffered serious erosion processes, the aspects of coastal management are not included in the priority list of the government until the 1980s. Before that, the responsibility for monitoring and implementing control works and mitigation coast was placed under the jurisdiction of the joint committee by the DID, district office, and local government. As early as the 1980s, efforts were initiated to strengthen aspects of coastal management. The government has established a Coastal Engineering Technical Centre in January 1987, and this center serves as technical advisor for any development projects involving coastal component. In addition, the agency is also responsible for performing the monitoring, control, and conservation of coastal zone with the primary focus on coastal areas experiencing critical erosion problems.

As stated above, most beaches on the east coast of Peninsular Malaysia experience critical erosion process. However, the opposite is detected in most of the west coast of Peninsular Malaysia. Most of the beaches in the east coast of Peninsular Malaysia are directly exposed to the northeastern monsoon. This monsoon capable to produce aggressive waves that could erode the coastal zone rapidly. However, the opposite is happening in the western part of Peninsular Malaysia where wind action and wave action are moderate and less aggressive and act as corrosion agents rather than agents of deposition.

In this case, the beaches in Selangor are not immune from both the natural process of coastal erosion and deposition processes. As one of the most developed states in Malaysia, Selangor experienced a very rapid development process. Land use changes occur rapidly and also removed the sediments eroded into the river system that flows directly into the Straits of Malacca. Removed sediment eroded into the river system is transported to the mouth of the river, and most of the muddy sediments were deposited directly into the mouth of the river. A serious deposition process occurs as the waves are not aggressive [5]. This situation is also assumed to occur at Carey Island-Morib coast. Preliminary observations in the field show that there are certain parts along the coast which suffered serious deposition process, while some others are experiencing erosion process. Therefore, this article seeks to clarify the changes of coastline along the Carey Island to Morib beach. The changes are based on the assumption of the presence of muddy material during low tide, and the velocity of waves during high tide is considered at high momentum. Both situations were observed during the first visit in the fields and these led to the study of coastline change along the Carey Island to Morib beach. Data on physical characteristics of the study area could be used as a baseline data to regulate and conserve resources in the coastal zone especially in the study area.

## 2 Methodology and Research Background

As stated above, the study of shoreline change is done in Carey Island Morib coast. The beach is located in the western part of Peninsular Malaysia 2° 36′32.64′ N 101° 41′05.68′ E and 2° 36′32.64′ N 101° 41′05.68′ E as shown in Fig. 1. The coastline is about 30 km long and characterized as muddy and sandy sediments. The Carey Island covered with Langat River in the north and south, and both are skirting out into the Straits of Malacca. The Straits of Malacca are one of the busiest sea routes in terms of ship movements and carrier trade. The beach is known for rest and recreation. Some beaches along the Carey Island to Morib are covered with mangrove and this mangrove is seen as a wave's buffer zone. Preliminary observations in the field show that certain parts of the mangrove clusters have collapsed and died, and this is an early indication of coastal erosion processes. Another part of the coast indicates the presence of a lot of precipitates, and these precipitates are an early indication of deposition process.

To track the changes to the coastline of Carey Island-Morib coast, edge geospatial technology is fully applied. A summary of the study methodology is shown in Fig. 2. In general, this article will only involve the measurement of

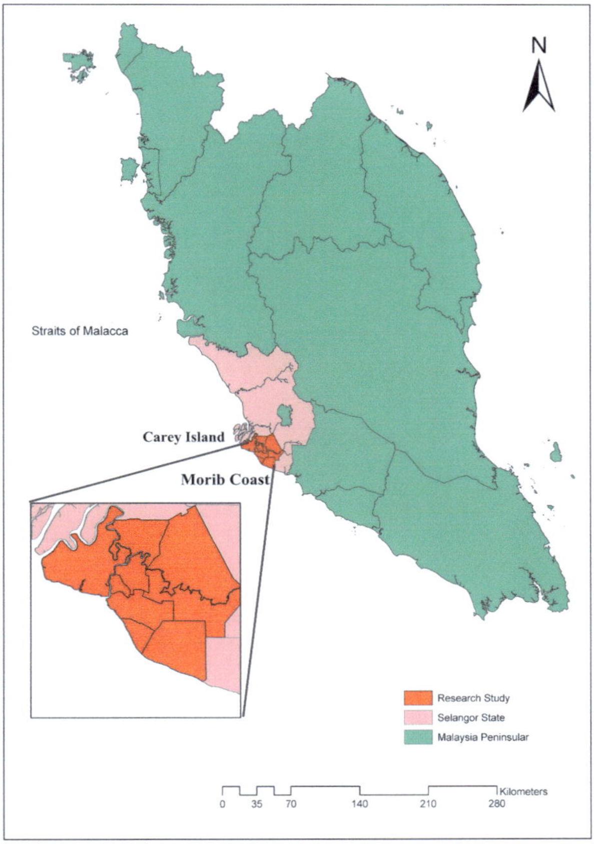


**Fig. 1** Carey Island-Morib coast. *Source*: Modified from Malaysia Peninsular Boundary map

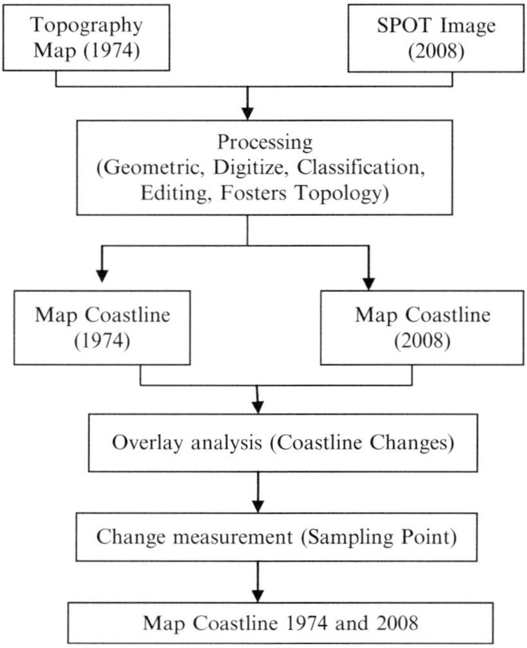


**Fig. 2** Flow chart of research methodology

shoreline change at the first stage of identifying the source of this change through map and source images. The second stage involves the measurement of the coastline in the field using the Global Positioning System (GPS), but this level will not be covered in this article.

The first process is digitizing the coastline of Morib coast and Carey Island (source topographic maps that represent the position of the shoreline in 1974). This process is done using geographical information system (GIS) in which a digital map layer is built into ArcGIS package. Meanwhile, the ERDAS package had been used to build the coastline of the study area. Both data have been through some geographic processes such as geometric, topological classification and thematic layer construction. In order to detect the changes of the Carey Island-Morib coastlines, all thematic maps were analyzed using simple overlay procedures. In addition, the measurement of distance change was based on 37 locations of field sampling points (Fig. 3).

## 3 Research Result and Discussion

Figure 4 shows the results of mapping Carey Island-Morib coast for 1974 and 2008. In general, both erosion and deposition processes can be traced along the coast. The north coast of Carey Island was experiencing serious erosion process while middle

**Fig. 3** Distribution of sampling points relative

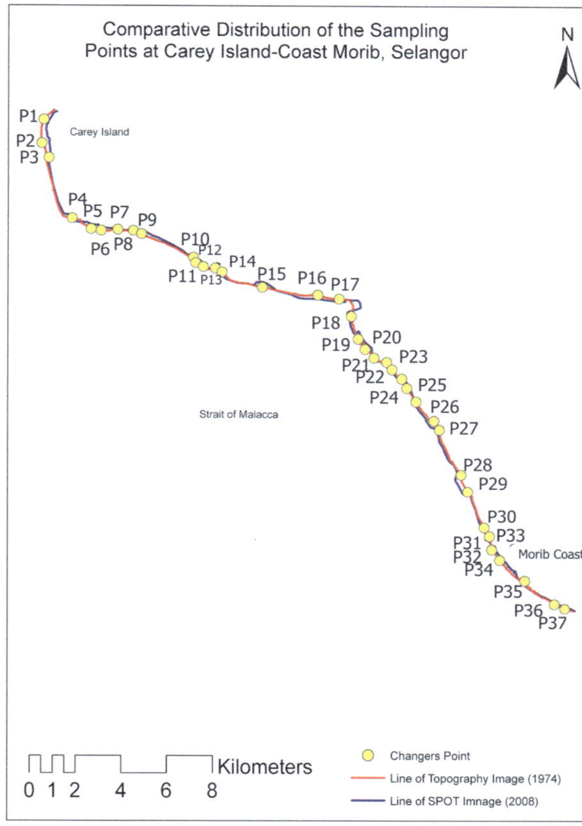


parts of Morib beach were suffered deposition process. Sg. Langat, south coast of Carey Island, is also found to undergo a process of erosion.

To clarify the picture of the erosion-deposition process that occured on both beaches, shoreline change measurement data are shown in Table 1. P1–P3 sampling locations that represent the area on the north coast of Carey Island recorded the average distance of highly eroded beach, about 261 m. Even the highest eroded beach distances are also recorded in this part of the sampling locations of P2 380.8 m. Meanwhile, the deposition process was detected seriously in the southern part of the Morib beach from P26 to P30 sampling points. The average value of deposited beach distance recorded for P26–P30 sampling locations is 214 m. The highest deposition distance was recorded at P27 and P29 sampling points of 294 and 299 m respectively. In addition, 20 sampling points experienced erosion process while another 17 sampling points experienced deposition process. The average value of eroded coastline was 170 m, and 157 m was the case of deposit coastline.

Based on these initial findings, it is clear that Carey Island-Morib coast experienced more erosion process than deposition process. As a comparison, Carey Island

**Fig. 4** Comparison of shoreline in 1974 and 2008

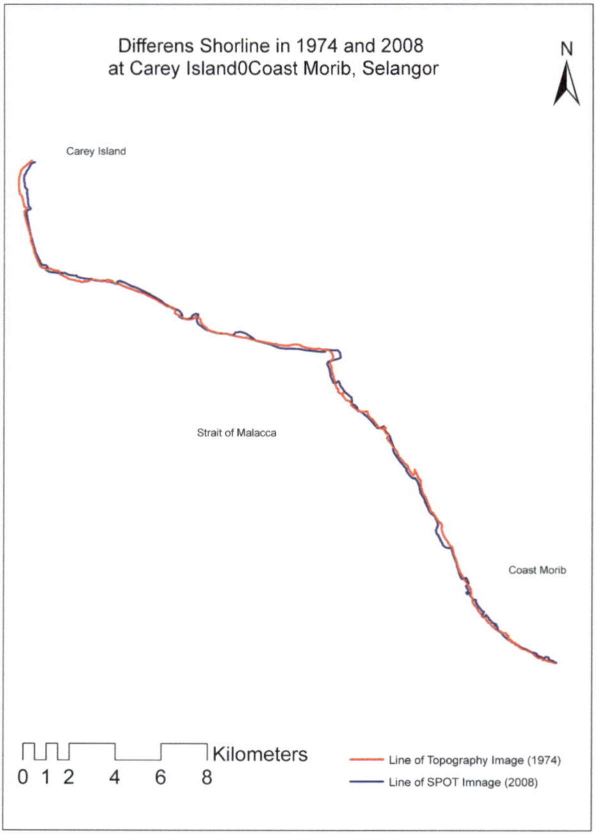


experienced more erosion process while Morib beach experienced more deposition process. Generally both processes are strongly influenced by the action of the dynamic agents of fluctuation which include coastal waves, currents, tides, and wind. All these agents act by affecting the energy intensity. The energy contained in the waves and currents is able to move the activity to alter the shape of the beach. Waves are the main cause of the changes taking place on the beach. Waves provide an important energy source for coastal construction, the production of materials on the basis of surface coastal sediments, and transporting the raw materials to the shore, offshore, and parallel to the beach. Strong waves with strong energy are capable to move beach materials, especially moderate and fine sands [6]. If the wave energy decreases, the precipitated material is deposited in the bottom and on the shore. The energy of wave is also can be reduced if the movement of the wave to the beach shore was hindered by any object such as shelf sand and huge beach rock. In addition, the energy of the wave could be increased when the wave reflected back to the sea [3]. Therefore, both of movement and reflected wave to and from the beach would affect the coastline profile along the Carey Island and Morib beach.

**Table 1** Coastline change based on erosion-deposition processes

| Sampling location | Coastline change (m) Erosion | Coastline change (m) Deposition |
|---|---|---|
| P1 | 236 | |
| P2 | 381 | |
| P3 | 166 | |
| P4 | | 127 |
| P5 | 179 | |
| P6 | 127 | |
| P7 | | 65 |
| P8 | | 72 |
| P9 | 140 | |
| P10 | | 53 |
| P11 | 245 | |
| P12 | | 143 |
| P13 | 140 | |
| P14 | | 157 |
| P15 | 214 | |
| P16 | | 173 |
| P17 | 130 | |
| P18 | | 200 |
| P19 | 100 | |
| P20 | 171 | |
| P21 | 94 | |
| P22 | | 158 |
| P23 | | 94 |
| P24 | 102 | |
| P25 | 130 | |
| P26 | | 113 |
| P27 | | 294 |
| P28 | | 123 |
| P29 | | 299 |
| P30 | | 245 |
| P31 | 150 | |
| P32 | | 218 |
| P33 | 216 | |
| P34 | 179 | |
| P35 | | 149 |
| P36 | 137 | |
| P37 | 171 | |

Currents in coastal area are created by the interference waves. This highly influential currents erode, transport and deposit sediment material. Currents also cause a rapid flow of water on the coast [2]. Wind action on the coastline occurs directly and indirectly. Currents and winds in both of these situations act as an agent of erosion or deposition. Wind indirectly creates waves and waves act as an agent of erosion and sedimentation in coastal areas. Direct action of the wave could be associated with its function in transferring fine materials to the berms while indirectly wind could create strong wave and current and eroding the coastal zone.

In addition, wind velocity along the study area could be decreased as the study area is barred by the Sumatera Island, Indonesia [7].

The period between high tide and low tide is called ebb tide. Tides in the ocean are more influenced by the moon than the sun. Time of high tide and low tide is closely related to place the month. Therefore, tides are also important in determining the length of the beach profile. Relatively, when the tides reached its highest level, the beach profile will be shorter, and the beach profile seems longer during the low level of tides. During low tide, the beach profile is longer and at high tide, it is shorter.

In addition, human factors such as the movement of large ships in the Straits of Malacca are believed to play an important role in influencing the process of erosion on the north coast of Carey Island. The north coast of the Carey Island is closed to the port activity known as Port Klang where many large ships moving daily within the port zone. Large vessels are capable of producing energy waves, and when the waves get to the beach, these act as an agent of erosion. Meanwhile, the process of muddy deposition in Morib could be related to changes in land use activity on the mainland. Nearby land to the sea along the Morib beach has been planted with oil palm. This plantation could lead to soil erosion as its soil surface does not cover with vegetation cover. Eroded soil, then has been transported into nearby streams and flowing down to the estuaries. During low tide, most of eroded sediment that reached the river mouth will be stranded and remains as muddy sediment.

## 4 Conclusion

This article has discussed the shoreline change in Carey Island-Morib coast. The findings clearly show that the study area experienced both erosion and deposition processes therefore affect the changes in Carey Island to Morib coastline. Geospatial technology involves the use of GIS and satellite imagery applied in this study. Based on the overlay analysis of thematic maps, the processes of erosion and deposition shown in the format of the line that representing changes in the coastline. However, coastal erosion processes were more dominant in Carey Island and more dominant deposition process in Morib coast. Apart from the action of natural agents, geomorphology, currents, tides, and wind, it is believed that the movement of large ships in the Straits of Malacca and Port Klang has resulted in beach erosion processes in Carey Island. Meanwhile, human activities, especially land use change are believed to affect the presence of serious muddy material deposition along the Morib beach. To confirm these initial findings, further studies need to be moved to include observations in the field to measure the physical characteristics of the beach. The combination of data obtained from the analysis of thematic maps and field data is convincing more interpretation of shoreline change at Carey Island-Morib coast that can be associated with erosion and sedimentation processes.

## References


1. Zakaria AS (1970) Geomorphology of Kelantan Delta. M. A. Thesis, University, New York
2. Komar PD (1976) Beach processes and sedimentation. Prentice-Hall, New Jersey
3. Sharifah Mastura SA (1988) Littoral environment observation study. No mimeographed series 12. National University of Malaysia, New York
4. Ahmad FA (2005) Effect of coastal development guidelines against stability in small islands of Langkawi Island. Faculty of civil engineering, Universiti Teknologi Malaysia, Johor
5. Sesli AF (2006) Mapping and monitoring temporal changes for coastline and coastal areas by using digital aerial photogrammetric and image data: a case study from Samsun, Turkey. Department of Geometrics Engineering, Ondokuz Mayiz University, Samsun
6. Marfai MA (2011) The hazards of coastal erosion in Central Java, Indonesia: an overview. Geography Faculty, Gadjah Mada University, Indonesia
7. Mustakim M (2012) Coast line detect changes using GIS Methods, Case Study-Kelanang Morib Beach. B. A. Academic Exercise, UKM, Bangi. Malaysia (Unpublished Academic Exercise)
8. Kasawani I, Shah AKA, Elder I, Hasmadi IM, Kamaruzaman J (2010) Coastal change detection using GIS in Setiu Lagoon, Terengganu, Malaysia. Faculty of Science and Technology, University Malaysia Terengganu, Malaysia
9. Basir N, Taufik M, Sukojo BM (2010) Coastal vulnerability model of sea water increases with harness interface remote sensing technology, Case study: Sentul Island. Remote Sensing Studies Program, Faculty of Civil Engineering and Planning Campus Sukolilo ITS, Surabaya
10. Smith MJ, Cromley RG (2012) Measuring historical coastal change using GIS and the change polygon approach. U.S. Geological Survey Troy, New York. Department of Geography University of Connecticut

# Assessing the Condition of Buried Pipe Using Ground-Penetrating Radar (GPR)

**S.W. Wahab, D.N. Chapman, C.D.F. Rogers, K.Y. Foo, S.W. Nawawi, and K.H. Abas**

## 1 Introduction

In the UK, most of the essential public service infrastructure was installed in the last 200 years, in which obviously the qualities of buried pipes have become degraded. To overcome this issue, approximately four million holes were dug every year in the UK roads in order to repair assets and new installations [1]. However, inaccurate asset management can lead to unnecessary holes dug in the wrong place, and this often causes third-party damage to other underground services. Those damages are potentially lengthy delays to construction works and increasing the direct costs of maintenance to the service provider and not least to the social disruption as well. It is therefore important to accurately locate these services which could minimise unnecessary holes dug for the repair and maintenance of existing assets.

The invention of ground-penetrating radar (GPR) technology has facilitated the possibility of detecting buried utilities and has been used primarily in civil engineering for detecting structural defects, such as voids and cavities in road pavements, slabs and bridge decks [2], but has not been used to assess the condition of

S.W. Wahab (✉)
School of Civil Engineering, University of Birmingham, Edgbaston, Birmingham B15 2TT, UK

Department of Survey and Mapping Malaysia, Wisma JUPEM, Jalan Semarak, 50578 Kuala Lumpur, Malaysia
e-mail: khairulhamimah@fke.utm.my

D.N. Chapman • C.D.F. Rogers
School of Civil Engineering, University of Birmingham, Edgbaston, Birmingham B15 2TT, UK

K.Y. Foo
Pulsar Process Measurement Ltd., Cardinal Building, Enigma Commercial Centre, Sandy's Road, Malvern, Worcestershire WR14 1JJ, UK

S.W. Nawawi • K.H. Abas
Universiti Teknologi Malaysia, 81310 Skudai, Johor, Malaysia

K. Ab. Hamid et al. (eds.), *The Malaysia-Japan Model on Technology Partnership: International Proceedings 2013 of Malaysia-Japan Academic Scholar Conference*, DOI 10.1007/978-4-431-54439-5_30, © Springer Japan 2015

buried pipes. Degraded pipes can be defined as pipes where, for example, cracking, differential deflection, missing bricks, collapses, holes, fractures and corrosion exist [2], [3], [4]. This GPR technology has become a favourable geophysical technique in detecting underground utilities. However, this technique still has certain limitations obviously depending on ground conditions, type and depth of objects. This feasibility research focuses on investigating whether off-the-shelf ground-penetrating radar (GPR) can be used for assessing the condition of pipes.

## 2 Basic Principles of GPR

GPR is a device used for noninvasive scanning which is able to record accurate depth reading and the signature of targets (radagram) for further properties interpretations and mainly to detect shallow or deep targets depending on the frequency of the antenna applied. The function of GPR wave transmission and reflection is shown in Fig. 1.

In physical brief, GPR is composed of a receiver and transmitter antenna, a control unit, battery supply and a survey cart. The control unit is the main part of the GPR because it controls the whole system. Generally, when the survey cart moves on the site surface, the transmitting antenna sends polarised, high-frequency electromagnetic (EM) waves into the ground. Because of different existing inhomogeneities in the ground, e.g. soil layers, underground utilities, stones, gravel, cavities and other anomalies, a proportion of the EM waves are reflected from the dielectric boundary between different materials and the rest is refracted and continues to deeper layers. The process is repeated until the EM waves become too weak. Reflection of the EM waves from the dielectric boundary is the consequence of differences in the electric and magnetic properties of materials of infrastructural objects and soil layers [5].

As a conclusion, the GPR works by sending pulses of energy into the ground and recording the strength and the time for the return of any reflected signal. This series of pulses over a single area is called a scan. Reflection occurred whenever the energy pulses are transmitted through various materials on their way to the buried target feature, thus changing the velocity. Their velocity changes depending basically on two primary electrical properties of the subsurface: electrical conductivity ($\sigma$) and dielectric permittivity ($\varepsilon$). Reflections are produced by changes in the dielectric contrast due to changes in the subsurface materials [6]. The greater the contrast between two materials at subsurface interface, the stronger the reflected signal, resulting in a higher-amplitude reflected wave [7]. Therefore, it is necessary to understand those characteristics of the materials, which affect both the velocity of propagation and attenuation.

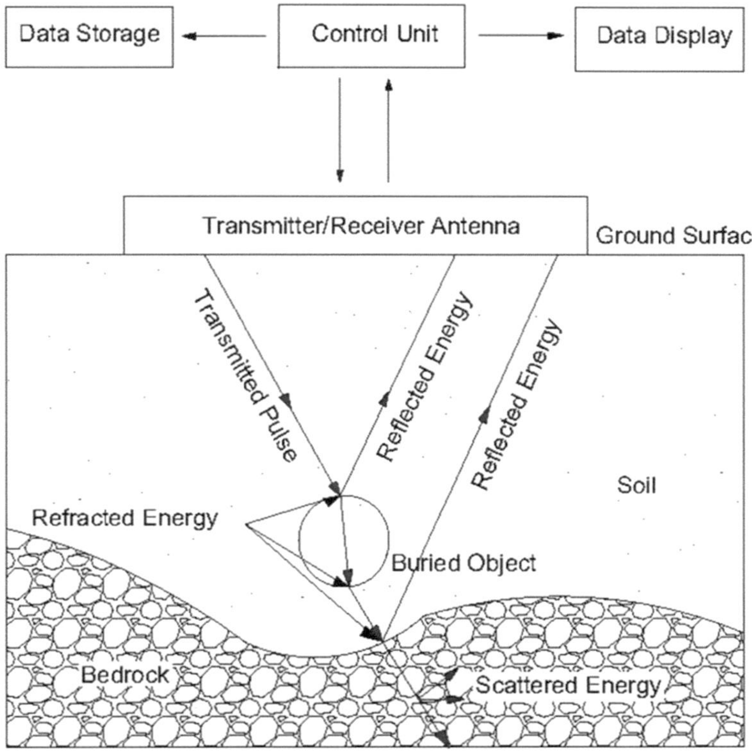


**Fig. 1** Functional diagram of GPR equipment [7]

## 3 Research Methodology

In this research, a laboratory test facility consisting of a box of dimensions 2.4 m (length) × 2.2 m (width) × 1.2 m (height) constructed from structural insulated panel material is used. The size of the box was determined as the minimum requirement for the GPR unit based on beam width antenna calculations so as to avoid signal reflection from the edges or base of the box and to ensure the complete shape of the hyperbolic trace from the targets to be captured. In these experiments, only a plastic pipe was chosen as the main pipe material. The experiments involve burying 0.2-m-diameter plastic pipes in the box (Fig. 2). One of these pipes is in a 'good' state (undamaged) and the other has a defect (i.e. break or hole), i.e. damaged pipe.

In these experiments, the depth of buried pipe was fixed at 0.5 m from the soil surface. Meanwhile, the position of the buried pipe was laid at 1 m gap between each other and 0.4 m from the wall. This depth and position was chosen as an ideal orientation, which could minimise those signals that are unrelated to the target (clutter). Clutter can be caused by breakthrough between the transmitter and

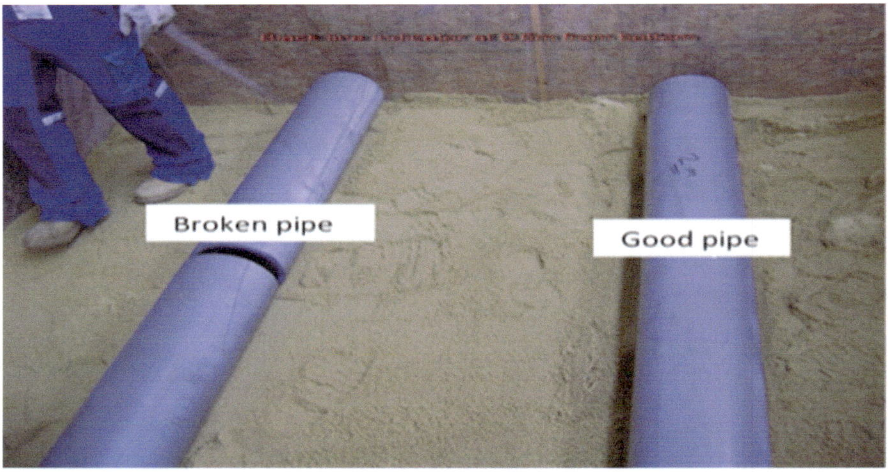


**Fig. 2** Arrangement of the two plastic pipes in the test box prior to burial

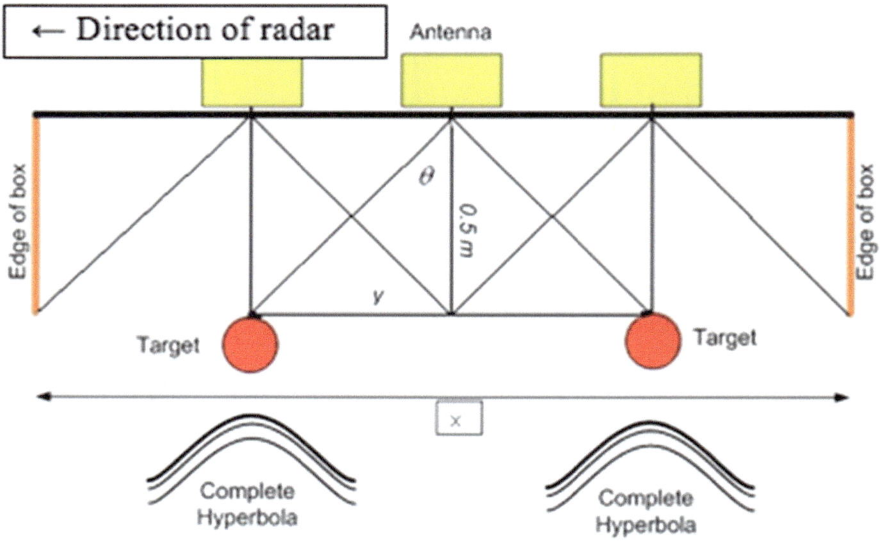


**Fig. 3** Minimum size of test facility

receiver antennae as well as multiple reflections between the antenna and the ground surface. Generally, clutter is more significant at short-range times (shallow target) and decreases at longer times (deep target). In theory, in order to get the complete shape of hyperbola from the targets (Fig. 3), Eq. (1) had been applied in which the width of the test facility must be at least $X = 4y$:

$$y = \tan^{-1}\theta\,(0.5) \tag{1}$$

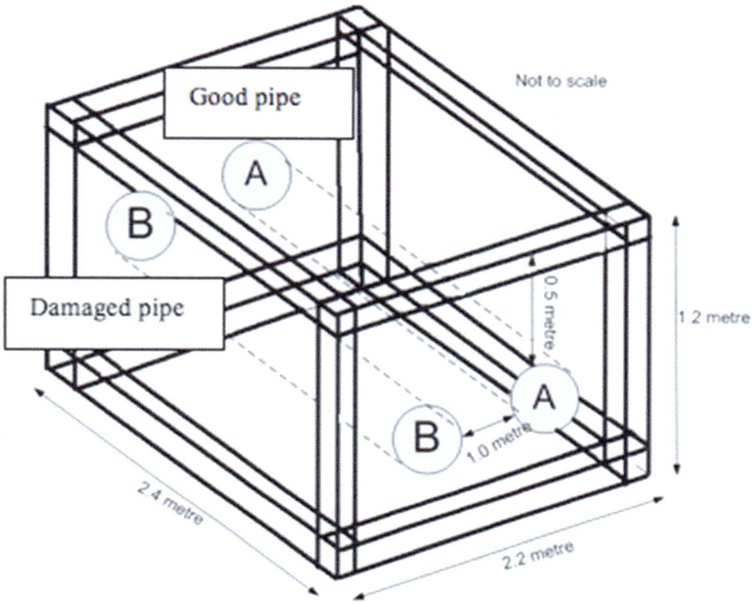


**Fig. 4** A schematic model of the test facility indicating the pipe position

A schematic model of the test facility indicates the pipe position and how it was constructed as shown in Fig. 4. A series of laboratory experiments were conducted to determine the validity and effectiveness of the GPR technology in assessing the condition of buried utilities with common plastic pipe damages. Several damages in plastic pipe were tested with respect to different GPR antenna frequencies. Three configurations of equipment setting were conducted during these experiments. The GPR surveys were carried out in order to obtain signal signatures from damaged and undamaged pipes. These surveys were organised through grid pattern across a test facility (Fig. 5). Ultimately, the information of all the radar signals were extracted and were further examined and analysed.

## 4 Results and Discussion

From the image shown in Fig. 6, the propagation velocity estimation (red line) is $5.8 \times 10^7$ ms$^{-1}$ on the basis of the shape of a hyperbola. This value can be compared with the velocity value as stated in Table 1 and it is acceptable.

From the radar images for X direction (Figs. 7 and 8), it is shown that there is an image distortion at survey grid X3 (damaged pipe) compared to survey grid X1 or X5 (undamaged pipe). Even though the image resolution is not quite clear to interpret the condition of the pipe, doing very high-dense grid might increase its survey in a particular area. The result from Fig. 8 is much clearer (sharp) compared

Fig. 5 Survey grid lines used for each experiment

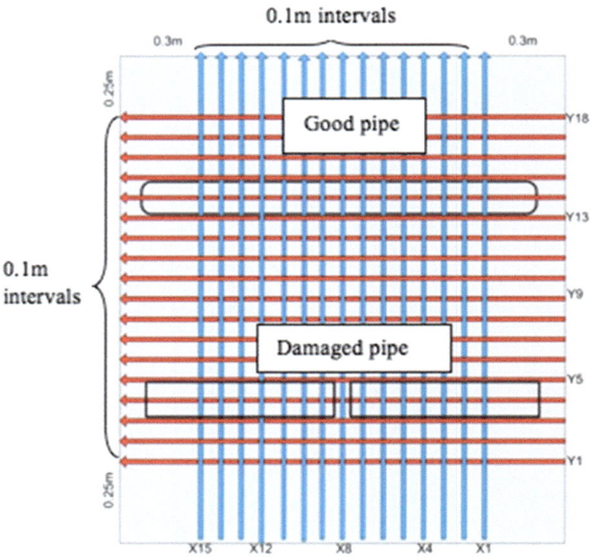


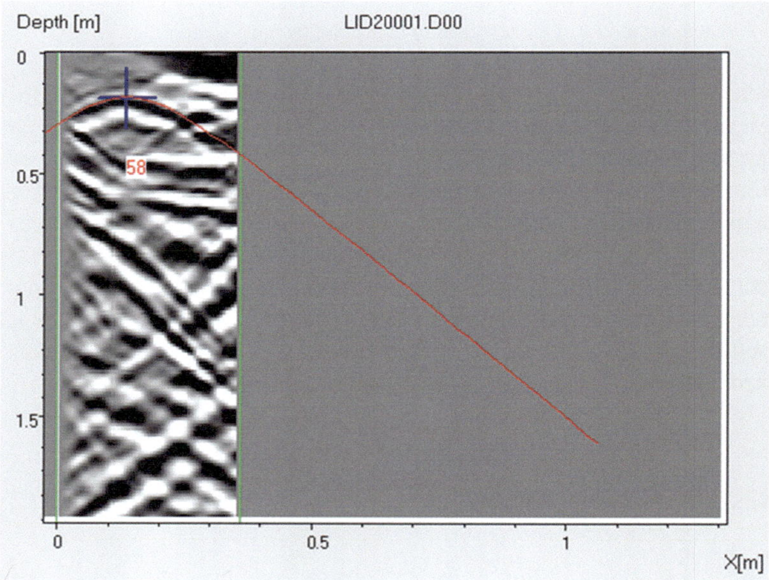


**Fig. 6** Propagation velocity estimation

**Table 1** Dielectric value for common materials (Daniels [6])

| Materials | Relative permittivity($\varepsilon_r$) | Velocity (mm/ns) |
|---|---|---|
| Dry sand | 2–6 | 120–170 |
| Wet sand | 10–30 | 55–60 |

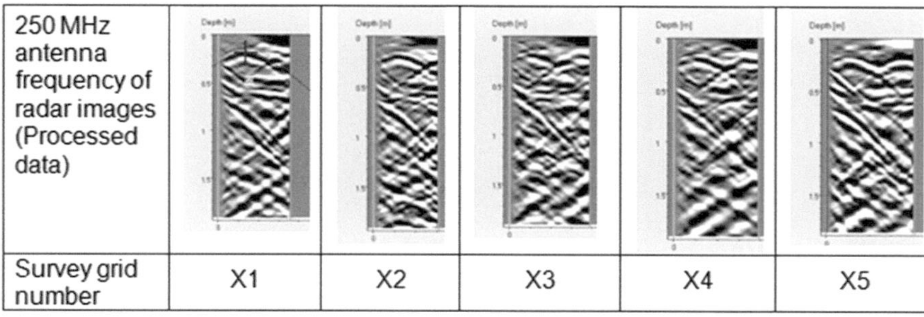


**Fig. 7** Processed data of X direction

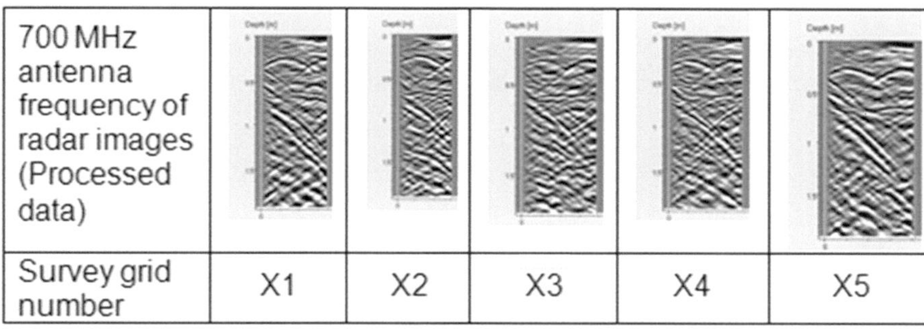


**Fig. 8** Processed data of X direction

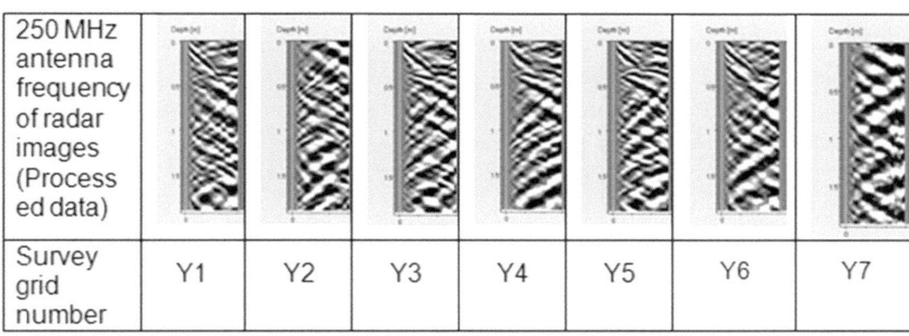


**Fig. 9** Processed data of Y direction

to Fig. 7. The resolution of high-frequency antenna is better than low-frequency antenna. On the other hand, this result also can be improved with some combinations of signal processing methods such as by GRED 3D signal processing or by using the Matlab to analyse any changes.

However, referring to radar images for Y direction (Figs. 9 and 10), it is quite hard to analyse the images because of high signal reflection from other sources. In this case, the condition of control environment is considered with no signal interference;

| | | | | | | | |
|---|---|---|---|---|---|---|---|
| 700 MHz antenna frequency of radar images (Processed data) | 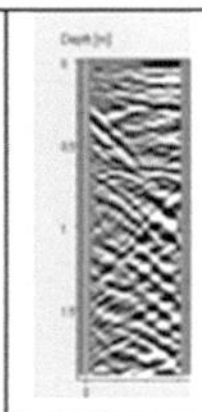 | 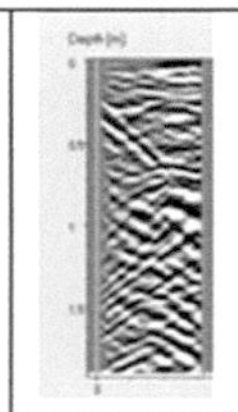 | 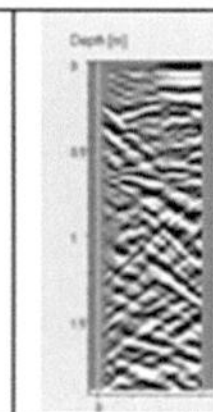 | 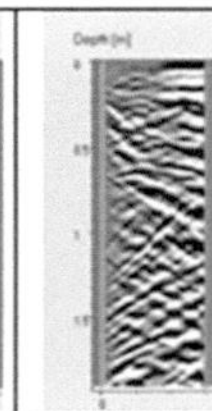 | 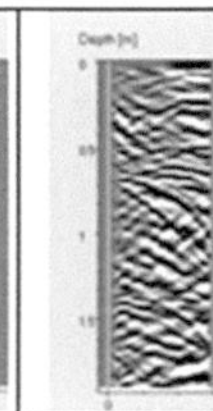 | 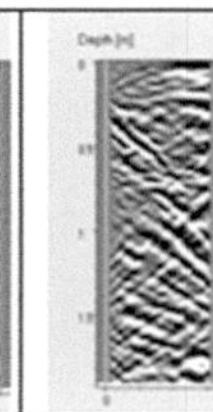 | 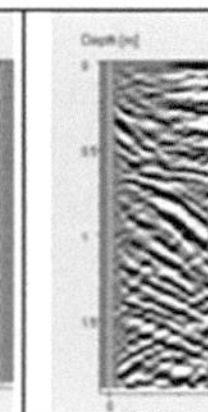 |
| Survey grid number | Y1 | Y2 | Y3 | Y4 | Y5 | Y6 | Y7 |

**Fig. 10** Processed data of Y direction

thus, the noise is assumed to be nearly zero. In this preliminary work, this condition is considered, but the authors feel that the results can be further enhanced by considering this noise in the analysis.

## 5 Conclusion

This work has successfully shown that both frequencies of 250 and 700 Mhz are capable of observing the defective regions in the pipe. At 700 Mhz frequency radar scan was sharper and brighter due to the signal resolution compared to 250 Mhz. However, the 250 Mhz antennae produced less signal interference compared to the 700 Mhz. Finally, pipe damage was observed easier when the survey was conducted perpendicular to the pipe, and it is necessary to established suitable grid spacing to identify and verify the damage region. Meanwhile, it is necessary to ensure the pipe alignments before radar scan was conducted.

## References

1. Beck AR, Fu G, Cohn AG, Bennett B, Stell JG (2008) A framework for utility data integration in the UK. In: Coors V, Rumor M, Fendel EM, Zlatanova S (eds) Urban and regional data management, Proceeding of the urban data management society symposium 2007. Taylor and Francis, London, pp 261–276
2. Koo DH, Ariaratnam ST (2006) Innovative method for assessment of underground sewer pipe condition. J Autom Construct 15:479–488
3. Sinha SK, Knight MA (2004) Intelligent system for condition monitoring of underground pipelines. Comput Aided Civ Infrastruct Eng 19:42–53
4. Silva DD, Davis P, Burn LS, Ferguson P, Massie D, Cull J, Eiswirth M, Heskee C (2002) Condition assessment of cast iron and asbestos cement pipes by in-pipe probes and selective sampling for estimation of remaining life. No Dig, Australia

5. Daniels DJ (2004) Ground penetrating radar, 2nd edn. The Institution of Engineering and Technology, London, pp 2–27
6. Eyuboglu S, Mahdi H, Al-shukri H, Rock L (2003) Detection of water leaks using ground penetrating radar
7. Kuo S, Zhoa L, Mahgoub H, Suarez P (2005) Investigation on ground penetrating radar for detection leaking pipelines under roadway pavements and development of fiber wrapping repair technique. Transportation, US

# Evaluation of Palm Decanter Cake as an Alternative Fish Feed

**Ying Ping Chang and Jun Hoeng Chuah**

## 1 Introduction

The oil palm industry is the key economy growth driver in Malaysia, providing thousands of jobs, and also one of the largest contributors to external trade. Solid waste generated from a palm oil mill includes empty fruit bunches, palm kernel cake, palm kernel shell, and decanter cake. Decanter cake is derived from the debris of the fresh fruit bunches and the mesocarp residue of fruitlets during crude palm oil processing. Generally, an average palm oil mill which processes 90 tonnes of fresh fruit bunches per hour can produce approximately 160–200 tonnes of decanter cake per month [1]. Thus, improper management and disposal of decanter cake can be disastrous to the ecology system. Palm decanter cake has been explored to be incorporated into bio-fertilizers [1] and ruminant feed [2], but the potential to incorporate it into fish feed is yet to be explored. Aquaculture is one of the fast-growing animal food-producing sectors and currently accounts for more than 60 % global fish production between year 2000 (32.4 million tonnes) and 2008 (52.5 million tonnes) [3]. Nile tilapia (*Oreochromis nilotcus*) and the catfish (*Clarias batrachus*) are widely farmed for human consumption in Malaysia. The cost of aquafeeds is crucial to ensure the development of commercial freshwater fish farming. There are researches to incorporate plant protein sources and spent yeast into selected freshwater fish diet [4, 5], to reduce the dependency on animal protein such as fish meal. Indeed, the composition of biomass governs its potential utilization. A viable alternative fish feed should contain low levels of fiber, starch, and antinutrients but relatively high protein and possess high nutrient digestibility. Digestibility studies are usually conducted on potential feed material to predict the nutritional quality of the feed and the availability of nutrients to the fed animals [6].

Y.P. Chang (✉) • J.H. Chuah
Department of Chemical Science, Faculty of Science, Universiti Tunku Abdul Rahman, Jalan Universiti, Bandar Barat, 31900 Kampar, Perak, Malaysia
e-mail: changyp@utar.edu.my

K. Ab. Hamid et al. (eds.), *The Malaysia-Japan Model on Technology Partnership: International Proceedings 2013 of Malaysia-Japan Academic Scholar Conference*,
DOI 10.1007/978-4-431-54439-5_31, 

In vitro protein digestibility studies using crude enzyme extract from fish intestine allow the observation of the dynamic breakdown of protein with little material used. The pH drop method had been used to study the protein digestibility of various aquatic feeds [7, 8]. To the best of the authors' knowledge, in vitro protein digestibility study on palm decanter cake to evaluate its suitability as an alternative feed for Nile tilapia and catfish has yet to be documented. Thus, the present study was undertaken to quantify nutrient content in palm decanter cake and evalute its protein digestibility using crude fish gut enzymes from Nile tilapia and catfish.

## 2 Materials and Methods

### 2.1 *Sample of Decanter Cake*

Fresh decanter cake was obtained from Tian Siang Oil Mill Sdn Bhd, Air Kuning, Perak, Malaysia. The decanter cake was stored in a freezer at −20 °C before further analysis.

### 2.2 *Crude Fish Gut Enzyme Preparation*

Three live speciments of mature *Oreochromis nilotcus* and *Clarias batrachus*, respectively, were collected from Kim Seng Fishery Co. at Temoh, Perak, Malaysia. The fishes were acclimatized for one week before they were subjected for enzyme extraction [9]. The alimentary tract was collected and weighed after the fishes were starved for 24 h. The alimentary tract was then grounded with a homogenizer and mixed with cold distilled water at a ratio of 1:10 before centrifugation at 10,000 (×*g*) for 10 min at 4 °C. The lipid layer was discarded and the supernatant was collected and stored at −20 °C.

### 2.3 *Proximate Analysis*

Decanter cake samples from 1.2.1 were analyzed in accordance to Weende proximate feed analysis scheme [10], in triplicate. Moisture, crude fat, crude protein, ash, and crude fiber contents of palm decanter cake were determined based on the standard AOAC method [11]. Moisture content was expressed in percentage of wet weight basis (wb) while crude fat, crude protein, and crude fiber contents were expressed in percentage of dry weight basis (db).

### 2.4 Determination of In Vitro Protein Digestibility

In vitro protein digestibility of decanter cake was carried out based on the pH drop method [12] with some modification. Fresh decanter cake sample from 2.1 that equivalent to the amount of ingredient that provides 160 mg of crude protein was weighed and soaked with 20 mL distilled water overnight at 4 °C. Then, the pH of the mixture was adjusted to pH 8, and 2 mL of thawed crude fish gut enzymes from 1.2.2 was added into the sample suspensions. The pH changes were recorded every minute up to the 10th minute with a pH meter. The same procedure was repeated for casein which was used as a reference of pH drop. The plots of pH versus time were constructed and the gradient represents the pH changes over time, $-\Delta$pH. The percentage of the relative digestibility of protein (RPD) was calculated as the percentage ratio of $-\Delta$pH of the sample and the $-\Delta$pH of casein as follows:

$$\mathrm{RPD}(\%) = \frac{-\Delta\mathrm{pH\,of\,sample}}{-\Delta\mathrm{pH\,of\,casein}} \times 100$$

## 3 Results and Discussion

### 3.1 Chemical Composition

Fresh decanter cake contained excessive water which was 76.5 ± 1.2 % wb. This result is comparable to those reported by Paepatung et al. [13] and Dewayanto [14]. High moisture content may induce microbial growth which is undesirable in fish feed pelleting. Thus the decanter cake should be kept dry for long-term storage and further processing. In this study, decanter cake samples were dried and estimated the moisture content simultaneously before storage in an air-tight container at room temperature for compositional analysis. Table 1 shows the composition of dried decanter cake.

The crude lipid content found in decanter cake exceeded the recommended amount of 6–8 % for Nile tilapia [15], but may be acceptable for catfish since lipid in catfish diet can be up to 16 % without detrimental effects [16]. The crude protein content of decanter cake was below the recommended amount of 24–30 % in order to be accepted as a suitable feed for broodstock [17, 18]. Hence, decanter cake can only replace partially commercial fish feed to prevent adverse effect on fish growth. The crude fiber content of decanter cake was slightly lower as compared with the amount in dried palm oil mill effluent (20.1 %) as reported by Wong and Wan Zahari [19]. However it was above the acceptable maximum levels of 6 % [16] for catfish and 5 % for Nile tilapia [15], respectively, which may affect its digestibility in these freshwater fishes.

**Table 1** Composition of dried decanter cake

| Content | Percentage (db) |
|---|---|
| Crude protein | 14 ± 0.8 |
| Crude lipid | 9.5 ± 1 |
| Crude fiber | 17 ± 0.9 |
| Ash | 8.7 ± 0.8 |

The results were expressed as means ± standard deviation. $n = 3$

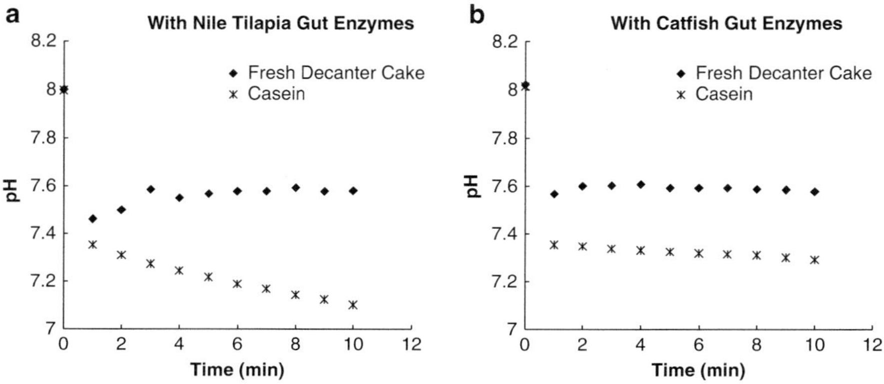


**Fig. 1** Plots of pH drop of fresh decanter cake and casein by enzymes from (**a**) Nile tilapia and (**b**) catfish

## 3.2 *In Vitro Protein Digestibility*

Figure 1a, b shows the pH drop of fresh decanter cake and casein using crude fish gut enzyme of Nile tilapia and catfish, respectively. This shows that the digestive enzymes in both types of fishes are able to hydrolyze proteins in decanter cake. However, both pH drops were not as steep as compared to the pH drop for casein probably due to the presence of high levels of crude fiber, such as non-starch polysaccharide in palm decanter cake that may decrease both in vivo and in vitro digestibility of dry matter in many fish species [20]. The pH drop of decanter cake was relatively lower if using fish gut enzymes of Nile tilapia as compared to those of catfish. Consequently, the RPD of fresh decanter cake using crude gut enzymes from Nile tilapia (23.5 %) was apparently lower than that of catfish (54.9 %). This may be because Nile tilapia is a type of herbivorous fish in nature with lower protein requirements [21]. Thus, the crude gut enzymes may not be as efficient to digest protein as compared to an omnivorous fish such as catfish.

## 4 Conclusions

The high levels of crude fiber and marginally low content of crude protein in decanter cake of palm oil mill make it unsuitable to completely replace the conventional fish feed. However, the in vitro protein digestibility result indicates that it is readily digested and may incorporate as part of the protein source in fish feed formulation.

**Acknowledgment** This study was supported by Universiti Tunku Abdul Rahman, Malaysia. Authors are thankful to Tian Siang Oil Mill Sdn Bhd and Kim Seng Fishery Co., located at Perak, Malaysia, for providing the palm decanter cake and the live specimens of freshwater fish, respectively.

## References

1. Khalid H, Ahmad TM, Rohaya MH, Ahmad KD (2008) Palm-based bio-fertilizer from decanter cake and boiler ash of palm oil mill. http://palmoilis.mpob.gov.my/publications/TOT/TT-412.pdf. Accessed 28 Jan 2013
2. Dashini DOG (2009) Oil recovery from palm oil solid wastes. Dissertation, University Malaysia Pahang
3. FAO Fisheries and Aquaculture Department (2011) World Aquaculture 2010. http://www.fao.org/docrep/014/ba0132e/ba0132e.pdf. Accessed 18 Dec 2012
4. El-Saidy DMSD, Gaber MMA (2003) Replacement of fish meal with a mixture of different plant protein sources in juvenile Nile tilapia, *Oreochromis niloticus* L. diets. Aquacult Res 34:1119–1127
5. Soltan MA, Hanafy MA, Wafa MIA (2008) Effect of replacing fish meal by a mixture of different plant protein sources in Nile tilapia (*Oreochromis niloticus* L.) diets. Global Vet 2:157–164
6. Boisen S, Eggum BO (1991) Critical evaluation of in vitro methods for estimating digestibility in simple-stomach animals. Nutr Res Rev 4:141–162
7. Dimes LE, Haard NF (1994) Estimation of protein digestibility – 1. Development of an in vitro method for estimating protein digestibility in salmonids. Comp Biochem Physiol 108:349–362
8. Lemos D, Ezquerra JM, Garcia-Carreño FL (1999) Protein digestion in penaeid shrimp: digestive proteinases, proteinase inhibitors and feed digestibility. Aquaculture 186:89–105
9. Walsh MG (1991) Evaluation of dietary carbohydrate utilization by captive sable fish (*Anoplopoma fimbria*). Dissertation, University of British Columbia
10. Tacon AGJ, Metian M, Hasan MR (2009) Feed ingredients and fertilizers for farmed aquatic animals: sources and composition. FAO Fisheries Aquacult Tech Paper 540:3–6
11. AOAC (2005) Official methods of analysis, 18th edn. AOAC International, Gaithersburg
12. Lazo JP (1994) Evaluation of several in vitro enzyme assay for estimating in vivo apparent protein digestibility by the Pacific white shrimp (*Penaeus cannamel*). Dissertation, Institute Technologic de Estudios Superiores de Monterrey
13. Paepatung N, Nopharatana A, Songkasiri W (2009) Bio-methane potential of biological solid materials and agricultural wastes. Asian J Energ Environ 10:19–27
14. Dewayanto N (2010) Waste to valuable by products: utilization of carbonized decanter cake from palm oil milling plant as an effective adsorbent for heavy metal ions in aqueous solution. Dissertation, University Malaysia Pahang

15. Mjoun K, Rosentrater KA, Brown ML (2010) Tilapia: environmental biology and nutritional requirements. South Dakota Cooperative Extension Service, USDA. http://pubstorage.sdstate.edu/AgBio_Publications/articles/FS963-02.pdf. Accessed 01 June 2013
16. Robinson EH, Li MH, Manning BB (2001) A practical guide to nutrition, feeds, and feeding of catfish. Bulletin 1113, Office of Agricultural Communication, Division of Agriculture, Forestry and Veterinary Medicine at Mississippi State University. http://www.aces.edu/dept/fisheries/aquaculture/documents/guidetocatfishfeeding.pdf. Accessed 01 June 2013
17. New MB, Tacon AGJ, Csavas I (eds) (1994) Farm-made aquafeeds. FAO Fisheries Tech Paper, 343, FAO, Rome. http://www.fao.org/docrep/003/v4430e/V4430E00.htm#TOC. Accessed 1 June 2013
18. Webster CD, Lim CE (eds) (2002) Nutrients requirements and feeding of finfish for aquaculture. CAB International, Oxon
19. Wan Zahari M, Alimon AR, Wong HK (2012) Utilization of oil palm co-products as feeds for livestock in Malaysia. In: Makkar HPS (ed) Biofuel co-products as livestock feed: opportunities and challenges. FAO, Rome, http://www.fao.org/docrep/016/i3009e/i3009e.pdf. Accessed 19 June 2013
20. Krogdahl Å, Hemre G-I, Mommsen TP (2005) Carbohydrates in fish nutrition: digestion and adsorption in postlarval stages. Aquaculture Nutr 11:103–122
21. Abowei JFN, Ekubo AT (2011) Some principles and requirements in fish nutrition. Br J Pharmacol Toxicol 2:163–178

# Human Head Pose Estimation and Its Application in Unmanned Aerial Vehicle Control

**Chun Fui Liew and Takehisa Yairi**

## 1 Introduction

Head pose estimation is a process of extracting 3D positional and 3D orientation information from a human face in digital images or videos. Figure 1 illustrates 3D orientation of human head pose in yaw, pitch, and roll angle format. The recent rise of interactive system design makes head pose estimation problem becoming increasingly important. In addition to human–computer interface systems [1, 2], human head pose estimation system also plays key roles in intelligent smart room systems [3, 4] and driving assistance systems [5, 6]. It is believed that head pose estimation system would have a profound impact on future interactive technologies.

Compared to face detection and face recognition, head pose estimation problem attracts fewer attentions from research communities. It has been reported that geometric approaches in head pose estimation problem have not reached their full potential, yet modern approaches can detect facial feature locations promptly and reliably [7]. In this chapter, we propose an automatic and fast method to achieve accurate head pose estimation from static image based on geometric approach, with the aim to offer a head pose estimation algorithm (based on geometric approach and monocular 2D image) that can be put to use in other new application easily. In our approach, we use Pose from Orthography and Scaling with Iterations (POSIT) algorithm for head pose estimation. Since POSIT algorithm estimates object pose by a set of 3D model points and corresponding 2D projected image points, we use a statistical 3D human face model pre-captured from Kinect device together with CLM-based facial feature tracking to effectively locate facial feature locations on 2D camera image. After training, CLM-based facial feature tracking has the key advantage of fitting a new person's facial features more than 15 fps in an Intel

C.F. Liew (✉) • T. Yairi
University of Tokyo, 4-6-1 Komaba, Meguro-ku, Tokyo 153-8904, Japan
e-mail: liew@space.rcast.u-tokyo.ac.jp

K. Ab. Hamid et al. (eds.), *The Malaysia-Japan Model on Technology Partnership: International Proceedings 2013 of Malaysia-Japan Academic Scholar Conference*, DOI 10.1007/978-4-431-54439-5_32, © Springer Japan 2015

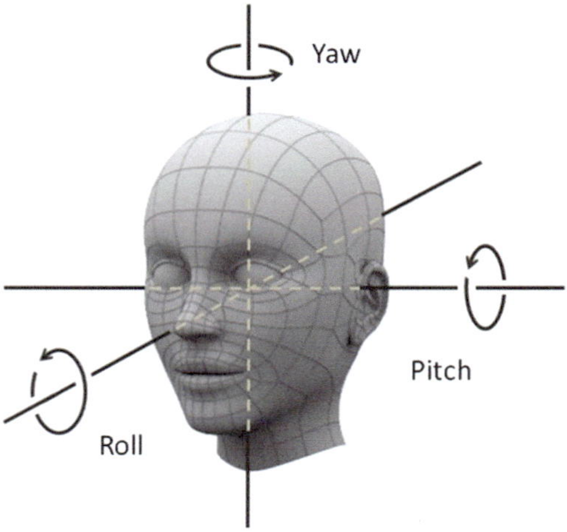


**Fig. 1** 3D orientation of head pose estimation

**Fig. 2** Our geometric approach in head pose estimation

i5-3230 CPU 3.0GHz processor. Figure 2 illustrates our geometric approach in head pose estimation problem.

In the following, Sect. 2 lists up common design considerations and our aims in building head pose estimation system. Section 3 outlines CLM-based facial feature tracking. Section 4 explains our process in capturing 3D human face model with Kinect device. We have discovered that the pre-captured 3D human face model is generic enough to be applied to other new persons. Therefore, users can use our system instantly without the need to capture their 3D personal face model with Kinect device. Section 5 presents our experiment results, and Sect. 6 describes a human–robot interactive application where we use human head pose to control an unmanned aerial vehicle (UAV). We explain some limitations of our current system and list up future works in Sect. 7.

## 2 Design Considerations

Geometric approach uses locations of facial features such as eyes, nose, and mouth to determine head pose from their relative positions. Compared to other machine learning approaches, where machine learning algorithms are trained with facial training images and corresponding head pose training labels, geometric approaches have not reached their full potential [7]. In order to satisfy the majority of other applications, we aim to achieve the following criteria with our head pose estimation system:

- Accuracy—tracking outcomes should have high tracking accuracy with average positional and orientation errors less than 2 cm and 5°, respectively.
- Autonomous—the system should be fully automatic. No manual initialization and additional inputs are required.
- Monocular—the system should be able to perform head pose estimation with a single 2D camera. Monocular image processing is more common among users, even though stereo image processing and 3D image processing might improve the head pose estimation accuracy and tracking range.
- Continuous head motion—the system should provide continuous tracking outcomes rather than limited and discrete pose estimation outcomes.

## 3 CLM-Based Facial Feature Tracking

CLM-based facial feature tracking has two major processes, where a local exhaustive search is first performed to obtain response maps at each facial feature locations and then optimization is performed over these response maps. Algorithm 1 and Algorithm 2 in Fig. 3 illustrate the overall process of CLM-based facial feature tracking. In order to realize real-time tracking, we avoid using complex algorithms. Logistic regression is used for local exhaustive search, and mean shift regularization is used for optimization of response maps. Despite its simplicity and low computation cost, it has been shown that mean shift regularization has superior performance over isotropic Gaussian estimate (ISO), Gaussian mixture model (GMM), and kernel density estimation (KDE) in CLM-based fitting problem of human face [8].

Logistic regression is the only part that needs to be trained in our method. With ground truth labels from frontal face database [9], facial features' image patches at eye corners, center of nose tip, mouth corners, and face shape corners are first extracted from original frontal images. Combined with image patches with negative labels, these image patches at each facial feature locations are supplied to logistic

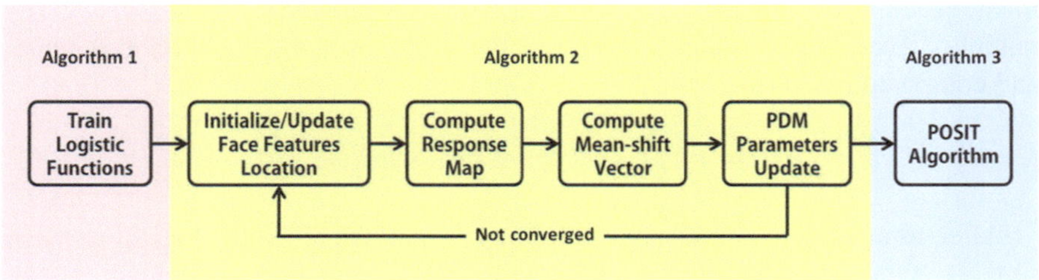


**Fig. 3** Training and update process flow of CLM-based facial feature tracking

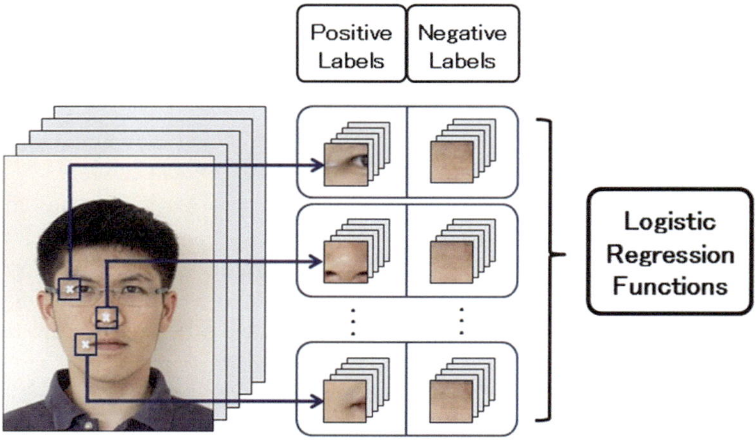


**Fig. 4** Logistic regression training

regression functions for training. Figure 4 illustrates the process of logistic training. After logistic training, one would gain a close-to-one logistic output if a visually similar image patch is passed to trained logistic function. Mathematically, the logistic output

$$C_j(\mathbf{x}_j, \mathbf{I}) = \mathbf{w}_j^T P(\mathbf{W}(\mathbf{x}_j, \mathbf{I})) + b_j \tag{1}$$

where $j$ denotes index of facial features, $\mathbf{W}(\mathbf{x}_j, \mathbf{I})$ denotes facial feature's image patch $j$ at location $\mathbf{x}_j$ of image $\mathbf{I}$, $P$ denotes normalization, and $\{\mathbf{w}_j, b_j\}$ denotes trained gain and bias of logistic regression function.

During the tracking process, the starting locations of facial features are initialized with the average locations of all training images. From Eq. (1), logistic outputs at each facial feature location can be computed in sequence. For illustration purpose, response maps with size of $9 \times 9$ and logistic function input image patch with size of $3 \times 3$ are shown in Fig. 5. It can be observed that logistic outputs of all 81 pixels inside the 9x9 response map window are computed in sequence. For instance, at pixel number 60, yellow plus mark in Fig. 5 represents facial feature's image patch location $\mathbf{x}_j$, and the yellow square represents $\mathbf{W}(\mathbf{x}_j, \mathbf{I})$ in Eq. (1). The output of every pixel inside the response maps is modeled as likelihood function and computed based on

$$p(l_j = 1 | \mathbf{x}_j, \mathbf{I}) = 1/(1 + \exp\{-l_j C_j(\mathbf{x}_j, \mathbf{I})\}) \tag{2}$$

After computing response maps with Eq. (2) at all facial feature locations, mean shift is computed over all response maps. Mean shift can be thought as the calculating center of gravity of response map with respect to the center of response

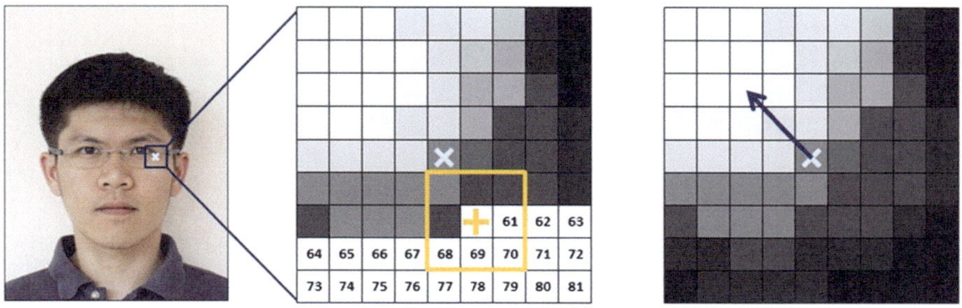


**Fig. 5** Search of local features' position

map, where pixel values inside the response map represent weight. For instance, in both response maps in Fig. 5, brighter pixels represent higher weight (closer to 1, which also means that these pixels are closer to the ground true facial feature location), while darker pixels represent lower weight (closer to 0, which also means that these pixels are further away from the ground true facial feature location). The right response map in Fig. 5 illustrates the computation of mean shift vector. After mean shift computation, one would obtain a two-dimensional vector that is pointing to the center of gravity, e.g., ground truth location in ideal case.

CLM-based facial feature tracking employs point distribution model (PDM), where deformable shape of a nonrigid object is assumed to be linear. PDM models nonrigid shape variations linearly and composes it with a global rigid transformation:

$$\mathbf{x}_i = s\mathbf{R}\left(\mathbf{x}_{i,mean} + \mathbf{\Phi}\mathbf{q}\right) + \mathbf{t} \tag{3}$$

where $\mathbf{x}_i$ denotes the 2D location of the PDM's $i$th landmark and $\mathbf{p} = \{s, \mathbf{R}, \mathbf{t}, \mathbf{q}\}$ denotes the PDM parameters, which consist of a global scaling $s$, a rotation $\mathbf{R}$, a translation $\mathbf{t}$, and a set of nonrigid parameters $\mathbf{q}$. $\mathbf{x}_{i,mean}$ denotes the mean location of $i$th PDM landmark in the reference frame, and $\mathbf{\Phi}$ denotes the location variance of training images (e.g., principal component of training images' facial feature locations). It has been shown by Cootes and Taylor that PDM is simple and efficient adequately to model the deformations of human face [10]. CLM-based facial feature tracking is generally posed as the search for the PDM parameters, $\mathbf{p}$, that jointly minimizes the misalignment error over all landmarks with appropriate regularization. With the mean shift vectors computed from response maps, we can update the facial feature locations with optimization technique that was originally proposed in [8]:

$$\Delta\mathbf{p} = -\left(\rho\mathbf{\Lambda}^{-1} + \mathbf{J}^T\mathbf{J}\right)^{-1}\left(\rho\mathbf{\Lambda}^{-1}\mathbf{p} + \mathbf{J}^T\mathbf{v}\right) \tag{4}$$

where $\rho$ represents arithmetic average eigenvalues of training images' facial feature locations, $\mathbf{\Lambda}$ represents the eigenvalues of training images' facial feature in diagonal

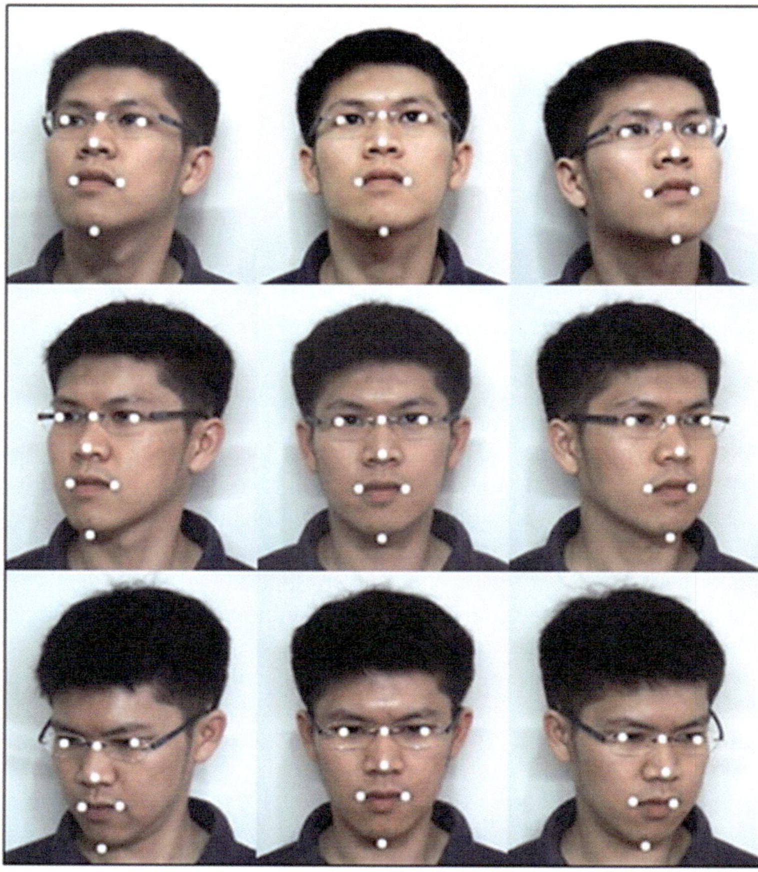

**Fig. 6** CLM-based facial feature tracking in nine directions

matrix form, **J** represents Jacobian matrix of image, **p** represents PDM parameters in Eq. (4) in vector form, and **v** represents mean shift vector. Figure 6 shows nine tracking results of eye corners, nose tip, mouth corners, and jaw locations after computation of logistic functions, mean shift vectors, and optimization updates.

## 4 Actual Implementation

We perform head pose estimation with POSIT algorithm [11] by utilizing 3D face model points and facial feature locations tracked by CLM described in Sect. 3. Since we require 3D face model for POSIT algorithm, we scan a real human face with Kinect 3D sensor. We planned to capture multiple 3D human faces and take mean values as our 3D face model in the first place. However, we discovered that

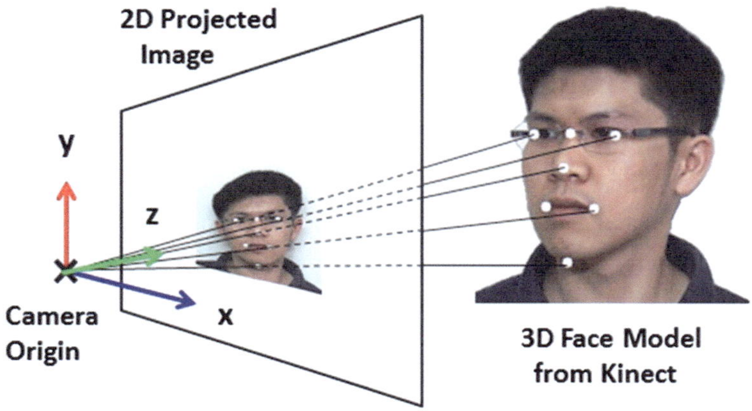


**Fig. 7** 3D head model projection for POSIT

the first captured 3D human face model is generic enough to be applied to other new persons without further adjustment. Therefore, we continue to use the first captured 3D human face as our final 3D face model for POSIT algorithm. Figure 7 shows the projection of our captured human face model on an image plane. The transformation relationship between the marked 3D facial feature points and their corresponding 2D projected image points is determined by using POSIT algorithm. Since both CLM-based facial feature tracking and POSIT algorithm are fast, our system achieves tracking speed higher than 15 fps. The finished program is fully automatic and only monocular images are required. No temporal filtering is performed in the meantime.

# 5 Experiment Results

As illustrated in Fig. 1, the head pose estimation results are represented by yaw, pitch, and roll angle format. Figure 8 displays some real-time head pose tracking results, where the first head pose is pointing at upright angle (hence with positive yaw angle of 23° and positive pitch angle of 15°) and the second head pose is pointing at slightly up-left angle (hence with negative yaw angle of 19° and positive angle of 9°). Figure 9 shows similar results under second user and verifies that our system is people invariant, even though only one generic 3D human face model is utilized now.

The real-time tracking results of yaw angle, pitch angle, roll angle, and depth (distance between camera center and user's nose tip) are summarized in Fig. 10. We evaluate our pose estimation accuracy with Kinect's 3D head pose estimation outcomes, in which we consider as ground truth since 3D head pose estimation is more reliable. We observe that 3D orientation can achieve absolute tracking error of less than 10°, while depth distance can achieve absolute tracking error of less than 5 cm.

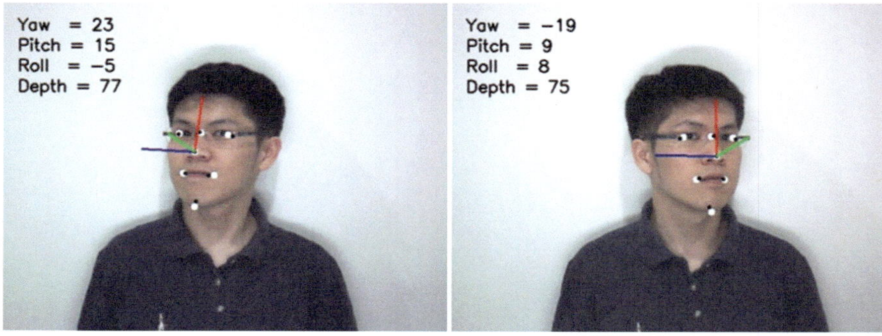


**Fig. 8** Real-time head pose tracking results (*first user*)

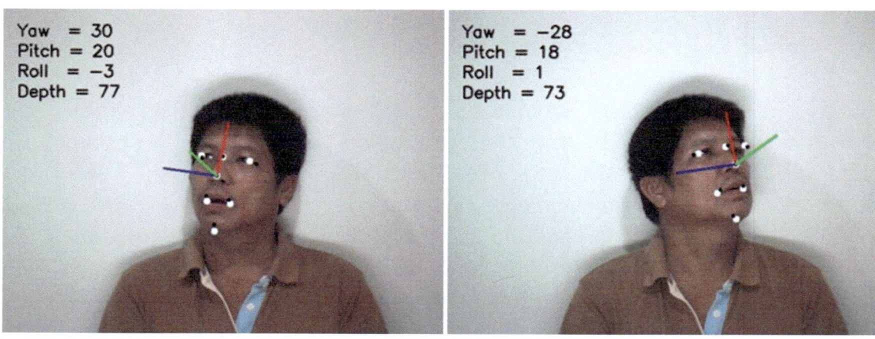


**Fig. 9** Real-time head pose tracking results (*second user*)

We also discover that accuracies are slightly affected by uneven light illumination. In both yaw and pitch experiments, the tracking errors increase when user turns head to the right-hand side and to the bottom, respectively. This is due to the uneven light illumination in the test room as the sunlight was coming from the user's right-hand side. The uneven light distribution affects the CLM-based facial feature tracking and subsequently affects the head pose estimation outcomes.

## 6 Applications

Head pose estimation has significant implications in human–computer and human–robot interaction fields. With estimated 3D head pose, robots can have a better understanding of human intentions and hence have a better interaction with human. Head pose information has also been utilized in smart room system to monitor occupants and measure their visual focus of attention [4]. Besides, head pose information creates a novel interface for human to interact with computer and robot. Some existing systems allow computer user to control mouse cursor by

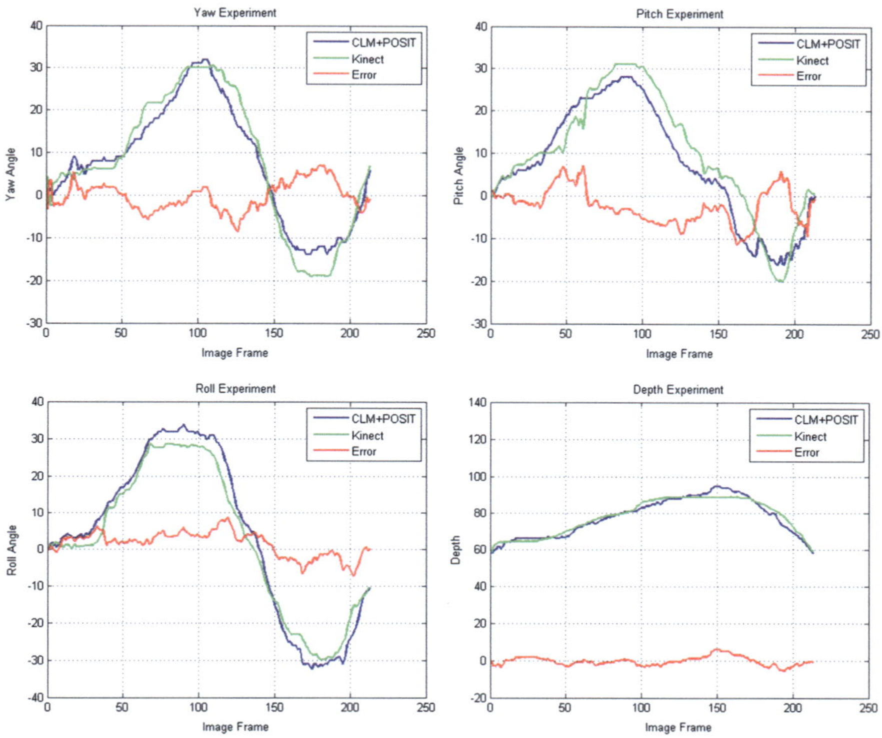


**Fig. 10** Tracking results in comparison with Kinect tracking

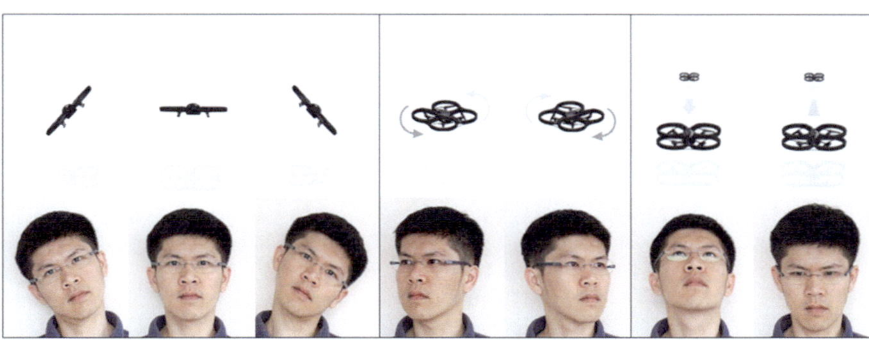

**Fig. 11** AR Drone control with head pose estimation

using head pose movements [1]. To our knowledge, no system has been developed to control aerial robot by using head pose movements. Figure 11 shows our application in controlling a UAV with only head pose movements, where roll, yaw, and pitch motions are used to control UAV's left-right drift, left-right turn, and forward–backward movements, respectively. After a few extensive tests, we find that it is not easy for user to control UAV with head pose while constantly maintaining visual tracking with UAV. We believe that hand gesture will be more suitable in this task and leave this as our future work.

## 7 Conclusion

We proposed to use head pose to control UAV, where the head pose estimation is achieved by using CLM-based facial feature tracking and POSIT algorithm. Since both CLM-based facial feature tracking and POSIT algorithm are faster than traditional methods such as Active Appearance Model-based facial feature tracking [9], our geometric approach can achieve real-time speed of higher than 15 fps. Nevertheless, current head pose estimation accuracy is slightly affected by uneven light illumination. Current system also has low tracking accuracy when yaw and pitch rotation angles are higher than 40°. We aim to improve these weak points and provide a better control method of UAV in our future work.

## References

1. Fu Y, Huang T (2007) hMouse: head tracking driven virtual computer mouse. In: IEEE workshop on applications of computer vision, Austin, TX, USA, 21–22 Feb 2007
2. Martins P, Batista J (2008) Single view head pose estimation. In: IEEE international conference on image processing, San Diego, CA, USA, 12–15 Oct 2008
3. Wu C, Aghajan H (2008) Head pose and trajectory recovery in uncalibrated camera networks – region of interest tracking in smart home applications. In: ACM/IEEE international conference on distributed smart cameras, Stanford University, CA, USA, 7–11 Sept 2008
4. Stiefelhagen R, Bernardin K, Ekenel H, Voit M (2008) Tracking identities and attention in smart environments—contributions and progress in the CHIL Project. In: IEEE international conference on automatic face and gesture recognition, Amsterdam, The Netherlands, 17–19 Sept 2008
5. Murphy-Chutorian E, Doshi A, Trivedi M (2007) Head pose estimation for driver assistance systems: a robust algorithm and experimental evaluation. In: IEEE intelligent transportation systems conference, Seattle, WA, USA, 30 Sept–3 Oct 2007
6. Ghaffari A, Rezvan M, Khodayari A, Vahidi-Shams A (2011) A robust head pose tracking and estimating approach for driver assistant system. In: IEEE international conference on vehicular electronics and safety, Beijing, China, 10–12 July 2011
7. Murphy-Chutorian E, Trivedi MM (2009) Head pose estimation in computer vision: a survey. Trans Pattern Anal Mach Intell 31(4):607–626
8. Saragih JM, Lucey S, Cohn JF (2011) Deformable model fitting by regularized landmark mean-shift. Int J Comput Vis 91(2):200–215
9. Gross R, Matthews I, Cohn J, Kanade T, Baker S (2008) Multi-PIE. In: IEEE international conference on automatic face and gesture recognition, Amsterdam, The Netherlands, 17–19 Sept 2008
10. Cootes TF (2001) Active appearance models. IEEE Trans Pattern Anal Mach Intell 23(6):681–685
11. Dementhon DF, Davis LS (1995) Model-based object pose in 25 lines of code. Int J Comput Vis 15(1–2):123–141

# Japan and US Smart Grid Effort? A Case Study

**Amy Poh Ai Ling**

## 1 Introduction

The Japanese version of the "smart city" is envisaged for the post-fossil fuel world. Alternative energy sources are harnessed in mass quantities [1]. In Japan, "smart grid" implies energy transmission and distribution to promote the stability of the electric power supply, by using information and communication technology while introducing a high level of renewable energy [2]. The focus will be on how to stabilize power supplies nationwide as large amounts of wind and solar power start entering the grid. This is because, unlike conventional power sources, such as hydro, thermal, and nuclear power, solar and wind energies are prone to the vagaries of the weather. People in Japan are still not familiar with the smart grid concept because the system has yet to gain currency. According to a nationwide survey released in December 2010 by the advertising agency Hakuhodo Inc., only 36.4 % of about 400 respondents aged from 20 to 70 years said they understood or had heard of a smart grid [3]. It was believed that a low-carbon society would not be realized without a fundamental shift in energy source use. The Japanese smart grid concept aims to make the best use of local renewable energy with a view to maximizing total efficiency. The Japanese government is aiming to increase the reliability of the grid system by introducing sensor networks and to reduce opportunity losses by introducing smart meters. The introduction of the smart grid will promote the use of renewable energy by introducing a demand response system.

A.P.A. Ling (✉)
Department of Technology Management for Innovation, Graduate School of Engineering, University of Tokyo, 7-3-1 Hongo, Bunkyo-ku, Tokyo, Japan

Japan Society for the Promotion of Science, Kojimachi Business Centre Building, 5-3-1 Kojimachi, Chiyoda-ku, 102-0083 Tokyo, Japan

Meiji Institute for Advanced Study of Mathematical Sciences, Meiji University, 8 F High-Rise Wing, Nakano Campus, 4-21-1 Nakano, Nakanoku, 164-8525 Tokyo, Japan
e-mail: amypoh@tmi.t.u-tokyo.ac.jp

K. Ab. Hamid et al. (eds.), *The Malaysia-Japan Model on Technology Partnership: International Proceedings 2013 of Malaysia-Japan Academic Scholar Conference*,
DOI 10.1007/978-4-431-54439-5_33, 

By focusing on electric vehicle (EV) technology, Japan is moving toward introducing charging infrastructure for electric cars [10]. Recently, increasing numbers of PV and wind power plants have been installed across the country as clean energy sources that emit no CO2 [4].

# 2 Japan and US Smart Grid Movement

The energy sources in Japan and the USA differ greatly, and the implementation of the smart grid tends to differ between countries, as do the timing and adoption of these technologies [5]. Japan is pushing for advanced integrated control, including demand-side control, so as to be ready to combine unstable power (i.e., reliant on weather conditions), such as solar, with its strong foundation of a highly reliable grid, as shown in Table 1.

With regard to Japan's nuclear contribution to energy supply, as of 2010, Japan's currently operating 54 commercial nuclear reactors have a total generation capacity of 48,847 MW, and about 26 % of electricity comes from nuclear power [6].

**Table 1** Japan and US smart grid movement

| Description | Japan | USA |
|---|---|---|
| Nuclear as % of all energy sources (%) | 26 | 19 |
| No. of electric power companies | Ten electric power companies (all IOUs) | More than 3,000 traditional electric utilities (IOUs = 210, public = 2,009, coops = 883, federal = 9) |
| Design | Vertically integrated in each region | Interdependent infrastructure |
| Energy supply | 0.7–25 million customers | A few thousand to more than five million customers |
| Aim | A low-carbon society | Focus on business and infrastructure |
| Reliability | Japan already has a highly reliable grid | Need for highly reliable transmission and distribution networks |
| | Going for advanced integrated control, including demand-side control, to accommodate unstable renewable power | Need for demand response for peak shaving and need to avoid additional infrastructure |
| Smart grid focus | More than $100 billion investment in the 1990s to upgrade generation, transmission, and SCADA network | Little investment (approx. $30 billion) in the 1990s into grid |
| | Last mile and demand-side management (DSM) | Now working across entire grid for enhancements |
| | Home solar power | Last mile and DSM are also important |
| Cyber security research | Protect smart meters, mutual monitoring, privacy in cloud computing | Grid computer, cryptography security |

This compares with 104 licensed-to-operate nuclear power plants operating in 31 states in the USA (with 69 pressurized water reactors and 35 boiling water reactors), which generate about 19 % of US electric power [7]. Japan's ten electric power companies are monopolies, being electric giants vertically integrated in each region. By contrast, there are more than 3,000 traditional electric utilities established in the USA, each with an interdependent structure. For this reason, the utilities' supply of nuclear energy to customers differs greatly between Japan and the USA. In aiming toward a low-carbon society, Japan depends greatly on nuclear power as an energy source, whereas the USA, in implementing its smart grid, focuses more on business and infrastructure. In terms of reliability, Japan already has a highly reliable grid compared with the USA, which needs more reliable and distributed networks across the nation to develop its smart grid system. Japan is developing its smart grid at a steady pace and has already been investing in grid projects for almost 20 years; over this period, there have been many developments. With proper security controls, smart grids can prevent or minimize the negative impact of attacks by hackers and thus increase the reliability of the grid, thereby gaining the trust and meeting the satisfaction of users [8]. Since smart grid cyber security is significantly more complex than the traditional IT security world, Japan focuses on the areas of smart grid cyber security concerns beyond smart metering, such as mutual monitoring and privacy in cloud computing. On the other hand, the USA is enhancing its smart grid cyber security in the age of information warfare, especially in the area of grid computer and cryptography security.

In conclusion, the USA focuses on businesses and infrastructure development, whereas Japan is striving to move toward a low-carbon society by developing the smart grid system.

## 2.1 *Japan Photovoltaic System*

The Japanese government developed its new policy on PV systems in 2010 based on the principles of energy security, global warming prevention, and an efficient supply of energy to end users across the nation. Its goal, set to be achieved by 2030, is focused on increasing independent energy supplies from 38 % to about 70 % and increasing the proportion of electricity generation with zero emissions from 34 % to about 70 % [9]. This new policy is supported by the government's aim of becoming a leading nation in environmental and energy sustainability through green innovation.

Figure 1 illustrates the cumulative installation of PVs in Japan and four other countries, namely, Germany, the USA, Spain, and Italy, extracted from data on trends in PV applications. As of 2009, Japan clearly ranks third, lagging far behind Germany. The demand for PV systems in Germany has remained persistently high for full 2 years. Spain ranks second, and Spanish companies and research centers are taking the lead in the recent revival of concentrated solar power, with expansive

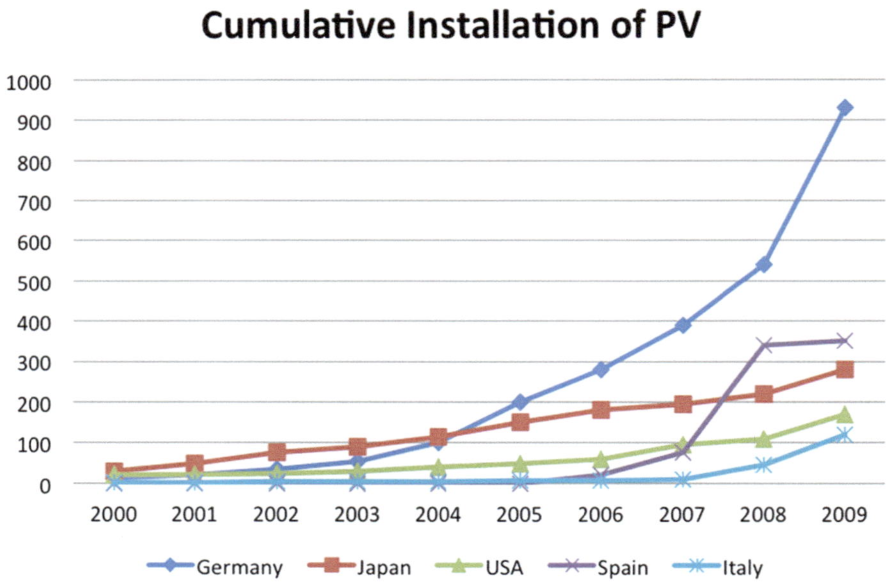


**Fig. 1** Cumulative installation of PVs in Japan, Germany, the USA, Spain, and Italy. *Source*: IEA PVPS (Data: Trends in photovoltaic applications/IEA/PVPS)

banks of solar mirrors being assembled around the country for concentrated solar plants.

The expanding installation of PV systems may increase the stability of extra-high-voltage transmission systems (Fig. 2). In the context of Japan's PV power generation development, the expected change in domestic electricity demand in Japan in relation to the expected installed PV system capacity by 2030, along with the actual cumulative installed PV system capacity as of 2007 is extremely huge. Domestic electricity demand is expected to increase sharply between 2010 and 2030. This is correlated with the expected installed PV system capacity, which experienced a sharp upturn in 2010, following the development of the detection of unintentional islanding and the development of technology that curtails the restriction of PV system output. Expanding installation of PV systems may increase the stability of extra-high-voltage transmission systems. At the final stage of PV system development, it is predicted that imbalances between output from PV systems and existing systems may influence frequency on utility systems.

NEDO and the European Commission (European Union) will jointly launch a project to develop concentrator PV cells, thus aiming to achieve a cell conversion efficiency of more than 45 %, which is the highest efficiency in the world [10]. However, the highest value seems to be ideal and might be still in research even at 2030 without any large breakthrough for practical use. On the other hand, Sunetric, Hawaii's largest locally owned and operated solar installer, has donated two solar PV systems to raise funds for two local charities assisting Japan following the tsunami that hit northeast Japan on March 11, 2011. The first is the American

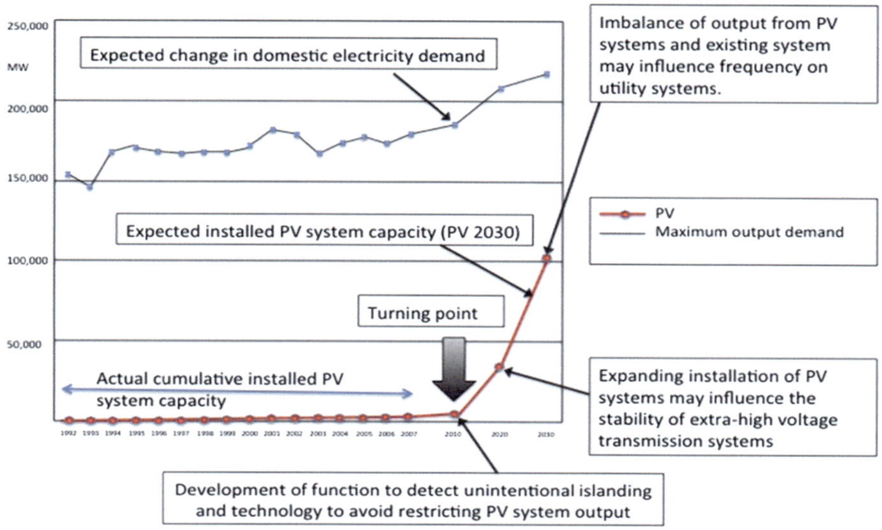


**Fig. 2** Estimation of Japanese PV penetration by 2030

Red Cross Hawaii State Chapter and the second is the "With Aloha" Foundation [11].

In summary, the Japanese government is encouraging further deployment of the conventional installation of residential PV systems for the sake of the PV community. Its current PV communities include Kasukabe and Yoshikawa in Saitama, Matsudo in Chiba, Kasugai in Aichi, Kobe in Hyogo, Tajiri in Osaka, Ota in Gunma, Wakkanai in Hokkaido, Shimonoseki in Yamaguchi, and Kitakyushu in Fukuoka. Although the national subsidy program for residential PV systems introduced by Japan's Ministry of Economy, Trade, and Industry (METI) was terminated, some local governments have programs for residential PV systems in the regions.

## 2.2 *Comparison of Japanese and US Energy Sources*

Japan lacks significant domestic sources of fossil fuels, except coal, and must import substantial amounts of crude oil, natural gas, and other energy resources, including uranium. In 1990, Japan's dependence on imports for primary energy stood at more than 85 %, and the country has a total energy requirement of 428.2 million tons of petroleum equivalent [12]. As for 2007, energy sources in Japan are coal 26 %, nuclear 26 %, natural gas 26 %, petroleum 13 %, and other renewables 1 %; energy sources in USA are coal 49 %, nuclear 19 %, natural gas 22 %, hydroelectricity 6 %, and petroleum 2 %. For Japan, the shares for coal, natural gas, and nuclear energy are similar, and the remaining share is split between hydroelectricity, petroleum, and other renewables. For the USA, the large share for coal is followed by natural gas and nuclear, with smaller shares for hydroelectricity and then petroleum. Japan is to be commended for having such a systematic and comprehensive energy planning process. While maintaining its goal of going green, Japan is utilizing more low-carbon energy sources such as geothermal, hydraulic, battery systems, and nuclear as major providers of energy.

## 3 Conclusion

Smart grid technology helps to convert the power grid from static infrastructure that is operated as designed to flexible and environmentally friendly infrastructure that is operated proactively. This is consistent with the Japanese government's goal of creating a low-carbon society, maximizing the use of renewable energy sources, such as photovoltaics and wind power. Nevertheless, public–private sector cooperation across various industries is necessary to establish smart communities. Japan uses renewable energy and promotes this measure through the development and employment of the natural grid. This chapter provided sufficient information for the reader to understand the Japanese concept of the smart grid and the government's associated strategy and the significance of the government's contribution to Japan's energy supply capacity, without documenting full case studies in detail. Japan is currently focusing on last mile and demand-side management and home solar power. Researchers have started to address challenges caused by large-scale solar power generation connected to the power grid as well as information security issues. Because the smart grid remains a novel field of study in Japan, it has great potential for further research.

Through this case study, the concept of Negawatt is considered to be another important factor in ameliorating the cost-efficiency of the power system in smart grid, although Negawatt's concept cannot be precisely measured, only theoretically proven.

**Acknowledgment** The author wish to thank the Japan Society for the Promotion of Science (JSPS) for research funding; Meiji Institute for Advanced Study of Mathematical Sciences (MIMS) and University of Tokyo, Graduate School of Engineering, Department of Technology Management for Innovation for mentoring; Professor Nawata Kazumitsu for supervising; and Professor Naotaka Uchitomi for valuable comments.

## References

1. Hosaka TA (2010) Japan creating 'Smart City' of the future. Associated Press Japan, Oct, in press
2. Shinkawa T (2010) Smart grid and beyond. BNEF Energy Smart Technologies Leadership Forum

3. Nakata H (2011) Smart grid pursuit slow off mark. Smart Grid, The Japan Times
4. Electricity Review Japan (2011) The federation of electric power companies of Japan. Electricity Review Japan, Japan
5. McGuire K (2010) The smart grid movement: Japan vs. U.S., TMCnet News, Technology Marketing Corporation. http://smart-grid.tmcnet.com/topics/smart-grid/articles/73301-smart-grid-movement-japan-vs-us.html
6. Nuclear Power Plants in Japan (2011) Nuclear power generation. Information Plaza of Electricity, The Federation of Electric Power Companies of Japan
7. Power Reactors (2011) Nuclear reactors. United States Nuclear Regulatory Commission (U.S. NRC), http://www.nrc.gov/reactors/power.html
8. Ling APA, Mukaidono M (2011) Selection of model in developing information security criteria on smart grid security system, Smart grid security and communications. In: The ninth international symposium on parallel and distributed processing with applications (ISPA), Korea
9. Yasui M (2010) Japanese policy on photovoltaic systems. ANRE, METI, Japan, pp 4–7
10. New Energy and Industrial Technology Development Organization, European Commission, European Union (2011) Aiming at developing the world's highest efficiency concentrator photovoltaic cells. New Energy and Industrial Technology Development Organization, European Commission, European Union. http://www.nedo.go.jp/content/100147758.pdf
11. Sunetric offers free solar PV systems to raise money for Japan. Cleantech News, April 2011
12. Williams M (2011) A legacy from the 1800s leaves Tokyo facing blackouts. Computerworld

# Threat to Individuals' Privacy in the Era of Social Networking: A Case Study of Facebook

**Chai Yeun Hoe**

*'While Internet Users May Feel Safe Behind Their Computers, They Have "Zero Privacy"'*

## 1 Introduction

Social networking sites are a type of virtual community that are primarily used for social interaction. On February 26, 2012, the web traffic data for Facebook, a free access site that allows users to connect and interact with other people, shows 845 million users worldwide, making it the largest of the social networking sites [1]. When Facebook first launched in 2004, it was restricted only to Harvard University students. On September 26, 2006, Facebook was opened to everyone of age 13 and above with a valid email address [2]. The exponential growth of Facebook has raised privacy and security concerns. With the increasing usage of Facebook as a tool for police searches and prosecution, people are becoming worried about their rights and potential privacy issues on this new and extremely popular site [3].

Australia is on its way to a comprehensive privacy law framework to protect individuals' information privacy [4]. The Australian Privacy Act [5] does affect Facebook given that it is a body corporate that satisfies the definition requirements in s 5B and does not fall within the exceptions in s 6 of the Act [6]. The relevant part of the Act is the National Privacy Principles (NPPs), which seek to regulate privacy practices of the private sector including online social networking services like Facebook.

C.Y. Hoe (✉)
Australian National University, Barry Dr, Acton ACT 0200, Australia
e-mail: chaihoe611@gmail.com

K. Ab. Hamid et al. (eds.), *The Malaysia-Japan Model on Technology Partnership: International Proceedings 2013 of Malaysia-Japan Academic Scholar Conference*,
DOI 10.1007/978-4-431-54439-5_34, 

# 2 Facebook's Privacy Policy and Its Interactions with the Australian Privacy Law

## 2.1 *Information We Receive and How It Is Used*

Facebook collects vast amounts of personal information, including the public information required upon registration: name, email address, date of birth and gender as well as any other information provided voluntarily by the users [7]. In addition, Facebook also collects personal information about an individual user from 'someone else' [8] such as third-party application developers, any websites that use Facebook features (e.g. Facebook Platform and social plug-in) and other third parties that 'help Facebook deliver ads, understand online activity and generally make Facebook better' [9].

Facebook's data collection policy has stirred controversy over the years. Facebook collects personal information beyond what was necessary for the purposes of the site, in contravention of NPP 1: collection requirements. For instance, information such as date of birth [10], data collected via the loading of Facebook social plug-ins on websites [11] and data received from the computer, mobile phone or other device used to access Facebook (e.g. IP address, location, the type of browser used, the pages visited by the user and user's GPS location) [12] are not necessary for the site's functions or activities. Moreover, the collection of personal information about an individual from third parties breaches NPP 1.5 as Facebook fails to take reasonable steps to ensure that the individual is aware of the matters set out in NPP 1.3 [13]. In some situations, individual users are not even aware that their information has been exposed to third parties. Furthermore, Facebook's policy provides inadequate notice of how Facebook would use and share the user's information or activities with affiliate websites or search engines.

## 2.2 *Sharing and Finding You on Facebook*

Although Facebook tries to allow its users to control over their profile and personal information, for instance, users can always edit their privacy setting to customize their profile and can 'select a specific audience, or even customize his or her audience' [14], users are not able to control the actions done by others that might infringe that user's privacy. The recent Usmanov case in which Usmanov shared his ex-girlfriend's nude pictures on Facebook provides a good example that people are powerless when it comes to information shared by other users on social networking sites [15]. By virtue of the fact that Facebook is not able to control the content posted by a user at the time of posting, Facebook's privacy policy fails to provide adequate protection to individuals' privacy pragmatically. Facebook warns its users to be careful of what they post on Facebook, and it disclaims responsibility for any subsequent use of information shared on Facebook by other parties [16].

## 2.3 Sharing with Other Websites and Applications

Facebook's privacy policy states that it will provide the user's ID (accompanied with user's public information and other information such as age range, locale and gender) and user's friend list to the game, application or websites that the user visits or connects using the Facebook Platform ('Applications') [17]. Although Facebook provides its users with some control of the applications they use, users are not able to prevent applications from accessing their personal information unless they turn all Platform applications off in which they 'no longer able to use any games, applications or websites through Facebook' [18]. Despite its privacy policy, Facebook has been critized for its lack of capacity and technical safeguards to restrict third-party developers from accessing the personal information of users and their friends [19]. Additionally, Facebook does not notify users of the extent of personal information that is disclosed and 'provides users misleading and unclear information about sharing with third-party application developers' [20].

## 2.4 How Advertising Works

Facebook is commendable for its efforts in managing the interests of advertisers and respect of users' privacy [21]. Facebook promises that it will 'remove user's name or any other personally identifying information from it, or have it combined with other people's data in way that it is no longer associated with the user' before providing its advertising partners data. However, Facebook's privacy policy notes the possibility that advertisers might be placing cookies on users' computers [22], which arguably might invade individuals' privacy. Additionally, users are not able to opt out *all* advertising as advertising revenues are required to run the site [23].

## 2.5 Some Other Things You Need to Know

Facebook's privacy policy states that it may share users' information in good faith in response to legal requests and preventing harm including response to 'legal requests from jurisdictions outside of the United States' [24]. This seems to mirror NPP 4: transborder data flow. While this has helped law enforcement bodies to search for evidence in a variety of offences [25], it inevitably brings some privacy concerns. Users might contend that they have legitimate expectation of privacy in social networking sites that their private information would not be revealed to anyone including law enforcement bodies.

Facebook does 'provide initial responses to access requests within a reasonable period of time', and users are able to keep a record of their data by downloading it from the site [26]. This complies with NPP 6: access and correction [27].

Facebook's privacy policy states that 'we do our best to keep your information secure, but we need your help', and it provides a link to the Facebook Security Page that provides detailed information about staying safe on Facebook. This complies with NPP 4.1 that requires an organization to take reasonable steps to protect the security of data. However, commentators [28] argue that indefinite retention of personal information by Facebook breaches NPP 4.2, which requires organizations take reasonable steps to destroy or permanently de-identify personal information it no longer has any lawful use for.

### *2.6 Conclusion*

Facebook has been justifiably critized for its weak and shifting privacy rules [29]. Its privacy policy, in some respects, contradicts NPPs. In addition to what has discussed above, Facebook's privacy policy fails to address the data quality requirement provided in NPP 3 [30]. Facebook is not making a reasonable enough effort, in accordance with Principle 5.1, to express *clearly* policies on its management of personal information. It is not making a reasonable effort to notify users clearly about its privacy policy [31]. By virtue of the fact that Facebook provides often confusing [31], incomplete, and sometimes misleading [31] information about its privacy practices and the fact that users' information can be exposed to third parties without meaningful consent, it can be concluded that Facebook's privacy policy does not adequately protect individuals' privacy.

## 3 Facebook Features: Privacy Risks

The explosive growth of Facebook in recent years has pushed its potential privacy risk into the public spotlight. Such concerns include 'inadvertent disclosure of personal information, damaged reputation due to rumors and gossip, unwanted contact and harassment or stalking, surveillance-like structures due to backtracking functions, use of personal date by third parties, and hacking and identity theft' [31]. Risks to privacy invasion were ascribed more to others than to the self. Facebook's privacy policy primarily concerns about privacy protection of the user that provides information on Facebook, but it fails to address the privacy risks done by users to others. For example, a user might post content about others on Facebook such as indecent pictures of others [32] as well as accusations or allegations about others that are extremely invasive to individuals' privacy. This section discusses some of the features of Facebook that present challenges to individuals' privacy despite its privacy policy.

### 3.1 Public Search

Public search engines are essentially noteworthy given considerable controversies surrounding this feature since it first launched. Facebook made limited public search listings available to people who are not logged into the Facebook website in 2007. Previously that function was only limited to members. 'These search listings expose members' names, profile pictures, the ability to send a message to a member, view his or her friends and request to add that member as a friend' [33]. Facebook had made these listings available on search engines such as Google, MSN Live and Yahoo. Although users can opt out of the public listings through Facebook's individualized privacy settings page [34], many users are unaware or lack understanding of the public search change. Facebook has announced through its blog that it does not have a policy of notifying users of changes to the site via email [35]. The personal public information available on public search listings might be used by sexual predators to lure teenagers as well as used by criminals to commit crimes such as burglary, theft and robbery (e.g. by knowing users' recent activities and location).

### 3.2 Timeline

Mandatory changeover to the new Timeline profile has raised privacy concern. Some Facebook users reported discontent with having many Facebook status updates and photos from the past easily visible on their homepage [36]. Timeline allows Facebook to show a detailed look at user's whole Facebook history including everything a user has ever done on Facebook along with pictures. This change outraged Facebook users, who felt exposed and deprived of their sense of control over their information.

## 4 Australian Privacy Laws and the Shortcomings in Social Networks' Privacy Practices

### 4.1 Federal Law: Privacy Act 1988

The Privacy Commissioner has the power to investigate breaches of the Privacy Act with or without someone making a complaint. The Commissioner can take action overseas to investigate complaints [37]. The Commissioner tries to enforce the Privacy Act by 'conducting audits, monitoring government data-matching and issuing guidance material to agencies and organizations to help them comply with the Privacy Act' [21]. Even the most laudable attempts made by the Privacy Commissioner to protect individuals' privacy, the applicability of the Act is limited to (*personal*) information [38] about Australian citizens or permanent residents [39].

## 4.2 National Privacy Principles

The office of the Privacy Commissioner claims that the Australian privacy law is not prescriptive; rather, it requires organizations to adhere to a set of fair information practices or principles [21]. However, Facebook may not be required by law to take sufficient efforts to adequately notify its users of intricate details of its privacy policy nor does the law require Facebook to make a reasonable effort to advise users all of the purposes for which and the extent to which their personal information is used and disclosed [40].

Moreover, organizations must obtain consent from customers for use of their personal information for marketing purposes *only* where it is 'practicable' [41]. Controls on the transfer of personal information *overseas* are limited. Organizations need only take 'reasonable steps' to ensure information will be protected or 'reasonably believe' that the information will be subject to similar protection as in Australia.

Further, the regulation of privacy practices of the private sector has some notable exceptions, including small businesses with an annual turnover of $3 million or less [42] and the media and political organizations. Certain acts or practices are exempted from NPP: performance of contractual obligations, disclosure to the National Archives of Australia and an act or practice outside Australia [43]. Moreover, the Commissioner has limited powers: it can only make a declaration (not binding) on the finding of any breaches of the Act [44]. This raises the question of potential gaps of the Act in privacy protection [45]. The board scope of exemption as well as the limited powers of the Commissioner and the lack of strong enforcement mechanisms put questions on the effectiveness of the NPP in addressing the shortcomings in social networks' privacy practices.

Despite privacy laws, the reality is, however, that individuals are less likely to challenge the Internet giant, Facebook or anyone that infringes their individuals' privacy. This might due to the complexities of the *Privacy Act* (with two distinct principles applying to public and private sectors), the lack of awareness about privacy rights, the expensive and incumbent process to bring action or even the fact that individuals do not value privacy at all. Furthermore, privacy laws are limited in scope in mitigating the shortcomings in social networks' privacy policy given the very nature of the Internet (decentralized, ungoverned and global nature) and the complexities associated with it.

## 5 Conclusion

Users have made a choice to upload their information voluntarily on Facebook; therefore, they are partly responsible for potential invasion of their privacy. Facebook, with millions of users around the globe, is limited in power to adequately and effectively protect individuals' privacy. At the end of the day, privacy is about 'what we think, what we believe and value, what we want and what we want to do' [45], but privacy cannot be absolute in the era of social networking and privacy law is all about trying to find the balance between a right to privacy and other human rights and social interests.

## References

1. Acquisti A, Gross R (2006) Imagined communities: awareness, information sharing, and privacy on the facebook (Lecture notes, privacy enhancing technologies, )
2. Haynes AW (2007) Online privacy policies: contracting away control over personal information? Penn State Law Rev 11:587
3. Moses A (2009) Privacy watchdog puts bite on Facebook. Sydney Morning Herald (Sydney)
4. Ortutay B (2009) Facebook to end Beacon tracking tool in settlement. USA Today (New York)
5. Debatin B et al (2009) Facebook and online privacy: attitudes, behaviors, and unintended consequence. J Comput Mediated Commun 17:83
6. Abam C (2006) Welcome to facebook, everyone on Facebook Blog http://blog.facebook.com/blog.php?post=2210227130
7. Dwyer C, Hiltz SR, Passerini K (2007) Trust and privacy concern with social networking sites: A comparison of Facebook and MySpace (Conference Paper, Association for Information Systems, 2007)
8. Denham E (2009) Privacy commissioner's findings in the case of CIPPIC against Facebook (Report of Findings, Privacy Commissioner of Canada, 16 July 2009)
9. Facebook (2011) Date Use Policy. http://www.facebook.com/full_data_use_policy.
10. Davis G (2011) Facebook Ireland Ltd' (Report of Audit, Date Protection Commissioner, 21 December 2011)
11. Hull G, Lipford HR, Latulipe C (2009) Contextual gaps: privacy issues on Facebook. Ethics Inform Tech 13:289
12. Health Aston (2012) Ex-lover punished for Facebook revenge, Sydney Morning Herald, Sydney
13. IB Times Staff Reporter (2011) Sue facebook for quick crash and fun: top ten most whimsical lawsuits against Facebook to date. International Business Times, United States
14. Stoddart J (2011) Privacy in the era of social networking: legal obligations of social media sites Saskat Law Rev 74:263
15. Semitsu JP (2011) From facebook to mug shot: how the dearth of social networking privacy rights revolutionized online government surveillance. Pace Law Rev 31:291
16. Regan K (2003) Online privacy is dead – what now? E-Commerce Times. http://www.ecommercetimes.com/story/20346.html
17. Perlman L (2007) Facebook Ads on Facebook Blog. http://blog.facebook.com/blog.php?post=6972252130
18. Story L (2008) To aim ads, web is keeping closer eye on you, The New York Times, New York
19. Hodge MJ (2006–2007) The fourth amendment and privacy issue on the "new" internet: Facebook.com and Myspace.com. Southern Illinois University Law J 31 S.III.U.L.J 95

20. Waters N, Pacific Privacy Pty Ltd, Convernor and Australian Privacy Charter Council (2001) The new Australian privacy landscape (Seminar Paper, UNSW Continuing Legal Education, 14 March 2001)
21. Office of the Privacy Commissioner, Applying the Privacy Act, Privacy Law http://www.privacy.gov.au/law/apply
22. Fernandes R (2011) Facebook second most accessed site, behind Google in the US Tech2 (India)
23. Sengupta S (2006) Risk and riches in user data for facebook, The New York Times, New York
24. Staff Reporter (2002) Facebook timeline will be compulsory, The Sun, United Kingdom
25. Govani T, Pashley H, Student awareness of the privacy implications when using facebook <http://lorrie.cranor.org/courses/fa05/tubzhlp.pdf>
26. Pilgrim T (2011) Privacy law reform-getting the balance right (Speech delivered to Communications and Media Law Association, Euro)
27. Hashemi Y (2009) Facebook's privacy policy and its third-party partnerships: lucrativity and liability. Boston University J Sci Tech Law (B.U.J.Sci. & Tech.L) 15:140
28. Moses A (2009) Privacy watchdog puts bite on Facebook. Sydney Morning Herald, Sydney
29. Semitsu JP (2011) From Facebook to mug shot: how the dearth of social networking privacy rights revolutionized online government surveillance, 31 Pace Law Review 291
30. Privacy Act 1988 (Cth) sch 3: an organization must take reasonable steps to make sure that the personal information it collects, uses or discloses is accurate, complete and up-to-date
31. Debatin B et al (2009) Facebook and online privacy: attitudes, behaviors, and unintended consequence. J Comput Mediat Commun 83–86
32. Aston H (2012) Ex-lover punished for Facebook revenge. Sydney Morning Herald, Sydney
33. Fung P (2007) Public search listings on Facebook'on Facebook blog. http://blog.facebook.com/blog.php?post=2963412130
34. Facebook, Data Use Policy (23 Sept 2011) Facebook. http://www.facebook.com/about/privacy/
35. Abam C (2006) Welcome to Facebook, everyone' on Facebook blog. http://blog.facebook.com/blog.php?post=2210227130
36. Staff Reporter (2002) Facebook timeline will be compulsory. The Sun, UK
37. Privacy Act 1988 s 5B
38. Privacy Act 1988 (Cth) s 6
39. Privacy Act 1988 (Cth) s 5(1)(a)
40. Hashemi Y (2009) Facebook's privacy policy and its third-party partnerships: lucrativity and liability. Boston University J Sci Tech Law (B.U.J.Sci. & Tech.L) 15:140–155
41. Privacy Act 1988 (Cth) sch 3 cl 2.1(c)
42. Privacy Act 1988 (Cth) sch 3 cl 2.1(c) Pilgrim T, Privacy law reform-getting the balance right (Speech delivered to Communications and Media Law Association, Euro, 6 Sept 2011)
43. Privacy Act 1988 (Cth) sch 3 cl 6A
44. Privacy Act 1988 (Cth) s 52
45. Pilgrim P (2011) Privacy law reform-getting the balance right (Speech delivered to Communications and Media Law Association, Euro, 6 Sept 2011)

# Part V
# Green Technology

# Modelling of the Output Power from a Grid-Connected Photovoltaic System Using the Artificial Neural Network

**Shahril Irwan Sulaiman, Sulaiman Shaari, and Ahmad Maliki Omar**

## 1 Introduction

The photovoltaic (PV) system is a mode of renewable energy power generation. Besides being used for rural electrification, PV systems are also applied to urban areas where conventional grid electricity is readily available. The PV systems which are connected to the conventional distributed grid network are called grid-connected photovoltaic (GCPV) systems. A GCPV system typically consists of a PV array to convert solar irradiation into DC electricity and an inverter to convert DC electricity into AC electricity which has similar characteristics to grid electricity.

GCPV systems offer several unique features as a source of electricity generation. First, the systems utilize solar energy which is freely available as a source for generating electricity. The solar energy is also harvestable in most parts of the Earth. Second, the systems can be implemented as a centralized power generation or a distributed power generation, unlike the conventional power generation which could only be implemented as centralized power generators. Third, the system operation promotes zero release of greenhouse gas emissions to the environment, thus making it a clean option of generating electricity.

Despite having many benefits, the operation and performance of GCPV systems are strongly dependent on the weather conditions which are commonly fluctuating throughout the day. The weather parameters which affect the performance of GCPV systems are typically solar irradiance and ambient temperature. The fluctuation of these parameters causes inconsistent power output from the system [1]. Therefore,

S.I. Sulaiman (✉) • A.M. Omar
Faculty of Electrical Engineering, Universiti Teknologi MARA, 40450 Shah Alam, Selangor, Malaysia
e-mail: shahril_irwan2004@yahoo.com

S. Shaari
Faculty of Applied Sciences, Universiti Teknologi MARA, 40450 Shah Alam, Selangor, Malaysia

K. Ab. Hamid et al. (eds.), *The Malaysia-Japan Model on Technology Partnership: International Proceedings 2013 of Malaysia-Japan Academic Scholar Conference*,
DOI 10.1007/978-4-431-54439-5_35, © Springer Japan 2015

understanding of the relationship between the weather parameters and the power output is crucial such that the output performance of the system can be modelled and estimated.

Several studies have been conducted to model the power output from the GCPV system. First, a GCPV system kW output had been modelled using the artificial neural network (ANN) [2]. Two ANN models were developed to predict the AC kW output from the inverter. The first model utilized solar irradiance and ambient temperature as the set of inputs, while the second model used solar irradiance as the sole input to the ANN. Although the first model was found to produce lower error than the second model, the contribution from PV module temperature was not investigated. The contribution of PV module temperature is significant as it is associated with the solar cell temperature which directly influences the output power of a PV module [3].

This paper presents the modelling of the GCPV system output using the ANN. The ANN is used to model the AC kW output from the system based on selected sets of input parameters such as solar irradiance, ambient temperature and PV module temperature.

## 2 Artificial Neural Network

Although there are different types of ANN architecture used for prediction, the MLFNN has been widely used in solving many engineering problems due to its good generalization capability and simplicity [4]. Apart from that, the MLFNN is also capable of performing nonlinear modelling with reasonable accuracy [5]. Hence, the MLFFNN was employed in this study.

### 2.1 *Multilayer Feedforward Neural Network*

A general MLFFNN architecture with a single hidden layer is shown in Fig. 1. The MLFNN basically consists of an input layer, a hidden layer and an output layer. Every neuron in a layer has similar number of inputs with the total number of inputs to that layer. In addition, each neuron would also have a single output.

Assuming that the input of a neuron in the input layer, the output of a neuron in the hidden layer and the output of a neuron in the output layer of the MLFNN are represented by $x_p$, $y_q$ and $z_r$, respectively, while the number of neurons in the input layer, hidden layer and output layer are $i$, $j$ and $k$, respectively, $y_q$ can be calculated using

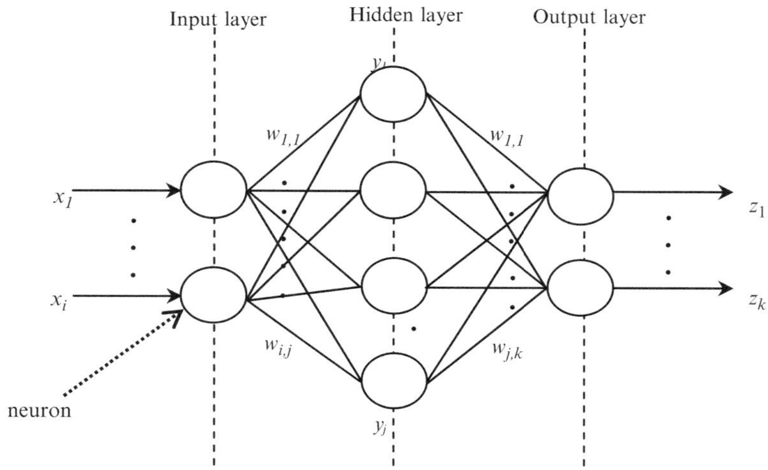


**Fig. 1** Basic structure of an MLFFNN with a single hidden layer

$$y_q = f_y\left(\sum_{p=1}^{i}\left(x_p w_{p,q}\right) - B_y\right) \tag{1}$$

where $B_y$ and $f_y$ are the bias and activation function for the neurons in the hidden layer, respectively. On the other hand, $z_r$ can be determined using

$$z_r = f_z\left(\sum_{q=1}^{j}\left(y_q w_{q,r}\right) - B_z\right) \tag{2}$$

where $B_z$ and $f_z$ are the bias and activation function for the neurons in the output layer, respectively. In the MLFNN, the data run rigidly from the input layer to the output layer via several processing units. Despite having no memory, the MLFNN output is heavily dependent on the input and the connection weight values in the network of nodes that would be modelled as fitting parameters for the MLFNN. The processing units, also known as the activation functions, can be linear and nonlinear, or a combination of both. However, the activation function for each neuron in the same layer has to be similar. Apart from that, although each neuron in a layer would have similar number of inputs, the connection weights could be different. In addition, as the neurons in the MLFNN are identical in structure, the information can only reside in the sets of weights and biases throughout the network. Nevertheless, a systematic method to update these weights and biases is called a learning algorithm or learning rule.

In this study, the search for the number of neurons, the type of activation function and the learning algorithm was performed heuristically such that the root mean square error (RMSE) of the model is minimized. The RMSE [6] of the model is determined using

$$\mathrm{RMSE} = \sqrt{\frac{\sum_{i=1}^{n} \left(z_{p,i} - z_{t,i}\right)^2}{n}} \tag{3}$$

where $n$ is the number of data patterns and $z_{p,i}$ and $z_{t,i}$ are the predicted and the target value of the MLFNN output, respectively. On the other hand, the coefficient of determination, $R^2$ [7], was calculated using

$$R^2 = 1 - \frac{\sum_{i=1}^{n} \left(z_{p,i} - z_{t,i}\right)^2}{\sum_{i=1}^{n} \left(z_{t,i} - z_{t\_avg}\right)^2} \tag{4}$$

where $z_{t\_avg}$ is the average value of the target MLFNN outputs and $R^2$ is a model fit commonly used in estimating the correlation between parameters in mathematical model. Nevertheless, the RMSE had been selected as the key performance indicator for the prediction, while $R^2$ was used to verify the validity of the RMSE of the network, i.e. $R^2$ of the prediction must be reasonably high although low RMSE was achieved by the network.

## 2.2 *Experimental Models*

In this study, the MLFNN comprises either two or three neurons at the input layer representing two or three inputs and one neuron at the output layer representing a single output. Three MLFNN models had been developed based on different types of inputs being assigned for the network. The three models are illustrated in Fig. 2. Model 1 consists of solar irradiance (SI) in W/m$^2$ and ambient temperature (AT) in °C as the inputs to the MLFNN. On the other hand, Model 2 consists of SI and PV module temperature (MT) in °C as the inputs. However, Model 3 utilizes all types of parameter, i.e. SI, AT and MT, as the inputs to the MLFNN. In all models, the output of the MLFNN was set to be the watt power output from the GCPV system. These models were evaluated to determine the best set of inputs which correlates with the output power of the system.

## 2.3 *Training and Testing*

The MLFNN is implemented in two stages, i.e. the training and testing [8]. The training is conducted by simulating the MLFNN models using the training data set after selecting the MFLNN design parameters such as the number of neurons in the

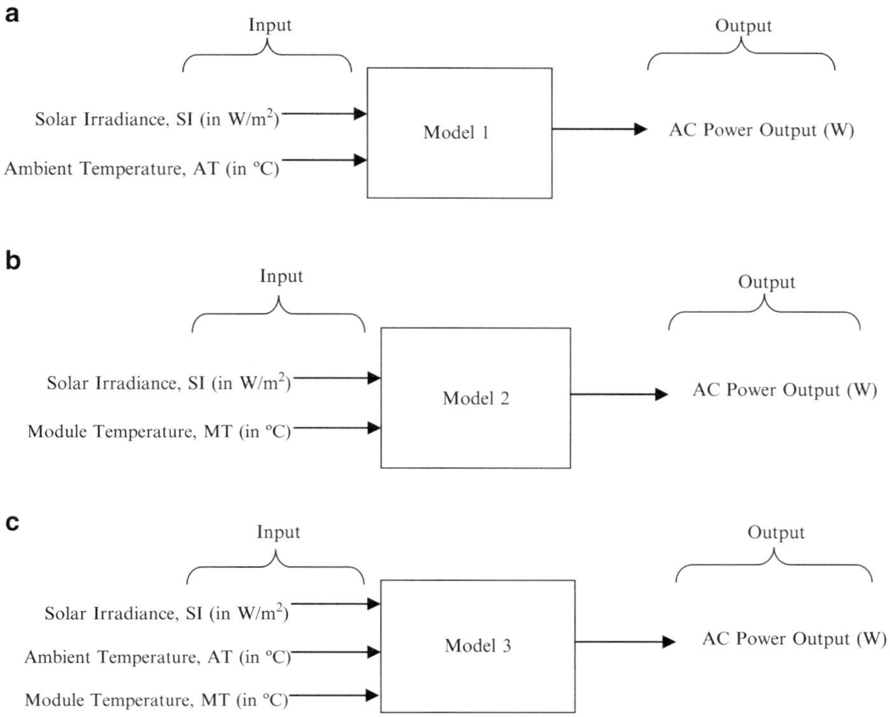


**Fig. 2** The three MLFNN models. Model 1 with SI and AT as inputs (**a**), Model 1 with SI and MT as inputs (**b**), and Model 1 with SI, AT and MT as inputs (**c**)

hidden layer, the type of activation function and the learning algorithm. Upon completion of training, testing is performed using the testing data set. If the testing produces the RMSE which is approximately similar to the RMSE obtained during training, the training process is not repeated. Otherwise, the training and testing process is repeated until satisfactory RMSE is obtained in both training and testing, respectively. Apart from that, $R^2$ is monitored during each training and testing to ensure high degree of fit for the models.

## 3 Data Collection

All data were obtained from a GCPV system located at Green Energy Research Centre (GERC), Universiti Teknologi MARA, Shah Alam, Selangor, Malaysia. The system comprises a retrofitted 10 kWp monocrystalline PV array (Yingli YL250C-30b) and an 8 kW inverter (Model: Sunny Tripower STP 8000TL-10). The irradiance and temperature sensors were connected to a built-in data logger inside the inverter while all data were recorded at 15 min interval. 50 % of the collected data or 2,556 data patterns were assigned for the training process, whereas the remaining 50 % of the collected data or 2,556 data patterns had been used for the testing process.

## 4 Results and Discussions

The MLFNN was executed using different models to identify the most suitable set of inputs for the MLFNN. Subsequently, the optimal types of parameter which strongly affect the output power of the GCPV system could be determined.

The results of the MLFNN training process are illustrated in Fig. 3. All models show good fit of data by having $R^2$ values very close to unity. However, Model 1 with SI and AT produces the lowest RMSE and also the highest $R^2$ when compared to Model 2 and Model 3. Model 1 yields approximately 13.76 and 18.58 % higher accuracy when compared to Model 2 and Model 3, respectively.

Upon completion of training, the MLFNN models underwent testing process. The results are shown in Fig. 4. All models show good fit of data by having $R^2$ values very close to unity and these values are commonly lower than the respective $R^2$ values obtained during training [9]. Similar to the training performance, Model 1 with SI and AT produces the lowest RMSE and also the highest $R^2$ when compared to Model 2 and Model 3. Model 1 yields approximately 5.40 and 17.80 % higher accuracy when compared to Model 2 and Model 3, respectively.

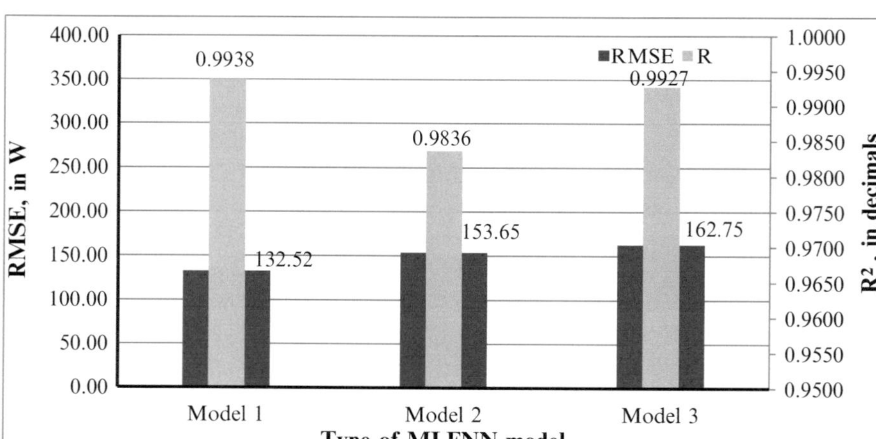


**Fig. 3** Training performance

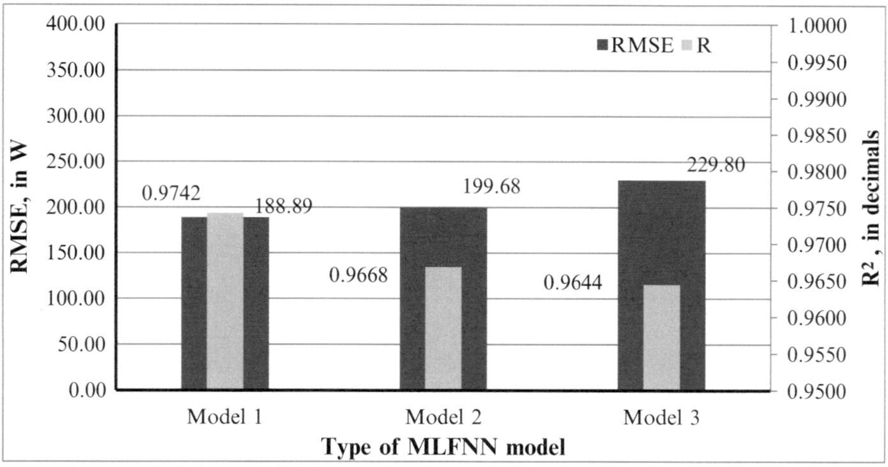


**Fig. 4** Testing performance

## 5 Conclusion

The MLFNN models were proven useful in modelling the output power from a GCPV system. However, the most accurate model was found to be Model 1 with SI and AT, as it yielded the lowest RMSE and highest $R^2$ during both training and testing.

## References


1. Tao C, Shanxu D, Changsong C (2010) Forecasting power output for grid-connected photovoltaic power system without using solar radiation measurement. In: 2nd International symposium on power electronics for distributed generation systems, HeFei, China, pp 773–777
2. Mellit A, Pavan AM (2010) Performance prediction of 20kWp grid-connected photovoltaic plant at Trieste (Italy) using artificial neural network. Energ Convers Manag 51:2431–2441
3. Skoplaki E, Palyvos JA (2009) Operating temperature of photovoltaic modules: a survey of pertinent correlations. Renew Energ 34:23–29
4. Gao DW, Wand P, Liang H (2007) Optimization of hidden nodes and training times in ANN-QSAR model. Environ Informat Arch 5:464–468
5. Rao KVS, Krishna MR, Babu DB (2009) Cryptanalysis of a feistel type block cipher by feedforward neural network using right sigmoidal analysis. Int J Soft Comput 4:131–135
6. Souza AP, Escobedo JF (2013) Estimates of hourly diffuse radiation on tilted surfaces in Southeast of Brazil. Int J Renew Energ Res 3:207–221
7. Zainuddin H, Shaari S, Omar AM, Sulaiman SI, Mahmud Z, Muhamad Darus F (2011) Prediction of module operating temperatures for free-standing (FS) photovoltaic (PV) system in Malaysia. Int Rev Model Simulat 4:3388–3394

8. Arafa M, Alqedra M (2011) Early stage cost estimation of buildings construction projects using artificial neural networks. J Artif Intell 1:63–75
9. Moghaddam MG, Ahmad F, Basri M, Abdul Rahman MB (2010) Artificial neural network modeling studies to predict the yield of enzymatic synthesis of betulinic acid ester. Electron J Biotechnol 13:1–12

# Basic Research on the $TiO_2$ Electrode and the Coloring Effect of Dye-Sensitized Solar Cell

**Arini Nuran, Matsutake Daiki, Mohamad Norsyafiq, and Akira Fujiki**

## 1 Introduction

Dye-sensitized solar cells (DSSCs) are a new type of solar cells developed by Gratzel et al. and composed of nanocrystalline porous semiconductor electrode-absorbed dye, a counter electrode and an electrolyte of iodide-triiodide ions [1]. It is a device utilized for the conversion of visible light into electricity, based on photosensitization produced by the dyes on wide bandgap mesoporous metal oxide semiconductors. This sensitization is due to the dye absorption part of the visible light spectrum. The sensitized dye works as a sunlight absorber which then converts it into electric energy.

Titanium dioxide ($TiO_2$) has attracted attention from researchers worldwide due to its potential applications in environmental protection and energy generation [2]. It has a high adsorption ability and good photocatalytic activity. Recently, $TiO_2$ has been applied largely in DSSC due to its nanocrystalline mesoporous nature that translates to high surface area for dye adsorption. The adsorbed dye molecules can then be excited by the solar energy to generate electron-hole pairs that are subsequently separated and transported within the lattice of $TiO_2$ [3].

Several metal complexes and organic dyes have been synthesized and used as sensitizers including porphyrins [4], platinum complex [5], and others. Ru-based

A. Nuran (✉) • A. Fujiki
Shibaura Institute of Technology (SIT), 307 Fukasaku, Minuma-ku, Saitama 337-8570, Japan
e-mail: md13004@shibaura-it.ac.jp

M. Daiki
Graduate School of SIT, 307 Fukasaku, Minuma-ku, Saitama 337-8570, Japan

Fujitsu Tokki Systems Limited, Tokyo, Japan

M. Norsyafiq
Shibaura Institute of Technology (SIT), 307 Fukasaku, Minuma-ku, Saitama 337-8570, Japan

Honda Malaysia Sdn Bhd, Alor Gajah, Malaysia

K. Ab. Hamid et al. (eds.), *The Malaysia-Japan Model on Technology Partnership: International Proceedings 2013 of Malaysia-Japan Academic Scholar Conference*,
DOI 10.1007/978-4-431-54439-5_36, 

complexes sensitizers have been widely used because they have better efficiency and high durability. However, these advantages are offset by their high cost, and they tend to undergo degradation in the presence of water [6].

In order to overcome those problems, we used natural pigments as sensitizing dyes. This is because they enhance the economical aspects besides having significant benefits from the environmental point of view. Natural pigments extracted from fruits and vegetables such as anthocyanins (blueberry) are proven to be suitable for DSSC.

# 2 Experimental Procedure

## 2.1 *Examined Materials*

For fabrication of DSSC, $TiO_2$ paste (Ishihara Sangyo Company, ST-01), acetic acid (Kanto Chemical Company), pure water, polyethylene glycol (Wako Pure Chemical Industries, 400), acetylacetone (Kanto Chemical Company, JIS K 8027), iodine electrolyte, graphite carbon pencil, FTO (fluorine-doped $SnO_2$ coated glasses), and blueberry were used in this research.

Experimental steps were done in three parts: preparation of the cathode electrode ($TiO_2$ paste), preparation of anode electrode (carbon), and preparation of dye-sensitized solution.

## 2.2 *Preparation Method*

### 2.2.1 Preparation of Cathode Electrode ($TiO_2$ Paste)

The $TiO_2$ films were made by spreading $TiO_2$ pastes on the fluorine-doped $SnO_2$ conducting glass (FTO) by the spin-coating technique. $TiO_2$ pastes were prepared by mixing-grounding the $TiO_2$ sample with distilled water, acetic acid, acetylacetone, and polyethylene glycol (PEG). Extensive stirring was proceeded to ensure complete dispersion of $TiO_2$ nanoparticles and to facilitate the spreading of the colloid on FTO glass [7]. Droplets of each paste were placed onto the FTO glass, on the turning table of a spin coater. Adhesive tapes were placed on the edges of FTO glass to form a guide for spreading the pastes at about 10 s. Then, air-dry for 10 min. After air-drying, the $TiO_2$ was heated at 100 °C for 30 s, followed by annealing at 450 °C for 60 min. The resulting electrode was to be cooled at 80 °C. Thickness of the $TiO_2$ film was controlled by multiple coating processes in which the coated substrates were subjected repeatedly to spin-coating and drying steps.

#### 2.2.2 Preparation of Anode Electrode (Carbon)

The FTO glass was wiped with ethanol. Then, the FTO glass surface was colored by using graphite carbon pencil. After that, the surface was checked to ensure that there was no space that the carbon did not cover.

#### 2.2.3 Preparation of Dye-Sensitized Solution

Anthocyanins (Blueberry)

Twenty frozen blueberries were left to defroze. No water was added into the solution. Cathode electrode was then immersed in the dye solution for 6 h, without the presence of light. After that, pure water was poured on top of the electrode to ensure that there was no blueberry stuck on top of the $TiO_2$ surface. The electrode was taken out of the dye solution, and left to dry for 15 min, at room temperature.

#### 2.2.4 DSSC Assembling

Cathode electrode and anode electrode were put together, overlapping each other, and a space at the end of each electrode was made. Next, fasten both electrodes using the double clip. Three drops of iodide solution were added at the end of the electrode. Let the solution spread over the entire electrode. Then, wipe off the remaining iodide solution using cotton swab soaked with alcohol. After that, a tester with crocodile clip was attached at both ends of the electrode, and the experiment was conducted under true light® (10 $W/m^2$).

True light has the same wave pattern as the sun. It is usually used in animal breeding.

## 3 Results and Discussion

### *3.1 Thickness*

The results produced show that as the thickness of $TiO_2$ increases, the power produced will also increase (Fig. 1). The sufficient film thickness will create large pore size and enough space that allows the more redox electrolyte to diffuse into the film.

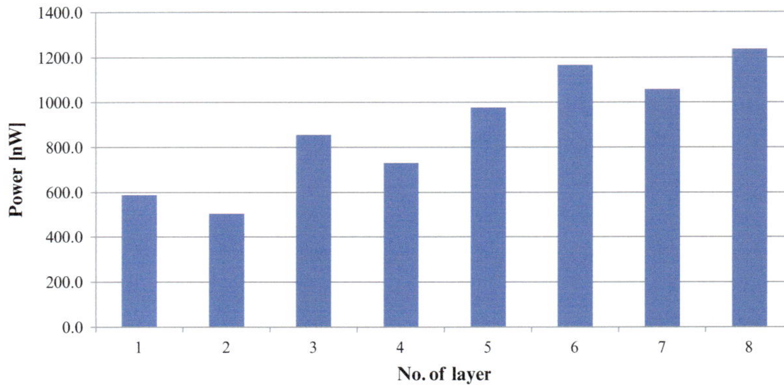


**Fig. 1** The energy produced with the number of layers of $TiO_2$

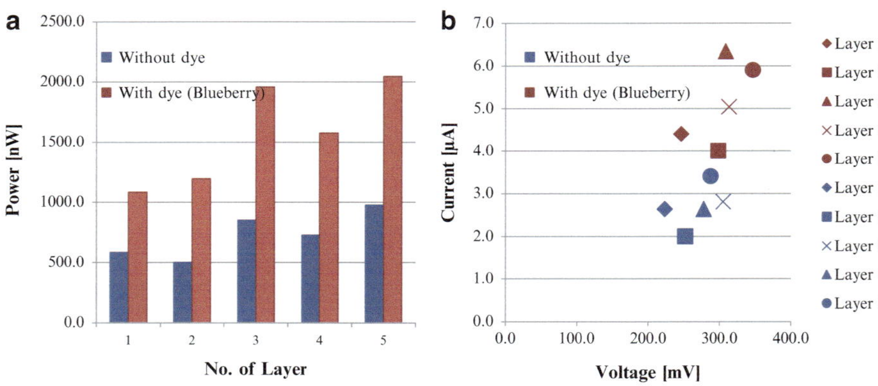


**Fig. 2** Number of layers of $TiO_2$ with and without dyeing. *Left figure* (**a**): The difference between $TiO_2$ with dye and without dye with the power produced. *Right figure* (**b**): The difference between $TiO_2$ with dye and without dye with the voltage and current produced

## *3.2 Thickness and Effect of Dye*

It is noted that the amount of dye absorbed for $TiO_2$ increases with the film thickness. The photocurrent generated by the solar cell is directly proportional to the amount of dye adsorbed on $TiO_2$ film [8]. This explains that more dye molecules are attached to the increased surface of $TiO_2$. Then, it can generate many electrons and improve the electron transport. The circuit voltage is almost the same, while the current density increases as the thickness of the $TiO_2$ layer increases (Fig. 2). This brings about the improvement of the efficiency.

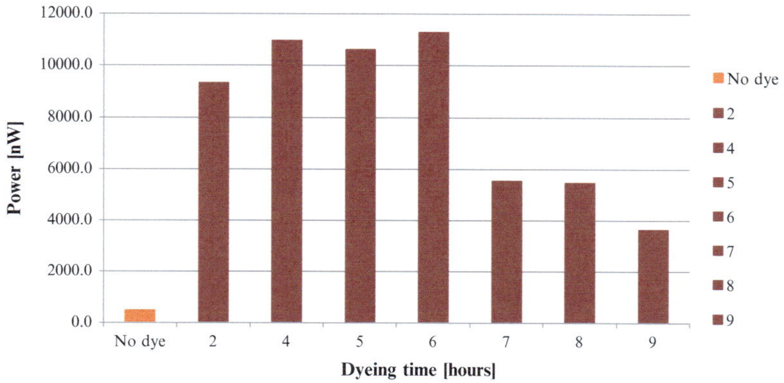


**Fig. 3** The bar graph shows the power produced according to the dyeing time

### 3.2.1 Suitable Condition For Dye (Blueberry) Absorption

Based on the graph below (Fig. 3), optimum dyeing time was 4–6 h. From 7 to 8 h, it shows the decrease in power produced. This is because the blueberries were used as the dye solution. Any time more than 6 h, the blueberries will start to rot. And this will change the dye solution. It might contain more acid than normal blueberries.

## 4 Conclusion

There are many factors that will influence the performance of DSSC, for example, the thickness of $TiO_2$, the nature of dye, the activity of redox electrolyte, etc. The criteria for making an effective $TiO_2$ working electrode include high surface area, appropriate bandgap, and appropriate dye solution. It is essential to increase the surface area of $TiO_2$ nanoparticles in order to increase the amount of dye adsorbed.

## References

1. Gratzel M (2005) Solar energy conversion by dye-sensitized photovoltaic cells. Inorg Chem 44:6841–6851
2. Saadoun L et al (2000) Synthesis and photocatalytic activity of mesoporous anatase prepared from tetrabutylammonium-titania composites. Mater Res Bull 35:193
3. O'Regan B, Gratzel M (1991) A low-cost, high-efficiency solar cell based on dye-sensitized colloidal $TiO_2$ films. Nature 353:737
4. Odobel F, Blart E, Lagrée M et al (2003) Porphyrin dyes for $TiO_2$ sensitization. J Mater Chem 13(3):502–510
5. Islam A, Sugihara H, Hara K et al (2001) Dye sensitization of nanocrystalline titanium dioxide with square planar platinum(II) diimine dithiolate complexes. Inorg Chem 40(21):5371–5380

6. Zhang D, Lanier SM, Downing JA, Avent JL, Lum J, McHale JL (2008) Betalain pigments for dye-sensitized solar cells. J Photochem Photobiol A 195(1):72–80
7. Wang H-H et al (2011) Preparation of nano-porous $TiO_2$ electrodes for dye-sensitized solar cells. J Nanomater 2011(2011):7
8. Ito S et al (1999) Low-temperature synthesis of nanometer-sized crystalline $TiO_2$ particles and their photoinduced decomposition of formic acid. J Colloid Interface Sci 216(1):59–64

# An Experimental Study on Bioclimatic Design of Vertical Greenery Systems in the Tropical Climate

**Badrulzaman Jaafar, Ismail Said, Mohd Nadzri Md Reba, and Mohd Hisham Rasidi**

## 1 Introduction

In the past, comfort is derived from natural resources. Now we tend to rely solely on mechanical systems, forgetting about the benefits of passive solar design and natural resources such as wind, water, or sun. Bioclimatic design aims to build in harmony with the weather rather than in opposition to it. Bioclimatic design is characterized by a willingness to adapt to the environment. This type of design allows substantial savings in air conditioning and lighting. Just by correctly orienting the house, using insulation to its best advantage, and correctly placing openings (windows and doors), you can save between 30 and 40 % on energy bills. There are natural or passive air conditioning systems, some of which have been around for hundreds of years and offer savings without generating extra costs. In a bioclimatic design, there are no set construction models. What works in one climate does not as a rule work in another. In warmer climates, the main facade is best oriented to the north and protected with shading elements. This condition is caused by the characteristics of the hot humid climate with summer temperatures (>20 °C) and moderate to winter (>0 °C). Annual precipitation and humidity are high with frequent rain showers. Freezing temperatures are uncommon, and relatively minor diurnal temperature fluctuations are typical [8]. Thus, site planning and design should seek to increase shade, cooling from evaporation, and breezes.

Living walls and green facades are very useful in warmer regions. They tend to be used on the western and southern parts of buildings that receive the most sunlight [2], thus preventing the sun's rays from hitting the dwelling directly. Moreover, the

B. Jaafar (✉) • I. Said • M.H. Rasidi
Department of Landscape Architecture, Faculty of Built Environment,
Universiti Teknologi Malaysia, 81310 Skudai, Johor, Malaysia
e-mail: badrazj@gmail.com

M.N.M. Reba
Institute of Geospatial Science and Technology (INSTeG), Faculty of Geoinformation and Real Estate, Universiti Teknologi Malaysia, 81310 Skudai, Johor, Malaysia

K. Ab. Hamid et al. (eds.), *The Malaysia-Japan Model on Technology Partnership: International Proceedings 2013 of Malaysia-Japan Academic Scholar Conference*,
DOI 10.1007/978-4-431-54439-5_37, 

distance between the plant layer and the constructed wall serves as the insulator: the vegetation cools the layer of air that is in contact with the wall of the house by means of evaporation of water on the surface of the leaves [12]. In this way, the surface temperature of the wall falls and less heat transfers to the dwelling. On the north facade which is the coolest since it does not face the sun, the microclimate obtained between the two walls is ideal for ventilating the home with cool air.

Rapid urbanization and the concern for climate change have led to a growing trend to bring nature into the cities [6]. Due to urbanization, urban heat island effects have become a serious problem. Greenery is expected to be an effective countermeasure, and much research is available regarding green roofs, green walls, street trees, and parks. In Malaysia, the use of vertical green system (VGS) is still new, needs more exposure to the best practice related to a method of increasing human comfort, and reduces energy consumption. Therefore, the approach of the Look East Policy adopted by Malaysia since 1982 is able to provide the best examples of VGS practice in Japan.

In Japan, vertical greenery system is widely used to reduce the urban heat island effect. This has been proved by the action of the Federal Government of Japan, and major cities offer incentives to developers that install green facades, living walls, and green roofs [4]. Tokyo experienced poor air quality and climate change in the early 1990s. The average temperature is now 5 °F (3 °C) higher than it was 30 years ago. On tall buildings, the ratio of wall surface to roof area is much higher. In the 1990s, Japanese municipalities and agencies of the Federal Government are aware that the concrete high-rise buildings are the dominant form of urban development, which contribute to the urban heat island. Therefore, the Japanese government has funded the installation work and research for Bio-Lung measuring 500′ long and 40′ high green walls. The results showed that 30 different green wall systems supplied by 18 different manufacturers in Japan have lowered the temperature of the building [4]. Results of the study, the act were drafted by the Tokyo Metropolitan Government (TMG) in April 2001. It requires all roof and wall surface greened in any new construction project with greater effect than 1,000 $m^2$ (250 $m^2$ to public utilities).

In this paper, experimental investigation of temperature and percentage of moisture at the surface of the VGS in front of the corridor is presented. The basic element of the study is a modular system and the cable system is installed next to each other. For this study, the ability to enhance the effect of high cooling by evaporation is achieved through a modular system.

## 2 Literature Review

The meaning of vertical greenery systems was collected from various research sources that produce a significant picture on terminology of vertical greenery system. Peck et al. [12] and Bass and Baskaran [3] define it as a way in which plants can be grown on, above, or in the walls of the building. This view is supported by Köhler [10] and Dunnett and Kingsbury [7] who state that vertical green system is the result of greening vertical the buildings with plants, either rooted in the ground, the wall material itself, or in modular panels attached to the facade to cover the building with plants and can be classified into the facade greening and living wall systems. Pérez et al. [13] have proposed a classification system called the green vertical system. This proposal is consistent with Wong et al. [20] and Cheng et al. [5] state that vertical greenery system (VGS) is comprised of the green facade and living wall. In this classification, a significant difference between the vertical greenery systems' detailed comparisons can be made. It is through the construction of different methods that plant species are grown, and the support structure and the materials used are dominant. Vertical greenery system concepts can be divided into two main categories such as green facades (support system) and living wall systems (carrier system) according to their growing method.

## 2.1 Temperature Reduction by Vegetation

Studies on the effect of the use of plants in vertical greenery system are different according to its purpose. It includes how they control the amount of solar radiation, reduce the cooling load, and improve internal and external thermal environment. Papadakis et al. [11] showed that the solar radiation incident on the shaded trees (100 W/m$^2$) is far lower than the area without shade (600 W/m$^2$). Stec et al. [16] reported that the temperature of the different layers of the double-skin facade is generally lower if plants are used instead of blinds in interior space. This is because the plants' surface temperature does not exceed 35 °C compared to blinds which have recorded temperatures above 55 °C. The installation of plants inside the double-skin facade can reduce the energy consumption of an air conditioning system energy up to 20 %. This study was strengthened by the "Bioshader" experiment conducted by Miller et al. (2007) at the University of Brighton, United Kingdom, where it is reported that there is a temperature difference if the double-skin facade was installed on the windows of offices, which have been compared with other offices without plants. Results show a reduction of internal temperature measurement at 3.5–5.6 °C. Another study by Wong et al. [19] on a freestanding wall in Hortpark (Singapore) with the type of vertical greening showed maximum reduction of 11.6 °C. This indicates that the greened façade less heat adsorbed not greened façade and reveals itself in less heat radiation during the afternoon and evening.

## 2.2 Humidity Increase by Evapotranspiration

Evapotranspiration process of plants that occurs through evaporative cooling of the leaves depends on the type of plant and exposure. It also includes the influence of climatic conditions. Dry environment or the effects of wind can increase the evapotranspiration of plants. In the case of vertical greenery systems, evaporative cooling from the substrate becomes important especially for the effect of substrate moisture [13]. Vertical greenery system can produce changes in ambient conditions (temperature and humidity) of green space between the screen and the wall of the building. This layer of air can produce a cooling effect on evaporation. Pérez et al. [13] verified that a microclimate between the wall of the building and the green facade is created, and it is characterized by slightly lower temperatures and higher relative humidity. As a consequence, especially in warmer climates, the cooling potential can lead to significant energy savings for air conditioning [1]. Sheweka and Mohamed [15] found that high humidity offsets the thermal comfort especially when temperatures are high and there is no wind to overcome thermal discomfort. Therefore, it is important to consider the location of the green wall in a closed place because it can affect the temperature and humidity.

Evapotranspiration is the process by which water is transferred from the land to the atmosphere through transpiration from plants. It shows that plants provide a cool environment where the land and the plants become perfect insulators [11, 18]. Vertical greenery system is an excellent way of saving on air conditioning, and their advantages do not end there. One of the solutions to improve the thermal performance of the vertical greenery system is to cover them with vegetation. Living walls and green facades enhance thermal comfort in a building environment both indoor and outdoor by reducing heat transfer to and from the building envelope [16]. They shade buildings from solar radiation, absorb solar radiation for photosynthesis and evapotranspiration, and reduce solar reflection and re-radiation to atmosphere [19].

## 3 Methodology

An experimental approach used in this study consists of two panels of the vertical greenery system characterized as green facades and living walls which are installed in front of a 5-storey office building corridor in Petaling Jaya, Selangor, Malaysia (Fig. 1). Two experiments were carried out to compare the air temperature and humidity in the internal corridor. The outdoor temperature and relative humidity were measured 30 cm from the facade using a HOBO U12-011 (Onset Computer Corporation, Massachusetts, USA) [9, 17] humidity and temperature data logger ($\pm 2.5$ % accuracy for RH and $\pm 0.35$ °C accuracy for temperature). Measurements of temperature and relative humidity are made on the surface of the vertical greenery system which is facing towards the north. It is measured at two different points (1,000, 2,000 mm from the bottom of the corridor) for each panel (living wall and green facade). From 27–31 May 2013, these measurements were recorded every 30 minutes with a data logger (Hobo U12-011 Temp\ RH, Onset Com, USA).

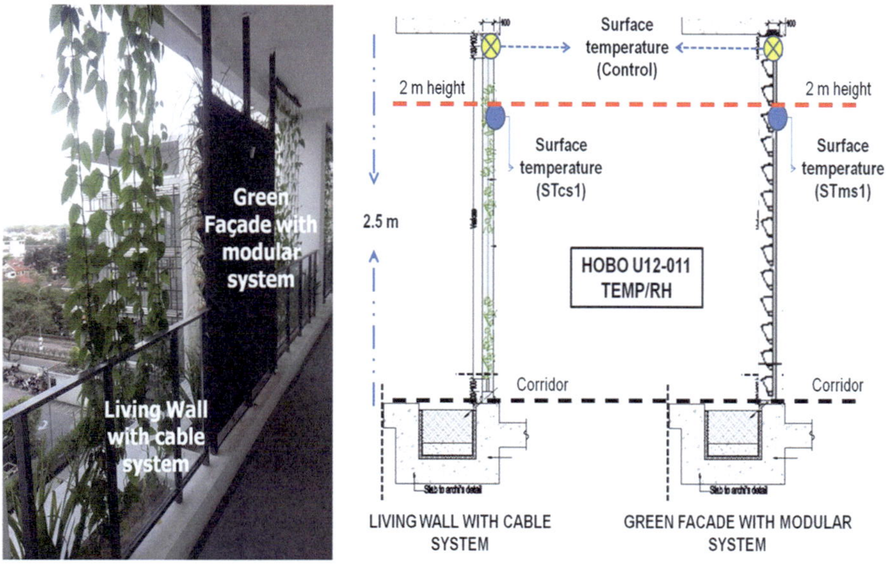


**Fig 1** Vertical greenery systems

## 4 Findings and Discussion

This paper presents the results of experimental studies conducted for 5 days as the reference. The focus of this study is to obtain data on the effect of living walls and green facade in the internal corridor. In this study, the effects of temperature and relative humidity on experimental plots at the site were evaluated.

### *4.1 Temperature*

Results in Table 1 show that the average temperature on the three plots is very different according to the VGS type used. Results in a modular system are lower (29.8 °C) and living wall (30 °C) compared to control (30.6 °C). These results suggest that a decrease in temperature occurs with the use of a vertical greenery system in the corridor. Maximum temperature taken from the living wall is 37.4 °C, which is higher than the green facade at 33.9 °C. This occurs because more sunlight can penetrate to the living wall with a cable system compared to the green facade's thickness and dense vegetation [5]. Thus, the use of green facade with a modular system is more effective in providing a cooling effect. This is because the highest temperature was 45.8 °C without any vegetation on the surface of the wall. It suggests that more heat energy into the non-shaded wall causes higher surface wall temperature [11].

**Table 1** Comparison of temperature by vertical greenery system

| Type of VGS | Max (°C) | Min (°C) | Mean (°C) | SD |
|---|---|---|---|---|
| Control | 45.814 | 24.171 | 30.628 | 5.948 |
| Living wall system | 37.453 | 25.965 | 30.014 | 2.793 |
| Modular system | 33.966 | 26.329 | 29.842 | 2.183 |

**Table 2** Comparison of relative humidity by vertical greenery system

| Type of VGS | Max (%) | Min (%) | Mean (%) | SD |
|---|---|---|---|---|
| Control | 96.791 | 28.548 | 70.635 | 19.718 |
| Living wall system | 86.797 | 49.759 | 70.397 | 11.562 |
| Modular system | 86.031 | 52.944 | 71.703 | 9.864 |

### 4.2 *Relative Humidity*

There are significant differences derived from the study site. Table 2 shows the average percentage of relative humidity on three plots including the control. The study found that the use of green facade with a modular system produce high humidity (71.7 %) compared to the living wall system (70.3 %). It happened because 5–30 % of the remaining solar radiation is passing through the leaves and is affecting the internal climate of buildings when it passes the façade or roof [14]. Both systems show similarities if referred to the maximum percentage of humidity, where the living walls (86.7 %) and green facades (86 %) are compared to the controls (96.7 %). However, due to heavy rain, the percentage of humidity in the control has been recorded as the highest reading. It is in line with studies by Sheweka and Mohamed [15], which state that high humidity compensates thermal comfort especially when temperatures are high and there is no wind to overcome thermal discomfort. Therefore, it is important to consider the location of the green wall in a closed place because it affects the temperature and humidity.

## 5 Conclusion

The vertical green system industry in Malaysia is in its infancy and it is important to ensure the opportunities available for knowledge transfer and exchange. Expertise and knowledge of the have proven effective within Japan. From the experimental results, it is clear that the temperature drop in the corridors occurs in a green facade compared to a living wall. This is because of the moisture available in a modular system due to the stagnant watering. This situation is contrary to the living wall system that uses a cable system that cannot absorb moisture. This research can be further developed to produce a more effective cooling method. It has the potential to be expanded for the validity of the temperature drop through the use of VGS. However, it can only be proven if the study period is longer than 1 month. This is due to the need to obtain patterns of temperature and humidity that influence the effectiveness of a bioclimatic design. It also indirectly will provide space for discussion and study on the species and morphology of plants suitable for use in this method. Other parameters that are proposed to be conducted in specific studies are aesthetics, the thickness profile, weight, ease of maintenance, ease of construction, type of plant, and cost.

**Acknowledgments** We would like to thank the Faculty of the Built Environment and Faculty of Geoinformation and Real Estate for providing us the materials and equipment to complete the experiment. We are also grateful to HashimDesign and Conlay Land Sdn. Bhd. for their assistance and necessary support in collecting the data.

# References

1. Alexandri E, Jones P (2008) Temperature decreases in an urban canyon due to green walls and green roofs in diverse climates. Build Environ 43(4):480–493
2. Ali-Toudert F, Mayer H (2007) Effects of asymmetry, galleries, overhanging façades and vegetation on thermal comfort in urban street canyons. Sol Energ 81(6):742–754. doi:10.1016/j.solener.2006.10.007
3. Bass B, Baskaran B (2003) Evaluating rooftop and vertical gardens as an adaptation strategy for urban areas. CCAF Impacts and Adaptation Progress Report, Toronto, p 111
4. Carpenter S (2008) Green roofs and vertical gardens. International Specialised Skills Institute, Melbourne (ISS Institute) Melbourne, Australia, p 48
5. Cheng CY, Cheung KKS, Chu LM (2010) Thermal performance of a vegetated cladding system on facade walls. Build Environ 45(8):1779–1787
6. Chiang K, Tan A (eds) (2009) Vertical greenery for the tropics, 1st edn. National Parks Board, Singapore
7. Dunnett N, Kingsbury N (2010) Planting green roofs and living walls, 2nd edn. Timber Press, Inc., Portland
8. Harris CW, Dines NT (eds) (1998) Time-saver standards for landscape architecture: design and construction data, 2nd edn. McGraw-Hill Publishing Company, Massachusetts
9. Jim CY, He H (2011) Estimating heat flux transmission of vertical greenery ecosystem. Ecol Eng 37(8):1112–1122
10. Köhler M (2008) Green facades – a view back and some visions. Urban Ecosyst 11(4):423–436
11. Papadakis G, Tsamis P, Kyritsis S (2001) An experimental investigation of the effect of shading with plants for solar control of buildings. Energ Build 33(8):831–836
12. Peck SW, Chris C, Kuhn ME, Bass B (1999) Greenbacks from green roofs: forging a new industry in Canada. CMHC, Toronto, p 78
13. Pérez G, Rincón L, Vila A, González JM, Cabeza LF (2011) Green vertical systems for buildings as passive systems for energy savings. Appl Energ 88(12):4854–4859
14. Perini K, Ottelé M, Haas EM, Raiteri R (2011) Greening the building envelope, façade greening and living wall systems. Open J Ecol 01(01):1–8
15. Sheweka SM, Mohamed NM (2012) Green facades as a new sustainable approach towards climate change. Energ Procedia 18:507–520. doi:10.1016/j.egypro.2012.05.062
16. Stec WJ, Van Paassen AHC, Maziarz A (2005) Modelling the double skin façade with plants. Energ Build 37(5):419–427. doi:10.1016/j.enbuild.2004.08.008
17. Sunakorn P, Yimprayoon C (2011) Thermal performance of biofacade with natural ventilation in the tropical climate. Procedia Eng 21:34–41. doi:10.1016/j.proeng.2011.11.1984

18. Wong NH, Chen Y, Ong LC, Sia A (2003) Investigation of thermal benefits of rooftop garden in the tropical environment. Build Environ 38:261–270
19. Wong NH, Tan AYK, Chen Y, Sekar K, Tan PY, Chan D, Chiang K et al (2010) Thermal evaluation of vertical greenery systems for building walls. Build Environ 45(3):663–672
20. Wong NH, Tan AYK, Tan PY, Wong NC (2009) Energy simulation of vertical greenery systems. Energ Build 41(12):1401–1408

# The Adoption of Green Manufacturing Practices in Furniture Firms

**Puteri Fadzline Tamyez, Norzanah Mat Nor, and Syed Jamal Abdul Nasir**

## 1 Introduction

According to Gomes and Shian [1], manufacturing is much related to the 3D syndrome—dirty, dangerous and degenerative. It is unfortunate to acknowledge the progress of Malaysia in green technology that is 25 years behind Europe particularly in green building technology, water and waste management and the transportation sector. Green technology is not merely described as technology or processes that are environmentally friendly and applied principles of sustainability. According to Lee et al. [2], green management comprises interaction between corporate and environment which further brings an impact in the field of business and environment. Nature or green technology is a current key theme which continues to conquer interior works and the furniture industry [3].

Furniture is an example of products that constitute to one of the major causes of environmental problems [4]. The limitation and overexploitation of natural resources have forced green design to be advocated as the mainstream of the future direction to protect the environment and resources [3, 5]. Most greenhouse gases emissions, nature resources consumption and other environmental problems are caused by manufacturing industries. References [6, 7] introduce 'green innovation' which aims to radically produce green products that are not only 'new' from distinctive competitors but also 'green' enough to create an influence and to minimise environmental impact [8]. The economic growth, industrialisation and growing population in Malaysia have forced companies to pursue to green manufacturing practices [9, 10]. The National Green Technology Policy (NGTP2009) has been developed by the government to encourage manufacturers to lessen environmental problems [10]. Green practices are also among the seven

P.F. Tamyez (✉) • N.M. Nor • S.J.A. Nasir
Faculty of Business Management, Universiti Teknologi Mara, Shah Alam, Malaysia
e-mail: 120rahmat@gmail.com

K. Ab. Hamid et al. (eds.), *The Malaysia-Japan Model on Technology Partnership: International Proceedings 2013 of Malaysia-Japan Academic Scholar Conference*,
DOI 10.1007/978-4-431-54439-5_38, © Springer Japan 2015

thrusts outlined in the National Timber Industry Policy as a policy direction in marketing and promotion [11].

## 2 Methodology

A total of 49 usable responses out of 403 samples completed the open-ended questions. Arnon and Reichel [12] stated the necessity to include open-ended questions in order to draw out personal thoughts about the topic of interest. Samples involved are the small-, medium- and large-sized furniture managers in Malaysia. The majority of the respondents (93.5 %) are Chinese who fell in the age ranged of 40–49 years old. Most of the respondents had 21–30 years of experience in the furniture industry where the majority of them are managers but not owners. In terms of the size of firm, the majority of respondents belong to the medium-sized firms which constitute 35.2 %, followed by small-sized and a minority of large-sized firms which constitute 10.3 %. Content analysis was chosen as a method of analysing the data resulting from the open-ended question. According to Ref. [13], an 'emergence of regularities' needs to be achieved after an iterative process of identifying units of similar characteristics.

## 3 Results and Discussion

### *3.1 Reluctance in Pursuing Green Processes*

A very large proportion of the participants (81 %) strongly claimed that they do not have any consideration to produce green products or green processes. This indicates a low possibility of adoption of green practices among the manufacturing firms. Lee [14] and Nawrocka [15] agreed to the notion and added that it is similar among SMEs that often have a narrow perspective on tackling green issues in an ad hoc manner.

This study is consistent with Valipoor and Ujang [16]'s study which revealed that most of their respondents are reluctant to sacrifice in changing their production facilities in order to pursue towards sustainable products. Factors that contribute to their hesitation are the lack of technology, proper facilities and appropriate suppliers. A similar study [17] added a lack of system of knowledge sharing and a large skill set of sustainable product design are some of the barriers to implement green products. This is also agreed by Ratnasingam et al.[18] and Ratnasingam and Wagner [19], and they stated that the lack of customer demand, no government regulation, lack of government incentives and lack of trained personnel to implement and maintain the system also contribute to the deterrents of the adoption of green manufacturing practices.

Not only that, Ratnasingam [20, 21], Ratnasingam et al. [18] and Ratnasingam and Wagner [22] investigated that ISO 14000 is less applied among the furniture firms and needs to be contemplated to evaluate whether the cost be reduced with the application of ISO 14000. Ratnasingam [21] added that ISO 14001 comprises important elements of an environmental management system which have been implemented by more than 130,000 organisations worldwide. A participant (No. 29) agreed to have consideration in producing green products depending on the market response and the additional cost that can be accepted by the customer. Another participant (No. 25) also had similar view and stated that their company has not put serious effort to implement this work or idea. The reason behind the reluctance of the retailers in pursuing green technology is their unwillingness in paying premium price for green furniture. Most buyers have the perception of purchasing cheaper furniture [1]. Some furniture buyers are not having any interest in purchasing sustainable products due to their unawareness on the benefits of the positive effects of green products and little realisation of sustainability to the public in Malaysia [16]. However, Adis et al. [23] argued that this is not the case for foreign buyers as they prefer certified environmentally friendly products from Malaysian firms. Noci and Verganti [24] revealed that advanced green manufacturing can only be carried out through implementation of high innovation capacity.

## *3.2 The Willingness of Firms in Pursuing Green Management*

Hence, it is worrying that there is only small proportion of participants that could take their products to a higher level by meeting requirements to produce green products, using water-based or nontoxic finishing materials and using plantation wood rather than tropical wood. Green management has become fashionable and a trend among large firms [25]. Gomes and Shian [1] agreed that there is an increasing number of manufacturers instituting green practices in response to the market that is becoming highly competitive. Ratnasingam [21] added that more export-oriented manufacturing firms showed significantly higher degree of green manufacturing. This is agreed by a research made by Ratnasingam and Wagner [19] in Malaysia which stated that only 54 % of the respondents practised some green manufacturing practices and only 8 % adopted the ISO 14000 environmental quality management system. This is also supported by communities [26] which stated that the use of nontoxic, sustainably produced or recycled materials in a process is one of the common principles of environmentally sustainable design. A study has been made by Kozak et al. [27] about the acceptance of customers in Canada on the market potential of wood certification. It is found that not only most participants had little knowledge on environmental labelling of wood products, but also they were willing to purchase and pay a small premium for these products, assuming equivalent design and quality.

The emergence of green innovation derives in response to environmental requirement, rising customers and corporate environmentalism which is described as the responsive way [20, 28, 29]. Goals of higher profitability and cost efficiency are other elements that led to the rise of green innovation, described as proactive way [30]. Several steps were taken by some firms such as using recycled materials (participant No. 13). Another participant uses only plantation wood as a step towards sustainability (participant No. 3). This is also agreed by Gomes and Shian [1] revealing certified sustainable wood is the easiest green step adopted by manufacturers. Forest plantation is described as wood that is harvested from forests that is sustainably managed environmentally [27]. Nonetheless, according to Ratnasingam [21], rubber wood is not considered a sustainable source as its social obligation is not met, and as a result there is difficulty for the plantation to be certified. Sri Lanka rubber wood however is all certified and largely used among manufacturing firms.

Other participants apply water-based finishing (participant No. 41), do not use toxic material which is harmful (participant No. 22), and comply with the FLEGT regulation (participant No. 34). Coating materials are very dependent on market requirements. Low-toxicity furniture results to low volatile organic compound (VOC) production although there is scarce application on recycled metal or plastic among manufacturing firms [21]. According to Ratnasingam [21], there are issues in sustainability that include adhesives, preservatives, wood certification, and packaging. Adhesives, for example, require a much lower volatile content and in packaging it is a matter of using more recycled fibre.

## 3.3 Guidance for Firms to Adopt Green Manufacturing Practices

It is a challenge for SMEs to gain the international standard [14] for their lack of price premium for green furniture, the cost of adoption and the absence of stringent legislations which are contributing factors to the limitation in the adoption of green practices [21]. The point of issue is how to bring sustainability and green technology into the furniture industry. In countries like USA, UK and Japan, green furniture has been the latest concern in terms of environmental issues [31]. Benefits of green manufacturing practices include energy reduction, better working environment and reduction in waste [21]. In addition, it is proven that manufacturers committing to ISO 14000 standards have reduced their energy use by up to 17 % as well as reduced their waste [1].

Therefore, firms must take the first step by being aware of the benefits of green manufacturing. Ratnasingam and Wagner [19] and Ratnasingam [21] supported this notion and added that it is an important marketing criterion and increasingly used as a strategic marketing tool. In fact, being green is considered as being distinctive from other competitors as better resource utilisation is achieved by reducing waste and pollution, as well as giving greater value to customers [32]. Additionally, extensive literature [33–36] reported that profitability of a firm could be enhanced by producing green innovative products.

Secondly, pursuing in owning certifications will allow firms to own their own green management. This is because certification plays an important role in determining the success of the wood industry, particularly in the wooden furniture products, which results to the willingness of consumers in purchasing value-added wood products [27, 37]. Certifications such as the International Standard Organization certification will increase a higher standard of promotion adoption [38]. However scholars argued that certification is unlikely to increase financial performance and only able to gain improved customer retention, satisfaction and positive public reputation [39].

Malaysian Timber Certification Council (MCTS) is an organisation that comprises two components: Forest Management Certification and Chain of Custody (CoC) Certification. This organisation assists in meeting demand for certified timber products [40]. Therefore, working closely with this organisation will hope to ensure market access for Malaysian timber products, particularly in improving environmentally sensitive markets. The increased consciousness on the environment in the global markets has led among stakeholders to take more responsibility and accountability to 'green the industry' [41]. Thus, more emphasis should be given in ensuring high quality and design on wooden furniture products, as well as sustainability which eventually gives a greater impact on business performance [42].

## 4 Conclusion

Marketing tools that are very commonly applied among manufacturing industries for waste reduction and process improvement initiatives are Total Quality Management (TQM) and Business Process Reengineering (BPR), Supply Chain Management (SCM) and Lean Manufacturing (LM). Designers also should take initiatives of making bold decisions on the application of resources, modes of consumption and the lifecycles of products and services [26].

According to Zhang and Wang [43], green product designs are undertaken through four stages: (1) process to improve technics, (2) waste recycling, (3) transformation products and (4) environmentally harmless in green product design. According to Callenbach et al. [44], technical and management skills of employees are a prerequisite in developing an innovation-focused environmental initiatives and programmes. Thus, consistent inspirations for employees need to be carried out to produce significant managerial implication. Moreover, an independent department needs to be established to ensure a smooth operation of green manufacturing practices adopted [14]. Green manufacturing practice is the way forward from producing the image to portray towards buyers and among competitors [23]. Low cost and differentiation are benefits that can be gained from green manufacturing practices [45–47]. This concurs with Handfield et al. [48] 's study which stated that design-oriented sustainability that is applied in the production process and related with the bulk of the cost of product will bring a more competitive edge in the market. Noci and Verganti [24] suggest that a combination of a firm's innovation with technological capability will pursue to a greater advancement in their green management practices.

## References


1. Gomes M, Shian YL (2012) Dr. Jegatheswaran Ratnasingam on green technology in furniture manufacturing. Furnish Now
2. Lee J, Park SY, Baek I, Le C-S (2008) The impact of the brand management system on brand performance in B–B and B–C environments. Ind Market Manag 37:848–855
3. Industry A o t G F (2013) Furniture trend report. FDM Asia, Shanghai New Eastern Media Co. Ltd, Shanghai
4. Datschefski E (2001) The total beauty of sustainable products. RotoVision SA, Switzerland
5. Mohamed S, Ibrahim ML (2007) Preliminary study on willingness to pay for environmentally certified wood products among consumers in Malaysia. J Appl Sci 7(9):1339–1342
6. Tseng(曾世昌) SC (2012) Green management approaches for manufacturing industries: framework and illustrative cases. Department of Business Administration, National Central University, China
7. Wong SK-S (2012) The influence of green product competitiveness on the success of green product innovation: Empirical evidence from the Chinese electrical and electronics industry. J Innovat Manag 15(4):468–490
8. Unruh G, Ettenson R (2010) Winning in the green frenzy. Harv Bus Rev 88(11):110–116
9. Eltayeb TK, Zailani S, Ramayah T (2010) Green supply chain initiatives among certified companies in Malaysia and environmental sustainability: investigating the outcomes. Resource, Conservation and Recycling
10. Savita KS, Dominic PDD, Ramayah T (2012) Eco-design strategy among ISO 14001 certified manufacturing firms in Malaysia: green drivers and its relationship to performance outcomes. IEEE 1:154–159
11. Ministry of Plantation Industries and Commodities Malaysia (2009) National Timber Industry Policy, Putrajaya
12. Arnon S, Reichel N (2009) Closed and open-ended question tools in a telephone survey about "The good teacher":An example of a mixed methods study. J Mixed Meth Res 3(2):172–196
13. Lincoln Y, Guba E (1986) Naturalistic inquiry. Sage, Newbury Park
14. Lee K-H (2008) Corporate environmental management and practices of SMEs: the case of Korean manufacturing industry. J Sustain Manag 8(1):73–86
15. Nawrocka D (2008) Environmental supply chain management, ISO 14001 and RoHS. How are small companies in the electronics sector managing? Corp Soc Responsibility Environ Manag 15(6):349–360
16. Valipoor S, Ujang B (2011) Challenges of sustainable design in Malaysian furniture industry. In: 2011 International conference on environment and industrial innovation, IACSIT Press, Singapore
17. Richardson JI, Tand Sherwin C (2005) Design and sustainability: a scoping report for the Sustainable Design Forum, published by The Design Council, 27 June
18. Ratnasingam J, TH M, Ioras F (2008) An assessment of Malaysian wooden furniture manufacturers' readiness to embrace chain of custody (COC) certification. Holz Roh Werkst 66:339–343

19. Ratnasingam J, Wagner K (2009) Green manufacturing practices among wooden furniture manufacturers in Malaysia. Eur J Wood Prod 67:485–486
20. Ratnasingam J (2006) Transforming the South East Asian furniture industry. Irama Persada, Kuala Lumpur
21. Ratnasingam J (2013) Status of green technology application in the ASEAN furniture industry. In: Industry seminar Malaysian international furniture fair 2013, Putra World Trade Centre, Kuala Lumpur
22. Ratnasingam J, Wagner K (2010) The impact of ISO 14001 on the operations management of wooden furniture manufacturers in Malaysia. Eur J Wood Prod 68:479–481
23. Adis AHAA, Sidin SM (2008) Impact of environmental factors as moderator on export marketing performance in wooden furniture industry. J Kemanusiaan 11:24–35
24. Noci G, Verganti R (1999) Managing green product innovation in small firms. R&D Manag 1:3–15
25. Alberti M, Caini M et al (2000) Evaluation of the costs and benefits of an environmental management system. Int J Prod Res 38(17):4455–4466
26. Communities E (2009) Design as a driver of user-centred innovation: Commission Staff Working Document, Brussels
27. Kozak RA, Cohen DH, Lerner J, Bull GQ (2003) Western Canadian consumer attitudes towards certified value-added wood products: An exploratory assessment. Forest Prod J 54:21–24
28. Johnstone Has˘c˘ic˘, I., N., & Popp, D. (2010). Renewable energy policies and technological innovation: evidence based on patent counts. Environmental and Resource Economics, 45(1), 133–155.
29. Porter M (1991) American green strategy. Sci Am 264(4):168–176
30. Rennings K, Rammer C (2009) Increasing energy and resource efficiency through innovation: an explorative analysis using innovation survey data. J Econ Finance 59(5):442–459
31. Chin Soon N (2011) Malaysian furniture. Kuala Lumpur, Malaysia Furniture Entrepreneur Association
32. Porter M, van der Linde C (1995) Towards a new conception of environment competitiveness relationship. J Econ Perspective 9(4):97–118
33. Lieberman MB, Montgomery DB (1988) First-mover advantages. Strat Manag J 9(1):41–58
34. Kash DE, Rycoft RW (2000) Patterns of innovating complex technologies: a framework for adaptive network strategies. Res Pol 29(7–8):819–831
35. Nassimbeni G (2003) Local manufacturing systems and global economy: are they compatible?: the case of the Italian eyewear district. J Oper Manag 21(2):151–171
36. Chiou TY, Chan HK, Lettice F, Chung SH (2011) The influence of greening the suppliers and green innovation on environmental performance and competitive advantage in Taiwan. Transport Res E 47(6)
37. Cavusgil ST, Zou S, Naidu GM (1993) Product and promotion adaptation in export ventures: An empirical investigation. J Int Bus Stud 24(3):479–506
38. Abdul-Aziz Z, Chan JFL, Metcalfe AV (2000) Quality practices in the manufacturing industry in the UK and Malaysia. Total Qual Manag 11(8):1053–1064
39. Owari T, Juslin H, Rummukainen A, Yoshimura T (2006) Strategies, functions and benefits of forest certification in wood products marketing: perspectives of finnish suppliers. Forest Pol Econ 9(4):380–391
40. Council MTC (2013) The Malaysian timber certification scheme. Kuala Lumpur, Retrieved from www.mtcc.com.my
41. Morris M, Dunne N (2003) Driving environmental certification: Its impact on the furniture and timber products value chain in South Africa. Geoforum 35(2):251–266
42. Lee CY, Zhou X (2000) Quality management and manufacturing strategies in China. Int J Qual Reliab Manag 17(8):875–898
43. Zhang N, Wang J (2009) Under the new industrialization model innovation product design thinking. IEEE

44. Callenbach E, Capra F, Goldman L, Lutz R, Marburg S (1993) Eco-management:the elmwood guide to ecological auditing and sustainable business. Berrett-Koehler, San Francisco
45. Christmann P (2000) Effects of best practices of environmental management on cost advantage: the role of complementary assets. Acad Manage J 43(4):663–680
46. Porter M (1985) Competitive strategy. Free Press, New York
47. Shrivastava P (1995) The role of corporations in achieving ecological sustainability. Acad Manag Rev 20(4):936–960
48. Handfield RB, Walton SV et al (1997) Green Value chain practices in the furniture industry. J Oper Manag 15:293–315

# Part VI
# Invited Papers

# Institute for Research, Development and Innovation (IRDI) of the International Medical University (IMU), Malaysia

**Wan-Loy Chu, Eng-Lai Tan, Stephen Ambu, Chee-Onn Leong, Vishna Devi Nadarajah, Patricia Kim-Chooi Lim, Shew-Fung Wong, Geok-Lin Khor, James Michael Menke, and Joon-Wah Mak**

## 1 Introduction

The International Medical University (IMU), Malaysia, which started as the International Medical College (IMC) in 1992 was granted university status in 1999. The IMU is based on an international partnership model, where the students can opt to do their preclinical years at the campus in Bukit Jalil before transferring to many Partner Medical Schools (PMS) in the UK, Australasia, Canada and the USA complete their MBBS programme (http://www.imu.edu.my/imu/). Students can also opt to complete the whole programme in Malaysia by continuing their clinical years at the IMU Clinical School in Seremban. The IMU also offers a wide range of health science programmes including Dentistry, Pharmacy, Chinese Medicine, Chiropractic, Nutrition and Dietetics, Nursing, Biomedical Science and Medical Biotechnology as well as postgraduate programmes by research and coursework. Research is given great emphasis by IMU as it is an important component of academic and scholarly activities that drive teaching and learning.

In the early years, research at IMU was supported mainly by internal grants as application for government funding was only open to public universities. Our researchers have successfully bid for external grants when government funding was open to private universities since 2002 (Fig. 1). The amounts of external funds have been increasing since then, with the Ministry of Education and Ministry of Science, Technology and Innovation being the major funding sources, while other funding comes from research institutions (e.g. Malaysian Palm Oil Board) and foundations (e.g. Malaysia Toray Science Foundation) and private sectors

W.-L. Chu (✉) • E.-L. Tan • S. Ambu • C.-O. Leong • V.D. Nadarajah • P.K.-C. Lim
S.-F. Wong • G.-L. Khor • J.M. Menke • J.-W. Mak
International Medical University, No 126, Jalan Jalil Perkasa 19, Bukit Jalil,
57000 Kuala Lumpur, Malaysia
e-mail: wanloy_chu@imu.edu.my

K. Ab. Hamid et al. (eds.), *The Malaysia-Japan Model on Technology Partnership: International Proceedings 2013 of Malaysia-Japan Academic Scholar Conference*,
DOI 10.1007/978-4-431-54439-5_39, © Springer Japan 2015

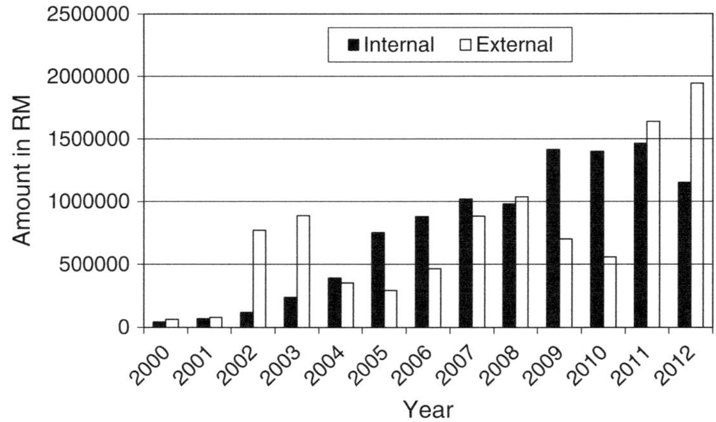


**Fig. 1** Total amounts of external and internal funding obtained by IMU between 2000 and 2012

(e.g. drug and food companies). There have also been substantial amounts of internal funds invested to support undergraduate as well as faculty and postgraduate research.

A credible environment and infrastructure, with a floor space of 26,000 square feet, has been established to support research activities at IMU. There is also an animal holding facility with state-of-the-art caging and maintenance system for experimental animals. An Environmental Health Laboratory equipped with an inductively coupled plasma mass spectrophotometry (ICP-MS) machine was set up for the testing of toxicants such as heavy metals in environmental and food samples. The facilities not only support research but also provide services to the industry, as part of the commercialisation activities.

There has been significant progress in research outputs considering that IMU is a relatively young university. The total number of publications from IMU has increased from 123 in 2008 to 149 in 2012, with an increase of total impact factor from 77.97 to 148.298 (Table 1). In terms of human capital training, IMU has produced a total of 12 Ph.D. and 31 M.Sc. graduates since 2004. In addition, two patent applications were filed in 2011.

Research activities of IMU at the initial phase were mainly driven by the interest of the principal investigators (PI) focusing on areas such as bioactive molecules, cancer biology and related stem cell research, pharmaceutics and drug delivery systems, nutraceuticals, environmental health research, clinical research and medical education research. These were later realigned into the following four main thrust areas: (1) environmental and population health, (2) cancer and stem cell research, (3) bioactive molecules and drug discovery and (4) health professional education research [3]. Specific areas of interest in these thrust areas include airborne organisms [2, 8, 11], nutrition [6, 10], bioactive molecules from various organisms [1, 4] and cancer biology [7, 9]. The major interests of health

**Table 1** IMU research publications and scientific presentations, 2008–2012

| | No. of presentations | | | No. of publications | | | Impact factor | |
|---|---|---|---|---|---|---|---|---|
| Year | Oral | Poster | Total | SCI/ISI journals | Others including books/chapters | Total | Total IF | IF/indexed paper |
| 2008 | 148 | 84 | 232 | 37 | 86 | 123 | 77.979 | 2.108 |
| 2009 | 149 | 80 | 229 | 48 | 66 | 114 | 122.159 | 2.545 |
| 2010 | 130 | 71 | 201 | 81 | 65 | 146 | 206.006 | 2.543 |
| 2011 | 180 | 96 | 276 | 116 | 96 | 212 | 274.029 | 2.362 |
| 2012 | 78 | 55 | 133 | 82 | 67 | 149 | 148.298 | 1.809 |

professional education research include problem-based learning (PBL), assessment and evaluation, curriculum development, ethics and professionalism and clinical training [5].

## 2 Institute for Research, Development and Innovation (IRDI): Roles and Functional Structure

The Institute for Research, Development and Innovation (IRDI) under the School of Postgraduate Studies and Research was established as part of the strategic plans of IMU to realign its research activities to bring research to a higher level. The IRDI was officially launched by the Chancellor, Dato' Sri Dr. T. Devaraj, on 18 September 2012. The mission of IRDI is to be recognised as a leading private research institute in Asia that creates value through integrating education, health care and research.

The primary aim of IRDI is to serve as a platform to support all research activities at IMU. The objectives of IRDI include the following: (1) to explore funding opportunities, (2) to liaise with industrial partners for contract research, (3) to provide supporting services to researchers such as statistical and editing services and advice on IP filing, (4) to coordinate the applications for both internal and external funding, (5) to coordinate the commercialisation of potential research products and services and (6) to coordinate corporate communication to publicise IMU research.

The IRDI is headed by a director who reports directly to the Vice-President, Research (Fig. 2). The director oversees the strategies and functionality of the Institute and provides leadership to the scientific community of the university. The Joint Committee on Research and Ethics is an important component of the Institute, playing a key role in evaluating research proposals for internal and external funding. In addition, there is a Scientific Advisory Board, consisting of distinguished scientists in the leading fields as well as representatives from the industry to advise on the research direction of the Institute, in alignment with its strategies.

There are two main arms under IRDI, namely, the Research and Development Arm and Commercialisation and Support Arm. The main functions of the Research

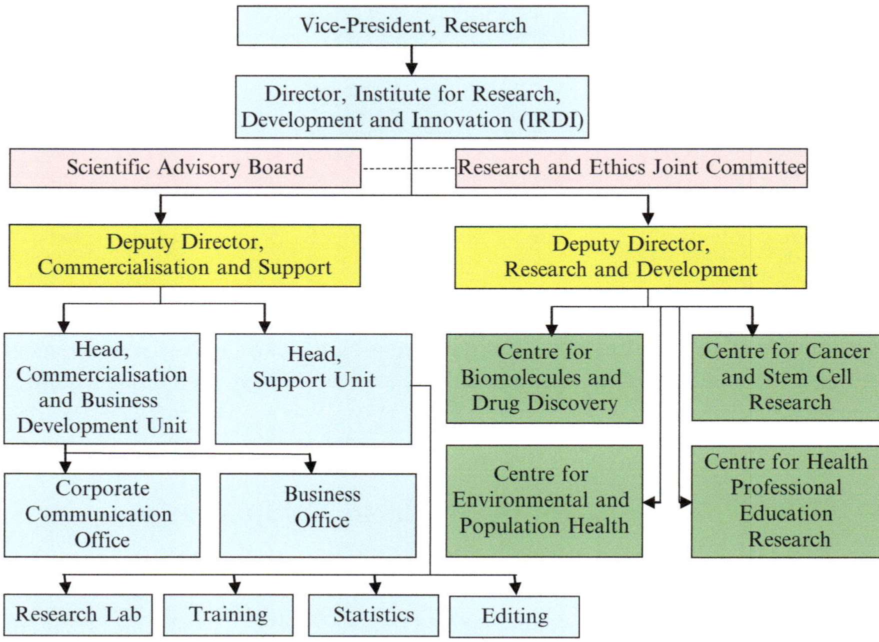


**Fig. 2** Organisational structure of the Institute for Research, Development and Innovation (IRDI), IMU

and Development Arm are as follows: (1) to drive research activities under the identified thrust areas; (2) to manage the resources for research and (3) to monitor and evaluate all research activities, including those conducted by undergraduate and postgraduate students.

Under the Research and Development Arm, there are four Centres of Excellence (CoE) based on the identified thrust areas, namely, (1) Centre for Bioactive Molecules and Drug Discovery, (2) Centre for Environment and Population Health, (3) Centre for Cancer and Stem Cell Research and (4) Centre for Health Professional Education Research.

## 2.1 *Centre for Bioactive Molecules and Drug Discovery (BMDD)*

The main aim of the Centre for Bioactive Molecules and Drug Discovery (BMDD) is to facilitate and to support a continuum of research and development activities from discovery and characterisation of bioactive molecules to clinical trials. Expertise in BMDD includes medicinal chemistry, cheminformatics, computational biology and chemistry, high-throughput and high-content screening and drug delivery. The expertise in chemistry includes organic and inorganic chemistry as well as analytical and synthetic chemistry. These experts have the capability to perform in silico drug design and discovery and synthesise small lead molecules. Our pharmaceutical chemists and technologists are adept in innovative drug formulation, employing the latest techniques in nanotechnology. We also aim to consolidate all our bioactive molecules in a library of compounds in order to facilitate research collaboration with our partner universities as well as external institutions through sharing of resources and expertise.

The Centre focuses on the identification of natural and synthetic compounds that have potential to develop into novel therapeutics or molecular probes. Our strategies are as follows: (1) to discover and develop novel and new compounds; (2) to resurrect drugs that failed in clinical trials and to discover new, beneficial activity of currently marketed drugs by developing new delivery platforms; (3) to rediscover new therapeutic targets of drugs with expired patent through high-throughput screening, cheminformatics and computation biology; (4) to discover therapeutics for the treatment of neglected diseases; (5) to offer research consultancy in generics and development of dosage forms for natural bioactive compounds; (6) to provide services in bioequivalence studies; (7) to conduct training workshops and seminars on advanced drug delivery and formulation and (8) to foster research collaboration through the provision of expertise in chemical synthesis and target validation.

## 2.2 Centre for Environmental and Population Health Research

The Centre for Environmental and Population Health Research (CEPHR) has a blend of clinicians, basic scientists and field researchers in its team, and it endeavours to work with local and international agencies to spearhead research in important areas of global health. The CEPHR identifies and shares its top priority objectives through its strategic plan by steering and aligning the current research on the following themes: (1) understand the epidemiology, biology and complex interactions between environmental agents and population health; (2) improve early detection and diagnosis; (3) accelerate progress in disease prevention and surveillance and health promotion and (4) develop effective and efficient treatments. These strategies could be strengthened by currently available tools and emerging fields in nanotechnology, genomics, proteomics and metabolomics.

The strategic plan for the CEPHR encompasses an integrated approach which aims to transform fundamental knowledge and research competency on environmental and population health to meet the needs of local and global communities for environmental sustainability. The more we know about environmental exposures and hazards to health, the better we can manage the risks in our efforts to promote human health and disease prevention activities. Environment, population and health are interrelated entities. Urbanisation and increased populations have directly or indirectly contributed to drastic changes in the global environment including air, water and soil contaminations, greenhouse effects, global warming and climate change. These environmental threats will eventually affect human health at different life stages (prenatal, pregnancy and old age). The incidence of disease and mortality in susceptible populations may increase due to emergence and re-emergence of communicable diseases, vector-borne diseases, nutritional diseases due to food insufficiency, respiratory diseases, non-communicable diseases and physiological and neurological disorders. Measures should be taken immediately to reduce the adverse health impacts of these environmental risk factors. These strategic themes and goals are also formulated to target and strengthen the early disease detection and prevention in addition to assessment and control of the environmental factors. Providing appropriate interventions and treatments in a timely manner will create a better health-support environment for the population.

## 2.3 Centre for Cancer and Stem Cell Research

Cancer is a complex, multistep process that can be interrupted at many stages, from initiation to disease progression to metastasis. While basic science exploration into the many factors that influence the cancer process has always been a major component of cancer research, it takes on increased importance today as accumulated knowledge and groundbreaking technologies such as imaging, nanotechnology and metabolomics are enabling us to interrogate in precise detail the tumour microenvironment and the larger biological system in which it exists.

At the Centre for Cancer and Stem Cell Research (CCSCR), we leverage our scientific expertise and technological forces with key partners to achieve our vision of reducing the burden and eliminate the adverse outcomes of cancer by leading an integrated effort to advance fundamental knowledge about cancer across a dynamic continuum of discovery, development and delivery.

To understand the machinery of cancer, we first have to focus on the biological nuts and bolts, that is, basic research—studying the molecular and cellular why's and how's of biology. The IMU Cancer Research team is populated with some of the top basic scientists, women and men, who explore the tumour cosmos within the confines of their laboratories.

The CCSCR has the following overarching research themes: (1) unravelling the complexities of the cancer process, normal cell growth and differentiation; (2) greater understanding of how cancers originate, progress and behave; (3) adapting discoveries to prevent or intervene the disease process at the early stage; (4) combining effective agents, including both conventional and molecularly targeted therapies, to develop less toxic and more effective treatments; (5) predict a patient's susceptibility to cancer and to use this knowledge to treat the disease on a personalised level and (6) advancing biomedical technologies, such as nanotechnology, genomics, imaging and proteomics, to identify human proteins and their biological functions, and creating instruments that deliver highly targeted treatments and predict therapeutic effectiveness.

## 2.4 Centre for Health Professional Education Research

The Centre for Health Professional Education (CPHE) Research encompasses research in the area of medical, dental, pharmacy and allied health professional education. Researchers supported by this Centre are faculty members from all IMU schools with an interest in HPE research. However, as with most health professional schools worldwide, faculty members are trained primarily as professionals in their respective fields (doctors, nurses, pharmacists and scientists) and not as educationists; hence one of the main aims of the Centre is to ensure that faculty have access to training in education research. For this reason, the Centre is strategically affiliated with the IMU Centre of Education (ICE), as ICE supports the Centre activities by providing training in education research, running research consultancy clinics and fostering networking by hosting international education conferences. The Centre also aims to contribute and be recognised as experts in HPE; hence, another strategy of the Centre is to develop international research collaborations with IMU's partner schools. Currently, there are research HPE collaborations with the University of Tasmania, University of Adelaide, University of Newcastle (Australia) and University of Otago.

The research areas supported by the Centre can be generally divided to teaching learning methods, assessments, clinical and communication skills, student preparedness and learning styles. For teaching learning methods, research is focused on problem-based learning, inter-professional learning and e-learning methods. Assessment research includes open and closed book assessments, key features assessment and formative assessment. Clinical and communication skills research includes peer teaching of clinical skills, innovations in teaching communication skills and clinical competencies. Research on student preparedness and learning styles covers aspects related to student preparedness to transfer to the clinical phase and student selection into health professional schools. In order to ensure that research projects are transformed into scholarly articles of international standards, the Centre also helps facilitate prepublication advice by having a panel of internal and external experts available for the researchers. The Centre will continue to review its initiatives and strategies within IRDI and IMU to improve and achieve its aims.

## 2.5 Commercialisation and Support Arm

The Commercialisation and Support Arm consists of two major units, namely the Commercialisation and Business Development Unit and Support Unit. The main functions of this Arm are as follows: (1) to coordinate the commercialisation of potential research products, findings and services; (2) to recommend and provide advice on patenting of research findings, products and processes; (3) to liaise with potential industrial partners for contract research; (4) to facilitate corporate communication for research with the aim of publicising IMU research to attract more funding; (5) to provide supporting services such as statistical advice, utilisation of bioinformatics tools, editing services for publications and grantsmanship training; (6) to ensure efficient utilisation and proper maintenance of all research facilities and equipment; (7) to enhance the quality of research through QC, QA and audit (e.g. ISO) and training (e.g. Good Clinical Practice Seminar) and (8) to enhance research culture within IMU.

Research entrepreneurship undertaken by IRDI will create systems to maximise returns on research efforts through planned synergistic collaboration with industrial partners, researchers, all stake holders and the community. It will promote an environment that will facilitate the processes that translate research results/products for downstream application. The outcomes expected will be creation of jobs, training of graduate researchers, financial returns, patents, publications and other research indicators, further research and generation of research funds. In line with IMU's efforts towards strategic partnerships and linkages, the Commercialisation Unit will seek to establish various research hubs with industrial partners in focused areas of interest. The "SRAS-IMU Herbal and Medicinal Research Hub" is the first research hub that was established in 2013 focusing on research and commercialisation of potential herbal and natural products. Other commercialisation activities include the provision of various laboratory services such as environmental testing services for both chemical and biological contaminants in air and water and consultancy and laboratory services for contract research in various fields (e.g. recombinant proteins as vaccine candidates). In addition, researchers can also access IMU's laboratory equipment on a rental basis.

Under this Arm, there is also an e-journal named International e-Journal of Science, Medicine and Education (IeJSME) which aims to provide a forum for disseminating articles of important scientific merit to international scientific community on three major areas, namely, basic science, medical science and medical education research (http://web.imu.edu.my/ejournal/default.asp). The journal was listed in the Malaysian Citation Index (MyCite) and ranked 76 of 112 Malaysian journals in 2012.

## 3 Concluding Remarks

With the realignment of research direction through IRDI, IMU aims to be more competitive in terms of securing external funding for its research and to provide support and training to enhance the capability of our researchers. We also aim to foster strategic partnerships with both local and foreign institutions to excel in the identified thrust areas. In addition, we aim to establish strategic linkages with the industry to promote commercialisation of our research products and service.

## References


1. Chan KK, Wong RSY, Mohamed SM, Ibrahim TAT, Abdullah M, Nadarajah VD (2012) *Bacillus thuringiensis* parasporal proteins induce cell-cycle arrest and caspase-dependant apoptotic cell death in leukemic cells. J Environ Pathol Toxicol Oncol 31(1):75–86
2. Chan LL, Mak JW, Low YT, Koh TT, Ithoi I, Mohamed SM (2011) Isolation and characterization of *Acanthamoeba* spp. from air-conditioners in Kuala Lumpur, Malaysia. Acta Trop 117:23–30
3. Chu WL (2013) Research at IMU: achievements, thrust areas and future challenges. IeJSME 7 (Suppl 1):S7–S18
4. Chu WL, Lim YW, Radhakrishnan AK, Lim PE (2010) Protective effect of aqueous extract from *Spirulina platensis* against cell death induced by free radicals. BMC Compl Alter Med 10:53
5. Hla YY, Babar MG, Mitha S, Zuhanariah Mohd Nordin (2012) Medical education research 2000–2011. International Medical University, 171 p
6. Khor GL, Chee WSS, Zalilah M, Shariff PBK, Mohan A, Rahman Jamalludin A, Theobald HE (2012) High prevalence of vitamin D insufficiency and its association with BMI-for-age among primary school children in Kuala Lumpur, Malaysia. BMC Pub Health 11:95–102
7. Low SY, Tan BS, Choo HL, Tiong KH, Khoo AS, Leong CO (2012) Suppression of BCL-2 synergizes cisplatin sensitivity in nasopharyngeal carcinoma cells. Cancer Lett 314:166–175
8. Ng EHP, Chu WL, Ambu S (2011) Occurrence of airborne algae within the township of Bukit Jalil in Kuala Lumpur, Malaysia. Grana 50:217–227
9. Poh YW, Gan SY, Tan EL (2012) Effects of Il-6, Il-10 and Tgf-B on the expression of survivin and apoptosis in nasopharyngeal carcinoma TwO1 cells. Exp Oncol (Online) 34(2):85–89
10. Voon PT, Ng TKW, Lee VKM, Nesaretnam K (2011) Diets rich in palmitic acid (16:0), lauric and myristic acids (12:0 + 14:0), or oleic acid (18:1) do not alter postprandial or fasting plasma homocysteine and inflammatory markers in healthy Malaysian adults. Am J Clin Nutr 94(6):1451–1457
11. Wong SF, Ling CA, Mak JW, Tan J, Ling SJ, Ho TM (2011) Molecular identification of house dust mites and storage mites. Exp Appl Acarol 55(2):123–133

# The Roles of the Institute of Nano Electronic Engineering (INEE), UniMAP in NCER: From Nanotechnology Perspectives

**U. Hashim, E.N. Elliazir, Jasni M. Ismail, A.R. Ruslinda, and M.K. Md. Arshad**

## 1 Introduction

Nanotechnology has been identified as new industrial revolution, and most countries are investing heavily in this market to assure their market competitiveness [1]. A forward-looking approach to the key technologies of the future is crucial to sustain an active and competitive industrial economy. This emerging technology offers rapid advances across many areas of science and engineering crucial to society [2]. The Malaysian government has taken a serious concern over the development of nanotechnology in the country. In the Third Industrial Master Plan (IMP3) that spans a 15-year period (2005–2020), nanotechnology is reported to be recognized as the new emerging field that can potentially generate technologies in a broad range of fields such as health, education, information technology, creation and usage of energy, safety, defense, biotechnology, food and agriculture, aerospace, manufacturing, and environmental rehabilitation. Researches in nanotechnology provide a fundamental understanding of phenomena and materials at "nano" scale and the ability to create and use structures to build devices and systems that have novel properties and functions. The advent of nanotechnology will be life changing. The development of nanotechnology is analogous to the development of electricity; it is a field of science and technology that has wide-reaching implications, not only allowing for the development of new products but also enhancing current products.

Nanotechnology R&D was started by the Malaysian government in 2001 and categorized as a strategic research (SR) program under IRPA in the Eighth Malaysia Plan (8MP) and funded by the Ministry of Science, Technology and Innovation (MOSTI). In the launching speech during the World Innovation Forum (WIF KL 2012), Datuk Seri Panglima Dr. Maximus Johnity Ongkili, Minister of Science,

U. Hashim (✉) • E.N. Elliazir • J.M. Ismail • A.R. Ruslinda • M.K. Md. Arshad
Institute of Nano Electronic Engineering, Universiti Malaysia Perlis,
Kangar Perlis 01000, Malaysia
e-mail: uda@unimap.edu.my

K. Ab. Hamid et al. (eds.), *The Malaysia-Japan Model on Technology Partnership: International Proceedings 2013 of Malaysia-Japan Academic Scholar Conference*,
DOI 10.1007/978-4-431-54439-5_40, © Springer Japan 2015

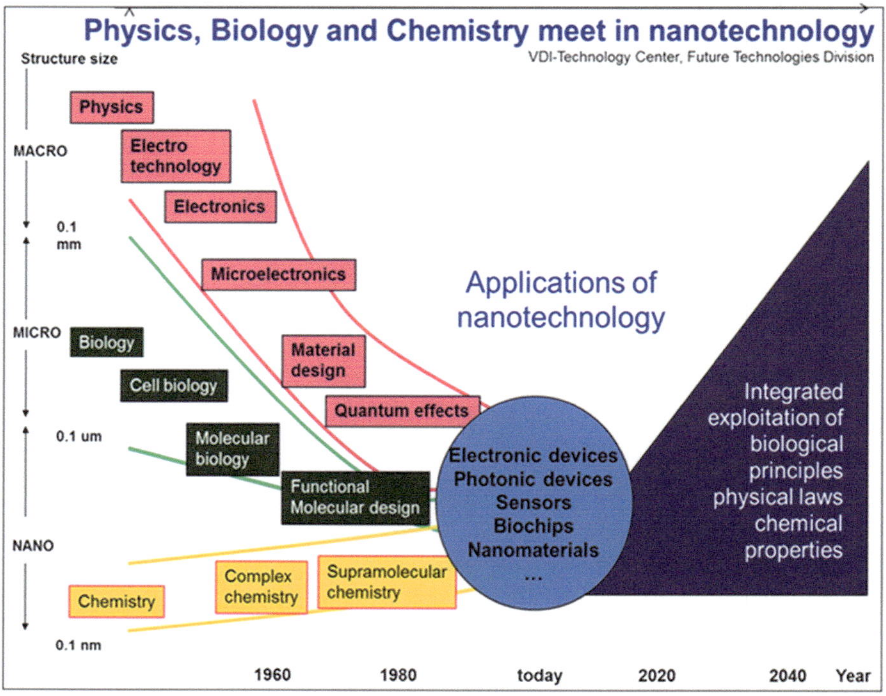


**Fig. 1** Harnessing convergence through nanotechnology

Technology and Innovation, said that in order for innovation to successfully transform the economy, inclusive and sustainable innovation has to be undertaken to ensure all sectors of the economy as well as the people benefit from a high-income economy. In order to remain competitive, nations need to embark on transforming their economy to a knowledge economy with innovation as its driver [3]. The Prime Minister in his 2013 Budget has allocated RM 600 million to five research universities to conduct high-impact research in strategic fields such as nanotechnology, automotive, biotechnology, and aerospace [4]. The multidisciplinary fields converge in nanotechnology as shown in Fig. 1.

Over the past years, there are many areas in which nanotechnology has been developed, emerging from the field of physics, biology, and chemistry, which include nanocrystalline, nanoparticles, nanocapsules, nanoporous, nanofibers, nanowires, fullerenes, nanotubes, dendrimers, molecular electronics, quantum dots, nanoelectomechanical and nanofluidic systems, nanotools, software, lithography, soft lithography, self-assembly, sol–gel technology, deposition, and manipulators.

All these technological areas provide the essential building blocks for the development of nanotechnology products and services. Countries like the USA, Japan, Korea, Germany, and the UK have invested huge financial capital to harness the potential of nanotechnology for their nations' future. Obviously, these investments are expected to bring high impact on socioeconomic for many decades ahead.

**Definition of Nanotechnology** Nanotechnology is the ability to control and restructure the matter at the atomic and molecular levels in the range of approximately 1–100 nm and exploiting the distinct properties and phenomena at that scale as compared to those associated with single atoms or molecules or bulk behavior. The aim is to create materials, devices, and systems with fundamentally new properties and functions by engineering their small structure. This is the ultimate frontier to economically change materials properties and the most efficient length scale for manufacturing molecular medicine. The same principles and tools are applicable to different areas of relevance and may help establish a unifying platform for science, engineering, and technology at the nanoscale. The transition from single atom or molecule behavior to collective behavior of atomic and molecular assemblies is encountered in nature, and nanotechnology exploits this natural threshold [5].

Nanotechnology is defined as the creation and utilization of materials, devices, and systems through the manipulation of matter at scales of less than 100 nm. In scientific terminology, the prefix "nano" means one-billionth, involving events that occur in the length range of a fraction of a nanometer to a hundred nanometer; that is the scale of distance in which atoms and molecules exist and interact [6].

## 2 Establishment History and Background

The Ninth Malaysia Plan has gazetted nanotechnology as a technology focus area and urged local tertiary education institutions to set up a center institution where technicians, engineers, and experts can be trained in microelectronic fabrication and nanotechnology.

In response to this, the Institute of Nano Electronic Engineering was established and approved by the Ministry of Higher Education (MoHE) in late 2008. INEE is a spin-off research cluster from the School of Microelectronic Engineering of the Universiti Malaysia Perlis, spearheaded by none other than Prof. Dr. Uda Hashim as the first and current director. The reason behind the inception of INEE is that UniMAP has a ready pool of active and experienced human capital in the field of micro- and nanoelectronic engineering in both teaching and research. The establishment of a research center of excellence is to support five complimentary modes of research: fundamental research, grand challenges, research centers of excellence, fabrication and user facilities for developing physical infrastructure, and education and societal implications [7].

INEE emphasizes on nanoelectronic engineering research activities that focus on nanobiochip, novel devices, memory devices, and nanophotonic devices. In the near future, INEE will expand its research activities to nanomaterials and nanobiotechnology. As a nanoelectronic research center, INEE's role is to encourage and embrace collaborative work and interdisciplinary researches among scientists and technologists. INEE strives to be a leading nanoelectronic research center in the region in line with Malaysia's goal of having world-class scientists. Working towards achieving that, INEE is in progress of accommodating state-of-the-art facilities and equipments, to be coherent with cutting-edge technologies for researchers to utilize and conduct comprehensive research without forgetting the educational segment. Innovative efforts, blending fundamental and applied sciences with technological advances in the field of nanofabrication, are the other combination factors to the creation of high-performance devices with greater speed, simplicity, and low cost for market needs in the twenty-first century. Through the realization of the abovementioned strong and important points, INEE aspires to make an impact on Malaysia's research communities, specifically in nanoelectronic research areas, and to strengthen Malaysia's economic capability over and above the well-being of the nation.

This ambition is in line with INEE's mission and vision, i.e., the mission is to serve as an excellent landmark and function as a reference institute in nanoelectronic engineering with performance beyond conventional devices which are based on the current industry standard, while the vision is to emerge as a leading frontier research institute across the spectrum of nanoelectronic engineering, from the life and physical sciences to engineering and medicine. The theme of INEE is "From Nano Structure to Systems."

## 3 Laboratory Facilities

INEE comprises of three laboratories with state-of-the-art scientific equipment and facilities to accommodate researchers involved in device design, fabrication, characterization, and testing, i.e., Micro Fabrication Cleanroom Laboratory (MFC), Nano Fabrication Cleanroom Laboratory (NFC), and Nano Biochip Laboratory. The Micro Fabrication Cleanroom is the first and largest purposely built for research and teaching program by a university in Malaysia [8]. The size of the cleanroom is approximately 115 $m^2$ with cleanliness class from ISO Class 5 to ISO Class 8. The cleanroom is currently equipped with ten-process modules comprising oxidation, diffusion, photolithography, physical vapor deposition, wet etching, wet cleaning, wafer test and characterization, scanning electron microscope, and e-beam lithography module.

The Nano Fabrication Cleanroom is housed in a combined cleanroom adjacent to the MFC. The NFC is facilitated with an electron-beam lithography (EBL) system for creating extreme fine nanopatterns for integrated circuits, inductively coupled plasma-reactive ion etching (dry etching) (ICP-RIE etcher) for anisotropic etches of all types of thin films, and a plasma-enhanced chemical vapor deposition (PECVD) unit for deposition of high-quality and stable interlayer dielectric film.

The Nano Biochip laboratory focuses on the testing and characterization of lab-scale silicon biosensor systems. The laboratory holds two clean booths of ISO

Class 10 known as the yellow and the white clean booths. The yellow clean booth is for testing and biology synthesizing under low-light conditions. The white clean booth is equipped with a high-range-frequency dielectric analyzer altogether with a probe station integrated oscilloscope for electrical characterization and lab-on-chip package. Currently, INEE's laboratories accommodate 24 equipments, 16 analyzers, and 8 facilities at a total value of RM15 million to assist researchers in pursuing their interests. This multimillion ringgit facility was set up to primarily aid in teaching and research purposes.

## 4 Postgraduate Programs and Research Focus Areas

Towards becoming a full-fledged tertiary education institution for both teaching and research purposes recognized by the Malaysian Ministry, INEE is currently focusing in four major research areas as described in Table 1.

In order to stay ahead as a knowledge and research center, INEE offers postgraduate programs in three master's levels and one doctorate level as shown in Table 2. We plan to offer more postgraduate programs in future expansion to gather more students and researchers.

**Table 1** Research focus areas of INEE

| | |
|---|---|
| 1. Nanobiochip research | (a) Nanostructure-based biosensor<br>(b) Nanomaterial-based biosensor<br>(c) MOSFET-based biosensor<br>(d) Lab-on-chips |
| 2. Functionalized nanobiomolecules research | (a) Biomarker synthesis and probe design<br>(b) Nanostructure surface modifications<br>(c) DNA immobilization and hybridization<br>(d) Simulation and modeling of nanostructure biomolecule interaction |
| 3. Functionalized nanomaterial research | (a) Nanotemplate<br>(b) Nanocomposites<br>(c) Nanoparticle<br>(d) Nanobiomaterial |

**Table 2** Postgraduate programs available at INEE at present (2012)

| |
|---|
| Doctor of Philosophy, Ph.D. |
| M.Sc. Nanoelectronic Engineering |
| M.Sc. Nanomaterial Engineering |
| M.Sc. Nanobiotechnology Engineering |

# 5 Research Achievement and Collaboration with National and International Partners

One of the paramount initiatives undertaken by INEE to promote and encourage intellectual publications, as well as internalization strategy, is to collaborate with both national and international partners. Table 3 describes in detail the research title and the collaboration teams involved.

**Table 3** Current major research achievement and collaboration with national and international partners

| | *Research title (national level)* | *Collaboration team* |
|---|---|---|
| 1. | Microfluidic for biomedical applications | UniMAP, UTM, Imperial College, University of Warwick |
| 2. | Nano-lab-on-chip for medical diagnostics | UniMAP, University of Hull |
| 3. | Development of highly sensitive carbon nanotube-doped biosensor chip for rapid screening of *Vibrio cholerae* | UniMAP, NPC, Islamabad, Pakistan |
| 4. | ZnO nanowire biosensor for biomedical application | UniMAP, University Engineering & Technology, Karachi, Pakistan |
| 5. | ZnO nanowire for nano e-nose applications | UniMAP, Linköping University, Sweden |
| 6. | Nano-lab-on-chip for medical diagnostics (EU grant application) | UniMAP, University of Nottingham |
| 7. | Fabrication of surface acoustic wave delay line for corrosion inspection | UTP, UniMAP |
| 8. | Nano DNA chips for carcinogen detection and profiling | UniMAP, AMREC, USM |
| 9. | Nanobiosensor for tropical disease detection | AIMST, UniMAP |
| 10. | SAW device on CMOS for resonator | UIA, UniMAP |
| 11. | Nanostructure-based biosensor for arsenic in seafood detection | Mardi, UniMAP, UPM |
| 12. | SAW device for *E. coli* detection | UniMAP, MARDI |
| 13. | Nano-lab-on-chip for medical diagnostics system | UniMAP, UMP, UTHM |
| 14. | Leptospirosis from rat urine | UniMAP, UPSI |
| 15. | Biomedical product–nanosilver coating on cannula surface | UniMAP, B.Brown Germany |
| | *Research title (international level)* | *Collaboration team* |
| 1. | Microfluidic for biomedical applications | UniMAP, UTM, Imperial College, University of Warwick |
| 2. | Nano-lab-on-chip for medical diagnostics | UniMAP, University of Hull |
| 3. | Development of highly sensitive carbon nanotube-doped biosensor chip for rapid screening of *Vibrio cholerae* | UniMAP, NPC, Islamabad, Pakistan |
| 4. | ZnO nanowire biosensor for biomedical application | UniMAP, University of Engineering & Technology, Karachi, Pakistan |
| 5. | ZnO nanowire for nano e-nose applications | UniMAP, Linköping University, Sweden |
| 6. | Nano-lab-on-chip for medical diagnostics (EU grant application) | UniMAP, University of Nottingham |
| 7. | Solar base of nanomaterial | UniMAP, Wayamba University of Sri Lanka |

The list above is part of the evidence justifying the recognition awarded by the MOSTI as a NanoMalaysia Centre of Excellence in June 2011. The recognition has placed INEE as a major nanoresearch hub in the northern region and especially in the NCER.

Based on excellent progress and achievement in research, in 2009, INEE received BioNexus status for Nano Fabrication Laboratory from Biotech Corporation and, recently in 2011, obtained the NanoMalaysia Centre of Excellence recognition from the MOSTI. Attributable to this recognition, a total of RM 1 million per year as a research grant is secured from both agencies for the period of 5 years to support research activities.

# 6 Impact to NCER

## *6.1 Society*

The existence of INEE as part of the NCER can help create public awareness on nanotechnology and road map to achieve as a parallel and consistent technology developer and provider. This effort will mark INEE as a reference to the public that nanotechnology does exist in the region corridor, as well as an introduction to the public on how NCER supports the advanced technology such as nanotechnology, and can be considered as a noble effort to mankind. The public has been introduced to the phrase of "nano," but when being questioned back, none of the correct answer was given. Thus, NCER has the added advantage to place the responsibility for INEE to educate the public on awareness towards the nanotechnology know-how surroundings.

INEE has been materialized as a Centre of Excellence by the MOSTI in June 2011; thus, this recognition is promoting the NCER capabilities in the advanced technology icon. Nanotechnology has been the national agenda with the vast of grants and investments provided. This activity will attract social awareness of the cash value that can be consumed when developing the high-end technology. NCER should be able to roadmap INEE's existence and promote public awareness on the future technology.

Furthermore, INEE is also promoting through the surrounding community of this advanced technology through its website by tagging the NCER role and support, which can be an in-conjunction joint effort. This kind of promotion will lead to various numbers of applications in daily activities. Multiplying the efforts of NCER boosts the recognition of INEE in the international arena. Since NCER is part of the big national agenda, the activities of the nanotechnology-related program should synchronize with the international interest as a future cutting-edge technology.

### 6.2 Research and Education

INEE is a reference point owing to its capabilities in providing research information and can potentially be an icon to NCER. This is hugely attributable to the low-maintenance labs, relatively when comparing with other mostly high-cost research labs. Since the establishment of INEE in 2008, gradually it has been referred by many other national institutions as a research collaboration partner.

INEE has been known as a "publication generator" to its research and lab users due to very active research activities. This is due to the generous cross-references by fellow researchers currently affiliated with INEE. Numerous M.Sc. and Ph.D. students have benefited from INEE laboratories. Due to its generous state-of-the-art research facilities that accommodate the research demand, the students at their own convenience, without much hassle, would complete their assignment in a timely manner. The students who graduated from this institution are from local and international communities, contributing intellectual numbers that can be produced by NCER.

As a center of excellence (CoE), INEE has been organizing numerous talks/ seminars and workshops to support the key performance indicator (KPI) on training activities. The series of workshops have brought about keen interest in the UniMAP surroundings, inclusive of other national institutions, to realize the importance of nanotechnology research capabilities in the region. As part of INEE's commitment to be a role player in this human capital development, INEE will continue striving in exploring all research potentials within the research community.

To accelerate the research milestone at INEE, providing the "know-how" technique and the capabilities is one of the main agenda. This agendum has created a clear vision and focus to researchers who have applauded INEE on its seriousness and high commitment to be a major player in the nanotechnology research hub. INEE has successfully merged the research collaborations with national and international partners. Vast numbers of collaborations had been achieved with the established institution, MARDI, since the beginning of INEE's establishment. A number of expertise exchange and reference have been accomplished throughout the collaborations where NCER should consider being involved as a platform provider.

To strengthen the NCER participations is to merge inMoU/MoA. Since INEE has been part of the major national agenda, NCER does not want to miss out to be part of it, since INEE does exist in the region. In fact, NCER can be a very proactive partner to initiate any small steps to promote this research environment in the larger and wider vision.

### 6.3 Economy

A state-of-the-art, low-cost operation lab can be an income generator. INEE lab equipment lineup, probably or most of them, is the only one that is available in the region; thus, most of the advanced research subjects require this set of facilities to support their research activities.

Vendors and contractors of the chain supply system are generating the exchange of financial flow. When supply and demand exist, the deal and cash flow is happening, causing the business atmosphere and leading to the economic growth.

A number of grants that INEE has obtained from the local providers and some private sectors will add up to the NCER financial strength. These grants will be utilized to move the research goals, where experiments, material, and technology are required. INEE's major strength is probably the highest grant values that are held by the fellow researchers who contribute to its capabilities in sustaining the financial when managing its laboratory requirement. Combined with the "know-how" that INEE possesses, these aspects become a good package in generating a commercial environment.

The number of international lab users from abroad steadily grows, thus encouraging the foreign exchange to the NCER value. When the name is there (INEE in the NCER), a flock of researchers from abroad will approach to utilize the INEE lab facilities. This will bring about a certain amount of forex to the NCER.

### 6.4 Future Plan

The establishment of INEE as a converging hub for innovation, research, and expertise in Malaysia, particularly for the NCER, aims to continuously promote nanoelectronics expertise within the Malaysian market, introducing the best in micro- and nanofabrication practices from around the world to Malaysian society and building direct links between the technology and education sectors. INEE emphasizes on nanoelectronic engineering research activities that focus on nanobiochip, novel devices, memory devices, and nanophotonic devices. After the passive nanostructure, such as quantum dots, nanowire, nanogap, nanotube, and nanoporus, had been successfully developed in the year 2008, current works are focused on functionalizing the nanostructure for several types of sensors and transistors. In a few years time, various types of sensors will be successfully produced by the INEE Nano Biochip Research Group. Subsequently, these sensors will be combined with the computer-aided systems to produce a Nano Bio-System, the so-called lab-on-chip, as shown in Fig. 2.

One of the major INEE road maps of achievement is to be one of the HiCoEs (Higher Institution Centres of Excellence) of MOHE in 2014. This will be a major recognition at the national and international levels since there are only 6 in the nation. To be one of the HiCoEs, INEE will triple the effort in all aspects in

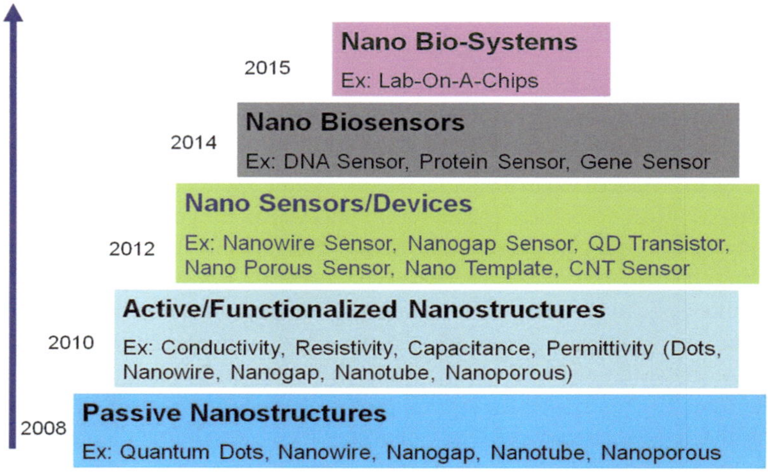


**Fig. 2** The INEE's nanostructures to system road map

accommodating the HiCoE criteria which are required by the MOHE. In the making of the HiCoE (which INEE is already on the track), NCER is a good platform to acclaim as one of its privileges to own the next HiCoE in its region, other than the Institute for Research in Molecular Medicine (INFORMM) and the Centre for Drug Research (CDR); both are from USM. Having a number of HiCoEs in the NCER will prove how committed the NCER is in observing the contribution to knowledge and technology activities. HiCoE recognition will be the ultimate challenge of international recognition by the national authorities (MOHE), and this will boost all the morale and pride of the northern community in having it here especially in UniMAP, Perlis, in the NCER.

Meanwhile, INEE will continue its growth within the next 4 years under the MOSTI's NanoMalaysia CoE grants and persist in observing its rights and roles in providing the low-cost, state-of-the-art lab-on-chip facilities to the surrounding and research communities, both domestic and foreign. INEE will be part of the NCER in contributing the necessary figure in both cash and research values by increasing its manpower of fellow researchers and lab facilities. The projection value is expected at RM23 million from its current value at RM18 million.

INEE is also making a new step in DNA know-how capabilities by opening its new research area in the INEE laboratory. This will be the new challenge in adapting the nanotechnology in DNA application research environment and multi-collaboration with other established research institutions at both national and international levels as required. The details of the projection plan have been acknowledged and are ready to move forward; this will be a major road map in the NCER where high expectations and attractions are expected.

Last but not least, INEE is planning to be one of the major commercial values for know-how technology transfer, where this figure will be the key of establishment of any research institutions; other than just producing publications and journals, INEE will set a new number of its commercial and income generator as the return of total investment.

## 7 Conclusion

The ability to fabricate nanodevices and nanostructures with the smallest resolution and fidelity is fundamental to all future research activities. We have demonstrated the application of e-beam and ICP etcher which are the core of today's nanofabrication technology using top-down approach. With the success of the current research in the nanoelectronics field, it is believed that the Nano Biochip Research Group will focus and lead their research towards nano-biosystem or lab-on-chip in the near future.

The NCER, as a body that monitors the activities that can contribute to the development of the socioeconomy, should work side by side with INEE. This partnership will create a systematic catalyst system to boost the portfolio as a technology–economy platform because the number or figures that have been produced by INEE in the financial scale are already on par with the rest of the national-level institution. The number should be elaborated more to show that the existing activities are actually already on track with the NCER commitment and goal. NCER too should take advantage of the achievement by INEE to increase the promotion in the region.

On the other hand, INEE will be marked in the NCER as one of the major players in the technology sector, as a main contributor to balance the NCER goal in agriculture, technology, and economic platform.

**Acknowledgment** The authors would like to express their deepest gratitude to INEE's team and UniMAP.

## References

1. Hashim U, Nadia E, Salleh S (2009) Nanotechnology development status in Malaysia: industrialization, strategy and practises. Int J Nanoelectron Mat 2(1):119–134
2. Roco MC, Bainbridge WS (2005) Societal implications of nanoscience and nanotechnology: maximising human benefits. J Nanoparticle Res 7:1–13
3. Datuk Seri Panglima Dr. Maximus Johnity Ongkili (2012) Minister of science, technology and innovation, launching speech, World innovation forum (WIF KL) http://www.wifkl.com/index.cfm?&menuid=112
4. Razak N (2013) Prime Minister of Malaysia, 2013 Budget Speech
5. Roco MC, Mirkin CA, Hersam MC (eds) (2010) Nanotechnology research directions for societal needs in 2020: retrospective and outlook. NSF/WTEC report Springer http://www.wtec.org/nano2/Nanotechnology_Research_Directions_to_2020/

6. MiGHT Report (2006) Identification of business and R&D opportunities in the application of nanotechnology in Malaysia
7. Roco MC (2003) Nanotechnology: convergence with modern biology and medicine. Curr Opin Biotechnol 14:337–376
8. Hashim U, Jamal ZAZ (2004) An undergraduate micro fabrication course at the Northern Malaysia University College of Engineering, Jurnal Penyelidikan dan Pendidikan Kejuruteraan, vol 1

# OER in Action: Case Study of Wawasan Open University

**Vighnarajah**

## 1 Introduction

The OER movement is still in its infancy in the Asian region despite the significant growth of the OER development and dissemination in other parts of the world. According to Menon and Ali [4], p. 1, Wawasan Open University is a pioneer institution in the region to act in response to the global sponsorship of this OER adoption. They further pointed to the OER-Asia initiative in 2010 under the advocacy of the university "for advocating and disseminating relevant information about OER as well as developing training materials and conducting research in this area." Aligned to the interests of the OER movement, joint and continuous dialogue among the OER Steering Committee, Management Board, Senate, and Board of Governors of the university results in the WOU-OER Policy declaration:

> WOU will promote and implement the creation, reuse, remix, repurpose and redistribution of Open Educational Resources (OER) within an Open Licensing framework.

The advent of high-speed Internet connection and cloud computing made it more possible to facilitate the development and dissemination of OERs. This is clearly evinced with the increasing numbers of OERs created and shared in various sites like YouTube, Facebook, Wikibooks, OER repositories, blogs, online reflective journals, and of such. At this juncture, it is interesting to note that the users may not have realized the inadvertent creation and sharing of instructional materials; but scrutiny of Facebook, for instance, may disclose how OERs are shared among users to aid each other in understanding a concept. On similar interests, Kanwar et al. [2] argue that effective utilization of OER promises assurance of minimizing the incongruities of formal and informal education.

Vighnarajah (✉)
School of Education, Languages and Communications, Wawasan Open University,
54 Jalan Sultan Ahmad Shah, 10050 Penang, Malaysia
e-mail: svighna@wou.edu.my; iolasus@gmail.com

K. Ab. Hamid et al. (eds.), *The Malaysia-Japan Model on Technology Partnership: International Proceedings 2013 of Malaysia-Japan Academic Scholar Conference*,
DOI 10.1007/978-4-431-54439-5_41, © Springer Japan 2015

Open educational resources (OERs) play a significant role in shaping the notion of open education movement in institutions of higher education. According to Johnson et al. [1], p. 6, the open education trend came to prominence as “a response to the rising costs of education, the desire for accessing learning in areas where such access is difficult, and an expression of student choice about when and how to learn.” This paper provides a practical measure for countries, such as Japan, with established policies on globalization of education to take an active role in developing institutional inertia and enriching the traditional practices of teaching and learning.

## 2 Adoption of OER in Course Development

The ICT in Education course is a core course offered in the Master of Education program in Wawasan Open University, and the development of this OER-based course pioneers the university’s policy in adopting OER for course development. The course development process was coordinated by the author alongside a team which consists of academic members, instructional designers, course writers, and external course assessor.

Although the development of OER-based courses is not a new idea and has been successfully attempted by other institutions such as the MA Teacher Education-International Programme by the Open University of Sri Lanka [3], the distinguishing factor of this OER-based course is the embedding of OER materials into the study package using open source eXe software (refer to Fig. 1). Among others, the OER materials used were audio, YouTube videos, open-access journal articles, and excerpts of open-access books.

Figure 2 presents an excerpt of Wikipedia for reading reference. This is very useful when reading requires referring to specific aspects of a text (in this case, the Wikipedia) without necessarily getting distracted or overwhelmed with bulks of text/information.

The development of this OER-based course has been completed and presented for the first time in the July 2012 semester. Feedback was gathered from students via interview survey forms on several aspects of the course and was used to improve the delivery of the course for the July 2013 semester.

With embedded integration of the OER materials, almost all students found the course to be more enriching and comprehensive in relation to other courses. One student in particular indicated that “It is more enriching in a way, yes, because there’s audio visual involved, plus a lot of hyperlinks and reading resources.” Figure 3 presents the list of text and media OERs used in the course.

This course is available in the public domain and can be accessed directly from Google Drive (https://drive.google.com/folderview?id=0B4IE5loqwltCWlRXcjh6Uy1rVFk&usp=sharing).

Users will be able to see two items when they click on the link provided above. One is the arrow icon for download of the course material; another item is the

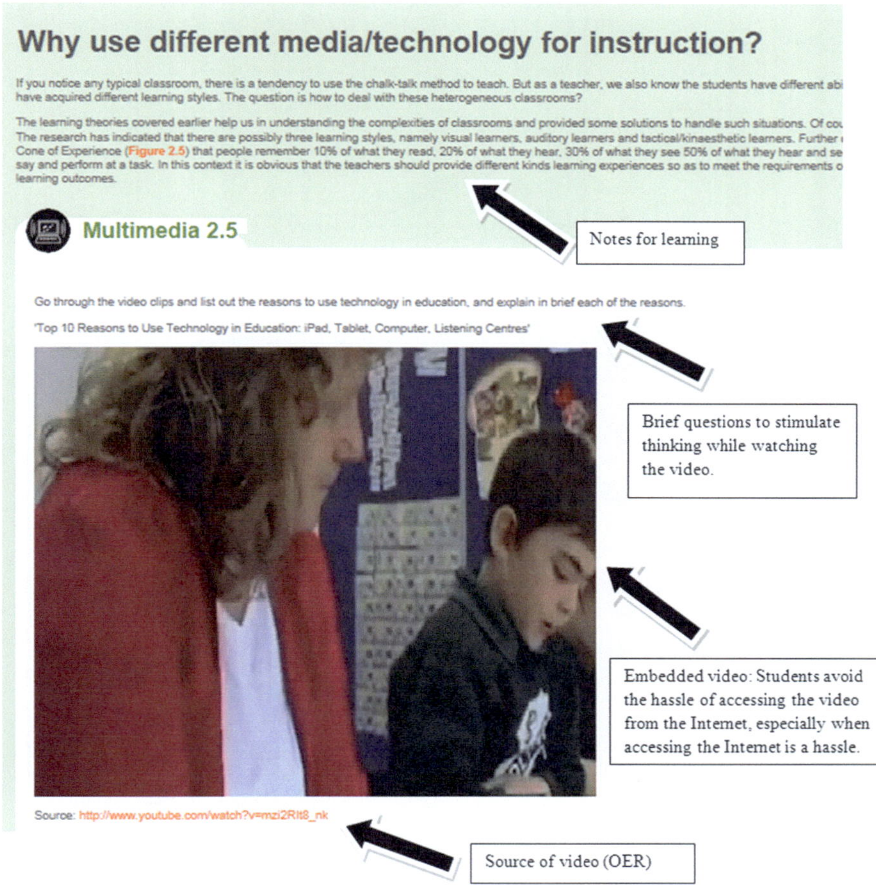


**Fig. 1** Embedded video in the course study package

Readme file. For users who wish to read the course material, they need to download the "eed50205.zip" file and follow the instructions shown in Readme. Pursuant to the first feedback, this course is still under revision until its presentation in the July 2013 semester.

## 3 Issues Surrounding OER Adoption for Course Development

The process of developing the OER-based course has been a very enlightening journey for the entire course development team. While the team members do have relevant experience in the creation and sharing of OERs, it is not until the hands-on work that we realize the complexity of embedding OER in the study package using

**Fig. 2** Embedded excerpt of Wikipedia for reading reference

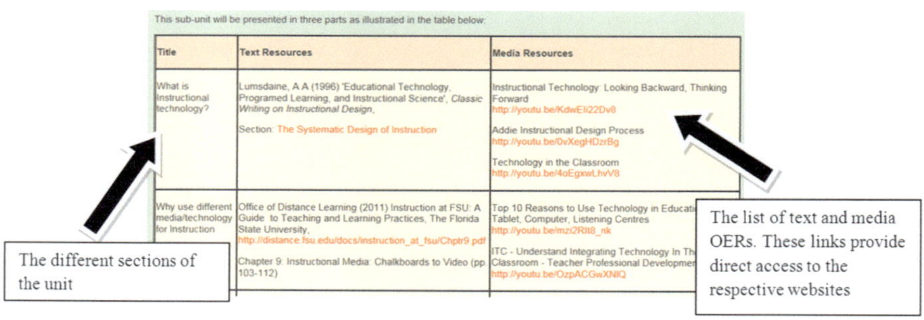


**Fig. 3** List of text and media OERs

the eXe platform. Several issues of OER integration came to light during the course development process.

One recurring issue that required constant monitoring was the use of identifying materials that qualifies to be regarded as an OER. The ethical and legal aspect of OER adoption in course development lies in that a material is not considered as an OER, despite being freely available online, unless it is presents one of the Creative Commons licenses. Readers may wish to check out other forms of open licenses. Hence, the discernment of these licenses is essential in ensuring effective and legally abiding integration of OER materials in the development of a course. The fundamental aspect that distinguishes an OER material from other instructional-based materials is the bearing of Creative Commons license that advocates intellectual property rights over the particular material. Referring to Table 1, the

**Table 1** Present abridged description of Creative Commons licenses (*Source*: http://creativecommons.org/licenses/)

| License | License attribution | Description of license |
|---|---|---|
| | Attribution CC BY | This license lets others distribute, remix, tweak, and build upon your work, even commercially, as long as they credit you for the original creation |
| | Attribution-ShareAlike CC BY-SA | This license lets others remix, tweak, and build upon your work even for commercial purposes, as long as they credit you and license their new creations under the identical terms |
| | Attribution-NoDerivs CC BY-ND | This license allows for redistribution, commercial and noncommercial, as long as it is passed along unchanged and in whole, with credit to you |
| | Attribution-NonCommercial CC BY-NC | This license lets others remix, tweak, and build upon your work noncommercially, and although their new works must also acknowledge you and be noncommercial, they don't have to license their derivative works on the same terms |
| | Attribution-NonCommercial-ShareAlike CC BY-NC-SA | This license lets others remix, tweak, and build upon your work noncommercially, as long as they credit you and license their new creations under the identical terms |
| | Attribution-NonCommercial-NoDerivs CC BY-NC-ND | This license is the most restrictive of our six main licenses, only allowing others to download your works and share them with others as long as they credit you, but they can't change them in any way or use them commercially |

Creative Commons licenses are presented on a continuum of one end that provides maximum dissemination and use of licensed materials to the other end which allows only restrictive use of the material. These licenses can be applied to a variety of OER materials including print materials, images, audio, video, learning objects, survey instruments, and open-access journal articles—basically, any form of instructional materials that the author wishes to share with others without the concerns of losing out on intellectual rights on the actual material or its derivation.

Another concern that the course team is currently looking into is the students' lack of awareness on the adoption of OER for course development. Students are still pretty much in the mindset that OER materials are "free" for use, and in that connection, the students assume that the OER materials are not of quality: "[N]ot much of quality is reflected [in the videos] due to the fact that most resources come under the CC license. Most, if not all of the videos are available free for download on the Youtube, and maybe the journal articles too. These are materials that anyone could download readily and freely from the Internet." In other cases, students were not aware of how to acknowledge OER materials. This is definitely a concern that the university is looking into as it has led to issues of (inadvertent) plagiarism.

## 4 Conclusion

This paper examined aspects pertinent to the adoption of OER for course development. This case study evinces that OERs can be embedded successfully into a single-study package as an alternative to presenting OER materials in a separate study guide as was attempted by other institutions. However, there still exist important concerns surrounding OER adoption in course development and especially how students perceive the use of OER in their open and distance learning courses. This study also calls for more case study research on the different ways of adopting OER in course development. In general, these are concerns that require the attention and collaboration of all relevant stakeholders.

**Acknowledgment** The author is currently a senior lecturer and program coordinator of the Master of Education (M. Ed.) program and coordinator of the Teaching Internship Placement. I would like to take this opportunity to thank Tan Sri Dato' Prof. Dr. Gajaraj (Raj) Dhanarajan, Prof. Dr. Dato' Ho Sinn Chye, Prof. Dr. Mohandas Menon, Prof. Dr. Phalachandra Bhandigadi, Mr. Ishan Abeywardena, and Ms. Patricia Toh Swet Ting of Wawasan Open University; Dr. Sanjaya Mishra of CEMCA; and Dr. Yuen Kin Sun of Open University of Hong Kong for their invaluable input in the development of this course.

## References

1. Johnson L, Levine A, Smith R, Stone S (2010) The 2010 Horizon report. Austin, TX
2. Kanwar A, Kodhandaraman B, Umar A (2010) Toward sustainable open education resources: a perspective from the global south. Am J Dist Educ 24(2):65–80
3. Menon M (2011) Integrating open educational resources in situated learning design for professional education. In: Proceeding of the presented at the 24th ICDE World Conference on open and distance learning: expanding Horizon-new approaches, 2–5 October, Bali
4. Menon M, Ali KM (2011) Learning from advocacy: a case of OER policy implementation in the Wawasan Open University. http://library.wou.edu.my/vertical/vf2012-23.pdf. Accessed 25 May 2013

# Lead Initiatives on Teaching and Learning in Malaysian Higher Education Institutions

**Rosna Awang-Hashim, Nena Padilla-Valdez, Norhafezah Yusof, and Tengku Faekah Tengku Ariffin**

## 1 Introduction

Higher educational institutions in Malaysia face dynamic change processes in light of local and international thrusts in the higher education landscape. This is clearly seen in the country's internationalization Agenda which has taken precedence over the other six thrusts under the National Higher Education Strategic Plan (NHESP) 2007–2020 [15]. At the instructional level, the continuous search for responsive engagement practices as crucial precursors for teaching and learning demands retooling teachers' skills and practices so as to enable their learners to assume greater independence and autonomy. Consequently, institutions of higher learning are expected to adopt the student-centred learning (SCL) approach. The focus of the present study is on Universiti Utara Malaysia where two scenario studies cum training programmes in learning engagement and motivation (LEM) and learner diversity (LD) were examined. The succeeding sections will discuss these two research studies which have inspired the current widespread interest in LEM and LD in the country.

R. Awang-Hashim (✉) • N. Padilla-Valdez • T.F. Tengku Ariffin
School of Education and Modern Languages, Universiti Utara Malaysia, 06010 Sintok, Kedah, Malaysia
e-mail: rosna@uum.edu.my; nena@uum.edu.my; faekah@uum.edu.my

N. Yusof
School of Multimedia Technology and Communication, Universiti Utara Malaysia, 06010 Sintok, Kedah, Malaysia
e-mail: norhafezah@uum.edu.my

K. Ab. Hamid et al. (eds.), *The Malaysia-Japan Model on Technology Partnership: International Proceedings 2013 of Malaysia-Japan Academic Scholar Conference*, DOI 10.1007/978-4-431-54439-5_42, 

## 2 Research Cases

There are two main researches involved here: firstly, Study One on learner diversity and, secondly, Study Two on learning engagement and motivation. The main methods employed in both studies were the survey and focus group.

### 2.1 Study One

It was argued that the governance of global education hubs necessitates a management perspective which is aligned to the tenets of diversity [10] and that the ability to manage institutional diversity required a multiple component mindset [8]. Gurin et al. [6] posited the twin concerns of global-mindedness [11] and diversity climate as constructs important in the study of intergroup and global understanding and, therefore, the need to appreciate the nature of the relationship between these two constructs. Moreover, Cushner and Mahon [3] hypothesized that an improvement in the global-mindedness of lecturers correlated with increased positivity towards diversity issues, which in turn improved instructional quality. Grounded on a similar view of the important role of these constructs, the present study examines the global-mindedness of lecturers. The design used qualitative and quantitative methods of inquiry. A questionnaire survey was distributed to 113 randomly selected lecturers from both private and public HEIs. A focus group method was used to analyse the lecturers' perspective on learner diversity. T he findings indicated that the majority of the respondents lacked awareness of global sensitivity, more specifically the ability to think within the framework of a global society.

### 2.2 Study Two

Teacher- and teaching-related factors, such as teacher autonomy support [7, 13, 17], teaching approaches [4] and quality assessment [1, 2, 19], were found to contribute towards student learning engagement and ultimately the learning outcome. Previous studies showed that students who were more engaged in learning were those who were more likely to adopt a deep learning approach [5, 16, 18]. The study population consisted of university/college students from public and private universities as well as community colleges and polytechnics in the country. Stratified random sampling was utilized in the selection of the student participants and a total of 838 students participated in this study. The findings indicated the significance of the teacher factor in influencing student engagement and this will in turn determine students' use of either a surface or deep learning approach.

## 3 Implications

Study One has given rise to some rather unique implications. Factors like inadequate conducive facilities and restrictive curriculum as well as the psychological aspects of the learners have attracted more attention from the government. As a result, there is the urgent need for it to address the issue of diversity and provide functional mechanisms that will encourage initiatives from the relevant stakeholders to enhance the quality of education with the inclusion of important areas of concern like learner diversity and global-mindedness. The data revealed that the lecturers needed to equip themselves with a greater knowledge of learner diversity, especially that which has been expanded to include other aspects such as learning styles, learning approach, motivation and expectations, social context of education and individual lifestyle beyond ethnicity, gender, religion, disability, culture and community.

In relation to Study Two, the research provides evidence to affirm the crucial roles of teachers in enhancing teaching and learning in the higher education context. In order to encourage students to learn in a meaningful way, teachers must think of teaching as helping students to learn, rather than just organizing activities or worse, merely teaching, period! The current findings support the link between teacher motivational strategy and effective learning (in this case, translated as student engagement) as noted by Jardine et al. [12]. At the same, the findings also lend further support to the claim that there is a close relationship between effective teaching and student success (in this case, students' uptake of the deep learning approach) [9]. Simply stated, good teachers (and/or teaching) can inspire students to be emotionally, behaviourally and cognitively engaged in the subject matter. When students find a class engaging, they will be more interested to learn the subject matter not only for the sake of getting good grades but more importantly for knowledge itself. Students who are engaged in learning are keen to do beyond what is required and enjoy learning at a deeper level.

## 4 Conclusion: Leading the Way Forward

The research team has been instrumental in conducting subsequent national and institutional trainings for future leaders and master trainers who will in turn be responsible for championing learning engagement and learner diversity approaches in their respective institutions. There is also, most importantly, the continuing effort to chart the progress of trainees and the effectiveness of the training programmes conducted. Moreover, a follow-up research which has incorporated the tenets of the Scholarship of Teaching and Learning (SoTL) has been initiated to gauge trainees' transferability of practice. This is done through the iterative analyses of evaluation feedbacks, post-mortem documents, and relevant workplace evidences. These efforts have been well supported by Universiti Utara Malaysia and the Higher Education Leadership Academy (AKEPT) of the Ministry of Education, Malaysia.

## References

1. Biggs JB, Tang C (2011) Teaching for quality learning at university. McGraw Hill, New York
2. Brennan J, Shah T (2000) Managing quality in higher education: an international perspective on institution assessment and change. High Educ 42:515–516
3. Cushner K, Mahon J (2002) Overseas student teaching: affecting personal, professional, and global competencies in an age of globalization. J Stud Int Educ 54:135–156
4. Entwistle NJ, Ramsden P (1983) Understanding student learning. Croom Helm, London
5. Floyd KS, Harrington SJ, Santigo J (2009) The effect of engagement and perceived course value on deep and surface learning strategies. Inform Sci 12:181–190
6. Gurin P, Dey EL, Hurtado S, Gurin G (2002) Diversity and higher education: theory on educational outcomes. Harv Educ Rev 1:330–366
7. Harde PL, Reeve J (2003) A motivational model of rural students' intentions to persist in versus drop out of high school. J Educ Psychol 95:347–356
8. Harvey C, Allard J (2002) Leading for diversity. Corwin Press, Inc., Thousand Oaks
9. Haskins R, Loeb S (2007) A plan to improve the quality of teaching in American schools [policy brief]. Future Child 17(1):3–14
10. Henze C et al (2002) Understanding and managing diversity. Prentice Hall, Upper Saddle River
11. Hett JE (1993) The development of an instrument to measure global-mindedness. Dissertation, University of San Diego
12. Jardine D, Friesen S, Clifford P (2008) Back to the basics of teaching and learning: thinking the world together, 2nd edn. Routledge, New York, NY
13. Liu NF, Littlewood W (1997) Why do many students appear reluctant to participate in classroom learning discourse? System 25:371–384
14. MOHE (2011) Pelan Strategik Pengajian Tinggi Negara Melangkaui Tahun 2020: Pelan Tindakan Pengajian Tinggi Negara 2011-2015. Putrajaya, MOHE
15. Robinson CC, Hullinger H (2008) New benchmarks in higher education: student engagement in online learning. J Educ Bus 84(2):101–108
16. Thaliah R, Awang-Hashim R (2008) Teacher's autonomy support and ESL classroom engagement: the road less travelled. The proceedings of the EDU-COM 2008 International Conference. Sustainability in higher education: direction for change. Edith Cowan University, Perth, WA, 19–21 November 2008
17. Trigwell K, Prosser M, Waterhouse F (1999) Relations between teachers' approaches to teaching and students' approaches to learning. High Educ 37:57–70
18. der Horst V, McDonald R (1997) OBE outcomes-based education: a teacher's manual. Kagiso, Pretoria

MIX
Papier aus verantwortungsvollen Quellen
Paper from responsible sources
FSC® C105338
FSC
www.fsc.org

If you have any concerns about our products,
you can contact us on
**ProductSafety@springernature.com**

In case Publisher is established outside the EU,
the EU authorized representative is:
**Springer Nature Customer Service Center GmbH**
**Europaplatz 3, 69115 Heidelberg, Germany**

Printed by Libri Plureos GmbH
in Hamburg, Germany